ALKALOIDS
A Treasury of Poisons and Medicines

ALKALOIDS

A Treasury of Poisons and Medicines

by

SHINJI FUNAYAMA

Professor, Nihon Pharmaceutical University

GEOFFREY A. CORDELL

President, Natural Products Inc.

ELSEVIER

AMSTERDAM • BOSTON • HEIDELBERG • LONDON
NEW YORK • OXFORD • PARIS • SAN DIEGO
SAN FRANCISCO • SINGAPORE • SYDNEY • TOKYO

Academic Press is an imprint of Elsevier

Academic Press is an imprint of Elsevier
32 Jamestown Road, London NW1 7BY, UK
525 B Street, Suite 1800, San Diego, CA 92101-4495, USA
225 Wyman Street, Waltham, MA 02451, USA
The Boulevard, Langford Lane, Kidlington, Oxford OX5 1GB, UK

Notices

Knowledge and best practice in this field are constantly changing. As new research and experience broaden our understanding, changes in research methods, professional practices, or medical treatment may become necessary.

Practitioners and researchers must always rely on their own experience and knowledge in evaluating and using any information, methods, compounds, or experiments described herein. In using such information or methods they should be mindful of their own safety and the safety of others, including parties for whom they have a professional responsibility.

To the fullest extent of the law, neither the Publisher nor the authors, contributors, or editors, assume any liability for any injury and/or damage to persons or property as a matter of products liability, negligence or otherwise, or from any use or operation of any methods, products, instructions, or ideas contained in the material herein.

Library of Congress Cataloging-in-Publication Data
A catalog record for this book is available from the Library of Congress

British Library Cataloguing in Publication Data
A catalogue record for this book is available from the British Library

For information on all Academic Press publications
visit our web site at http://store.elsevier.com/

This book has been manufactured using Print On Demand technology. Each copy is produced to order and is limited to black ink. The online version of this book will show color figures where appropriate.

ISBN: 978-0-12-417302-6

CONTENTS

FOREWORD

Professor Shinji Funayama (left) and Professor Geoffrey A. Cordell at the Medicinal Plant Garden, Nihon Pharmaceutical University, Saitama, Japan (April, 2013)

Since the earliest application of plants as medicinal and stimulant agents, alkaloids have been a part of the human experience. Over the millennia, numerous plants were introduced into systems of medicine around the world, and many of these have alkaloids as their active principles. When alkaloids were first isolated as bioactive agents from medicinal plants in the early part of the 19th century, it also became clear, as their structures began to be unraveled, that they offered tremendous challenges in terms of both complete structural and stereochemical assignment, and subsequently in synthesis. Indeed, some of the legendary organic syntheses of the 20th century are of complex alkaloids. When microorganisms were investigated for their bioactive principles, more novel groups of alkaloids were disclosed. The profound challenges to unraveling the exquisite processes for the formation of their amazing structural diversity continue to this day.

Their wide spectrum of biological activities provides numerous classes of useful medicinal agents, and yet, at the same time, they provide a global paradox. In spite of many health beneficent effects, two alkaloids, cocaine and morphine (and its derivative heroin), are the focus of a counter culture of illicit use of biological powerful alkaloids, and many alkaloids, because of their profound toxicity, are to be feared. On the other hand, the xanthine alkaloids, caffeine and threobromine, are an integral aspect of providing pleasure in the daily lives of most of the world in the consumption of tea, coffee, and chocolate.

Professor Shinji Funayama joined me at the University of Illinois at Chicago as a postdoctoral fellow studying the fascinating chemistry of acridone alkaloids. He returned to Japan to become a very prominent author of more than 11 single author books on various aspects of natural products and society. This volume was originally published in Japan in Kyoritsu-Shuppan with the title "アルカロイド–毒と薬の宝庫 (Alukaroido - Doku to Kusuri no Houko)". Following the translation, the name of the original volume has been retained, as "Alkaloids - A Treasury of Poisons and Medicines," a reflection of the paradox.

This is not a comprehensive treatise on alkaloids. It is a short volume, focused on some perspectives of the history and diverse biology of a selected group of alkaloids, with an emphasis towards their presence in the materials of Kampo medicines (Japanese Traditional Medicine) in Japan and their toxic potential. It is offered as an introduction, as a stimulant to further inquiry on these amazing natural products.

Geoffrey A. Cordell
Natural Products Inc., Evanston, IL

Introduction

Papaver somniferum (Papaveraceae)

Before describing the various groups of alkaloids individually, the place of alkaloids among the naturally occurring organic compounds, and in society will be discussed briefly in this chapter.

An understanding of what an alkaloid is will be mentioned in the first section, and in the second section, the aspects of classifying alkaloids on the basis of their biosynthetic origin will be presented. This method will be compared with a classification based on their carbon or heterocyclic skeleton, such as indole, isoquinoline alkaloids, or a classification of the alkaloids based on a chemotaxonomic approach, such as Rutaceae and Apocynaceae alkaloids. A short history of the study of alkaloids and the crude drugs as the origin of various alkaloids, and their relevance in the history of natural products chemistry will be described. Through these discussions, it will be established that the natural products classified as alkaloids relate to our life deeply, and on an everyday basis, as medicines, as dyestuffs, as flavors, as stimulants, and as toxic substances.

It is known that alkaloids show a broad range of biological activities. Among the biologically active compounds, there are especially many alkaloids which affect the central nervous system (CNS) and the autonomic nervous system. Some of the CNS stimulants will be described in the final sixth section.

Readers of this book will discover that alkaloids are a very important group of organic compounds which show a variety of highly significant, clinically and biologically useful properties. Among these activities of alkaloids are hypotensive, cardiotonic, hormone, pheromone, growth acceleration, antimalarial, antitumor, antiparasitic, sedative, analgesic, anti-Alzheimer's, and antimicrobial activities.

There are probably over 25,000 alkaloids derived from higher plants as presented in the *Dictionary of Alkaloids*. This modest volume offers a brief

Alkaloids
http://dx.doi.org/10.1016/B978-0-12-417302-6.00017-9

overview of the main alkaloid groups, their structures, their activities, and their basic biosynthetic pathways from a historical perspective.

1. THE DEFINITION OF AN ALKALOID

The word "alkaloid" was proposed in 1818 by K. F.W. Meissner (1792–1853), a pharmacist in Halle, Germany. The word alkaloid was coined from the word alkali (implying basicity), from "*al qali*" (referring to soda) in Arabic. The "-oid" suffix, meaning "like", derives from the Greek.

The definition of an alkaloid has changed significantly over the years, as more "alkaloids" have been structurally elucidated and the sources of alkaloids have broadened.

1. At the beginning, alkaloids were discovered only from higher plants, and those compounds showed basic properties and strong biological activities. Consequently, at that time, an alkaloid was defined as "the plant component which shows basic properties and strong biological effect." The basicity of alkaloids is derived from the presence of a nitrogen atom in the molecule in the form of an amine.

   Such a definition is no longer possible for the alkaloids. First of all, alkaloids are obtained from an extremely broad range of natural sources, not just the plant kingdom. For example, retronecine, danaidone, and hydroxydanaidal (derivatives of pyrrolizidine alkaloids) were isolated from the hair pencil of the male butterfly of the *Danaid* genus. Batrachotoxin, a poison arrow toxin component, was isolated from the skin of a frog. In addition, there are many examples of alkaloids of microbial, marine, and human origin, including a vast array of nitrogen-containing antibiotics.

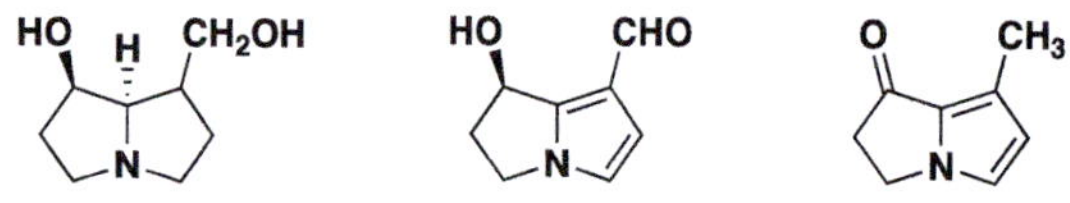

retronecine hydroxydanaidal danaidone

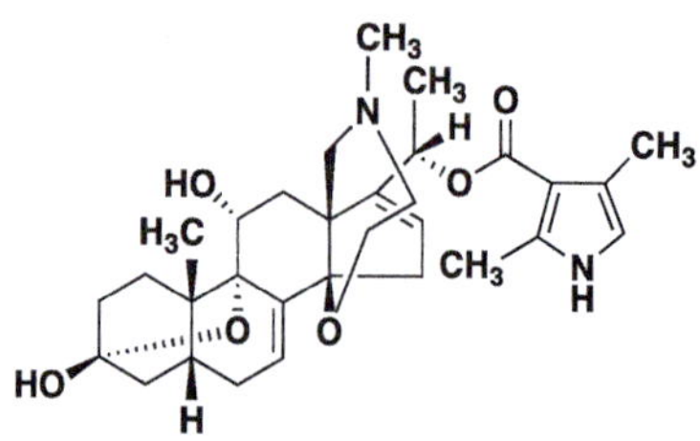

batrachotoxin

2. Alkaloids are not limited to those natural products which are basic in character. For example, colchicine isolated from *Colchicum autumnale* (Liliaceae) and used for the treatment of gout, etc. is not basic because the nitrogen atom in the molecule is present in a neutral amide group. However, the biosynthetic precursor of colchicine is autumnaline, a typical, basic phenethylisoquinoline alkaloid. Therefore any compound derived from such an intermediate should be classified as an alkaloid.

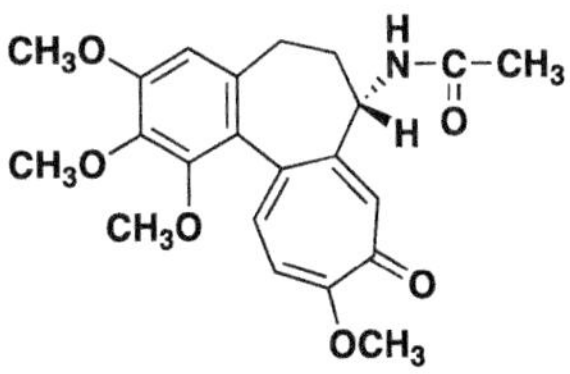

colchicine

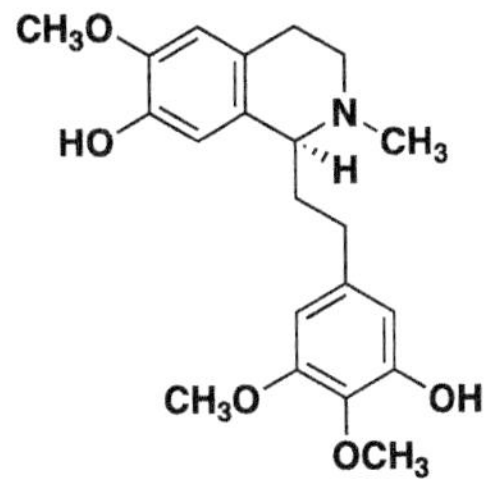

autumnaline

Another example pertains to the close structural isomers, pteleprenine and isopteleprenine, which were isolated from *Orixa japonica* (Rutaceae). In pteleprenine, the nitrogen is in the form of an amide, and therefore it lacks basicity. On the other hand, in isopteleprenine the nitrogen atom, being in a quinoline ring, is weakly basic. Given their common biosynthetic origin both compounds are classified as alkaloids.

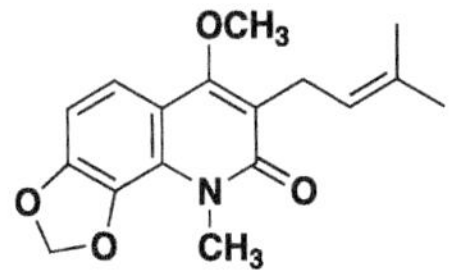

pteleprenine

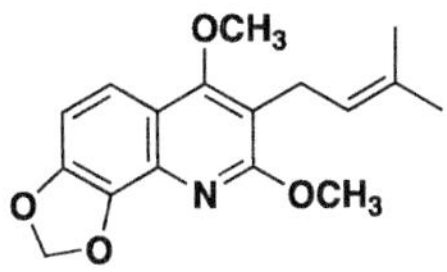

isopteleprenine

In the case of the phenanthrine derivative aristolochic acid-I, the skeleton is derived from an aporphine alkaloid precursor which has undergone oxidation, to the point where the nitrogen atom exists as a nitro group. Biosynthetically, this compound is also classed as an alkaloid.

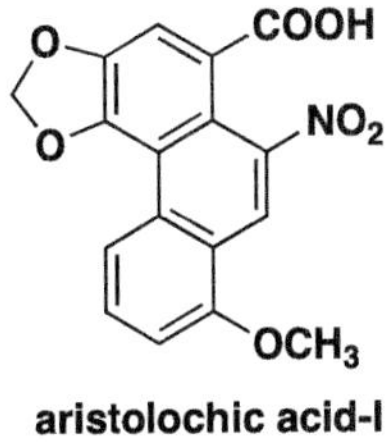

aristolochic acid-I

3. It is not appropriate to include the existence of biological activity in the definition for an alkaloid. Several years ago [1] it was shown that about 75% of known "alkaloids" had never been tested in a single bioassay. In addition, when an alkaloid with biological activity is isolated, compounds with a closely related chemical structure with no, or greatly diminished, biological activities will also be isolated. In such a case, all of these compounds are regarded as alkaloids, irrespective of whether they have a demonstrated biological activity.
4. Alkaloids are always compounds isolated from nature. However, there are many examples of alkaloid derivatives with a high profile which may be confused and sometimes classified as alkaloids. For example, **L**yserg **S**äure **D**iäthylamid (LSD) (LSD-25) is prepared by the amidation of lysergic acid, itself derived from the ergot alkaloids. Methamphetamine is prepared by the reduction of (−)-ephedrine, and heroin is prepared by the acetylation of morphine, and there are many derivatives of alkaloids which are pharmaceutical products. These semisynthetic compounds can be classed as alkaloid derivatives.

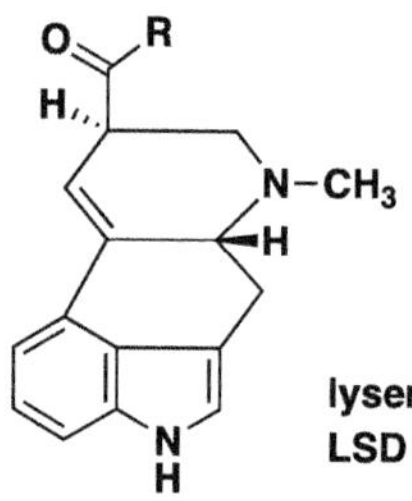

lysergic acid **R = OH**
LSD **R = $N(C_2H_5)_2$**

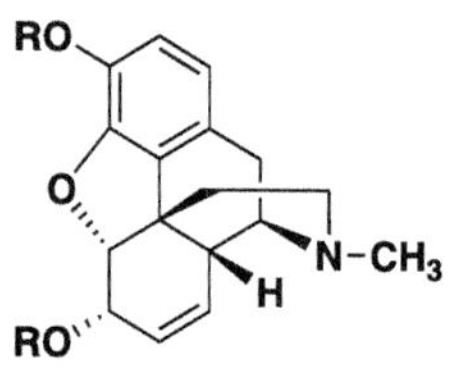

morphine **R = H**
heroin **R = $COCH_3$**

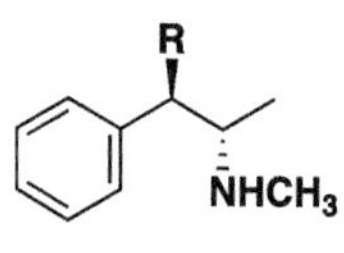

(-)-ephedrine **R = OH**
methamphetamine **R = H**

5. There are a significant number of unusual amino acids, simple peptides, pyrrole derivatives, and comparatively simple nitrogen-containing organic compounds, including purines and pyrimidines, which may be excluded from classification as alkaloids.

For example, L-α-kainic acid, obtained from the red algae *Digenea simplex* (Rhodomelaceae), might not be regarded as an alkaloid. However, this unusual amino acid is biosynthesized from L-glutamic acid and an hemiterpenoid unit, and therefore it is classified as an alkaloid rather than unusual amino acid.

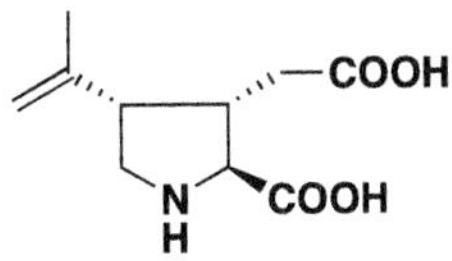

L-α-kainic acid

The pyrazolopyrimidines, such as caffeine of tea and coffee origin, are alkaloids, as is the phytohormone zeatin from maize.

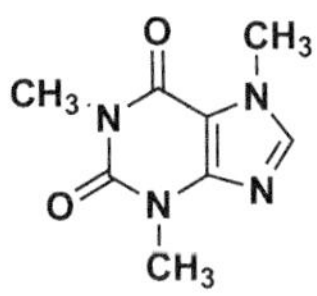

caffeine

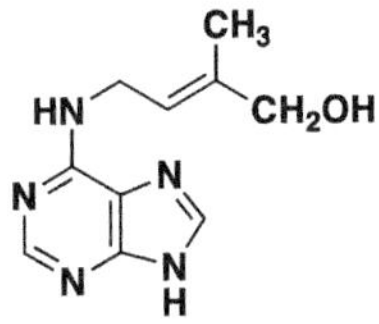

zeatin

Also, comparatively simple *N*-containing compounds (amines), such as serotonin and histamine, are alkaloids, as are mescaline, isolated from the peyote cactus, *Lophophora williamsii*, and (+)-coniine obtained from *Conium maculatum*.

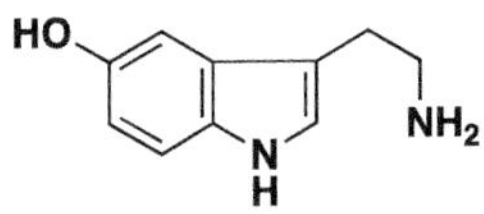

serotonin

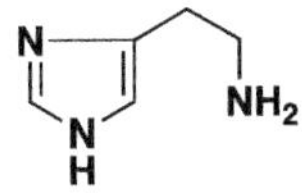

histamine

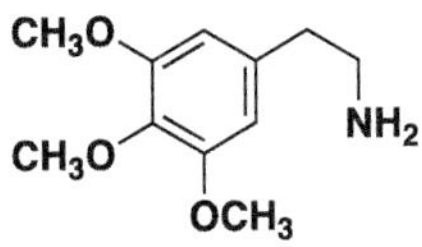

mescaline

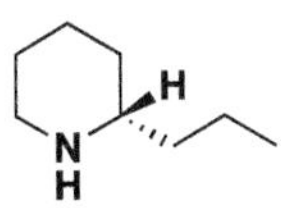

(+) -coniine

Based on the above discussion, the precise definition of an alkaloid is rather unclear. That is appropriate for such a broadly available, structurally diverse group of metabolites. It may be said that alkaloids are naturally occurring, nitrogen-containing compounds. The group excludes the amino acids of primary metabolism, complex peptides and proteins constructed from those amino acids, and nucleic acids. Thus we see alkaloids, not in terms of a comprehensive definition, but rather as a way to classify a large number of natural metabolites containing nitrogen possessing great structural diversity and derived from any natural source. As has been suggested previously, with experience, "you know one when you see one" [2]. Our task then is to classify them so that the breadth and depth of their molecular frameworks can be assimilated into the larger organization of natural product structures.

LITERATURE CITED

[1] G.A. Cordell, M.L. Quinn-Beattie, N.R. Farnsworth, Phytother. Res. 15 (2001) 183–205.

[2] G.A. Cordell, Introduction to Alkaloids. A Biogenetic Approach, Wiley Interscience, New York, NY, 1981. 1055 pages.

2. CLASSIFICATION OF ALKALOIDS

Several approaches to the classification of alkaloids are available, including, chemical, taxonomic, biological, and biosynthetic. At various times, each of these approaches has been used in terms of the presentation and discussion of alkaloid development. We will discuss these approaches in turn.

Alkaloids were often classified on the basis of their chemical structure. According to this system, alkaloids are organized based on a common, typically heterocyclic, nucleus, such as isoquinoline, indole, quinolone, quinazoline, pyrrolizidine, and tropane alkaloids, etc. Another method to classify the alkaloids, is to use their natural origin. So we could organize them based on a plant family, such as Amaryllidaceae, Solanaceae, and Rutaceae alkaloids, or based on a genus, such the *Catharanthus* alkaloids. Classification may use the name of a prototypical alkaloid of the group such as aconitine-type or morphine-type alkaloids. Frequently, this approach also follows a common biosynthetic or biogenetic origin.

Using this type of chemical classification is useful, and such a classification is used predominantly in this book. However, this approach also encounters some challenges. For example, autumnaline and (*S*)-reticuline are typical isoquinoline alkaloids, whereas colchicine and morphine, which are derived from these isoquinoline alkaloids no longer possess an isoquinoline moiety. So, morphine is sometimes classified as a member of the morphinan alkaloids, and colchicine is classified as a special alkaloid with other, structurally related alkaloids. Consequently, autumnaline and colchicine, and (*S*)-reticuline and morphine are not discussed in the same isoquinoline alkaloid category. On the other hand, by examining them from a biosynthetic perspective, their common origin from phenylalanine and tyrosine brings them together.

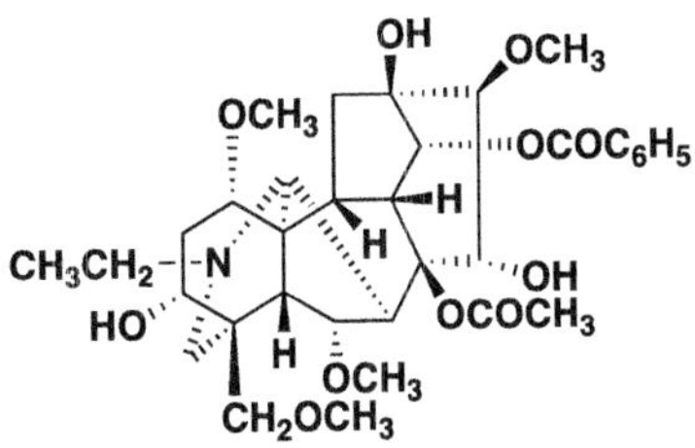

aconitine

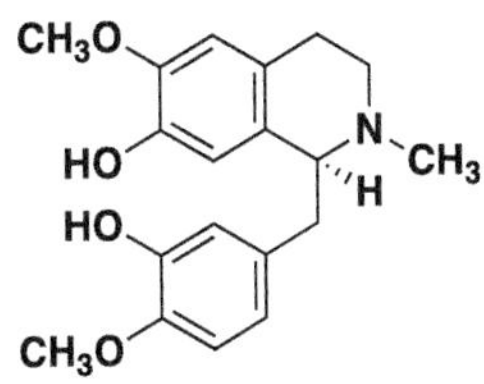

(*S*)-reticuline

Also, berberine is isolated as a main alkaloid of the Rutaceae plant, *Phellodendron amurense*, and the same alkaloid is also isolated from the Ranunculaceae plant, *Coptis japonica*. A classification, such as a Rutaceae or Ranunculaceae alkaloid is therefore not possible for berberine.

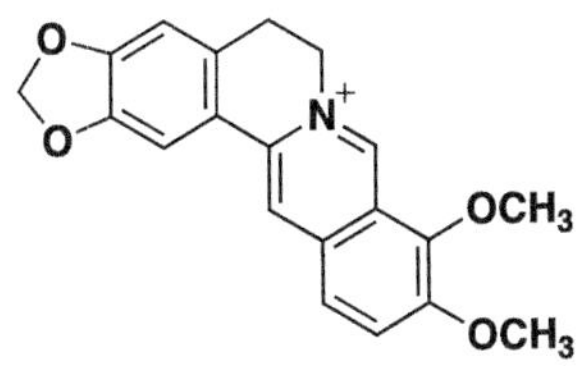

berberine

A large plant family may produce several groups of structurally diverse alkaloids. One example is the plant family the Solanaceae. Therefore, if a classification based on a Solanaceae alkaloid is used, solanine, isolated from the bud of potato (*Solanum tuberosum*), and the tropane alkaloids, are grouped with scopolamine from *Scopolia japonica*, and nicotine from tobacco leaves (*Nicotiana tabacum*). As a result, for an overall alkaloid classification, this is not an effective approach. These methods are sometimes convenient when selected aspects of alkaloid chemistry and biochemistry are discussed, and they are also used, in part, in this book.

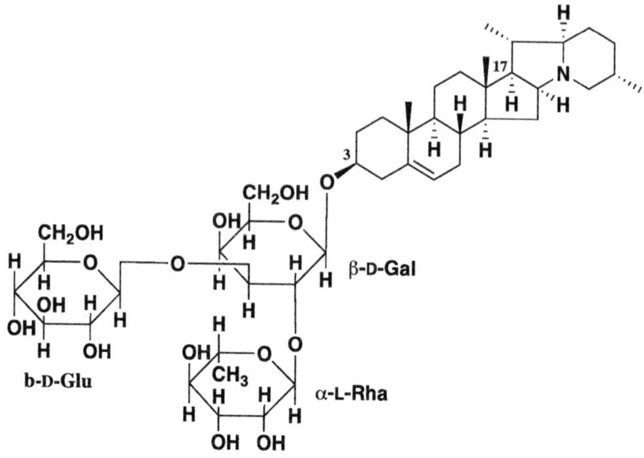

α-solanine

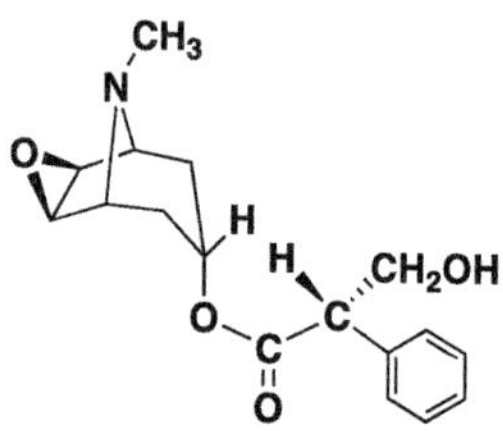

scopolamine

The system recently used is to classify the alkaloids based on their biosynthetic origins [1]. According to this approach, it is appropriate to discuss autumnaline and colchicine, and (*S*)-reticuline and morphine in the same section based on a common amino acid precursor. Similarly, (−)-hyoscyamine and scopolamine isolated from *Scopolia* sp. (Solanaceae) can be discussed in the same section with cocaine isolated from *Erythroxylon* (Erythroxylaceae).

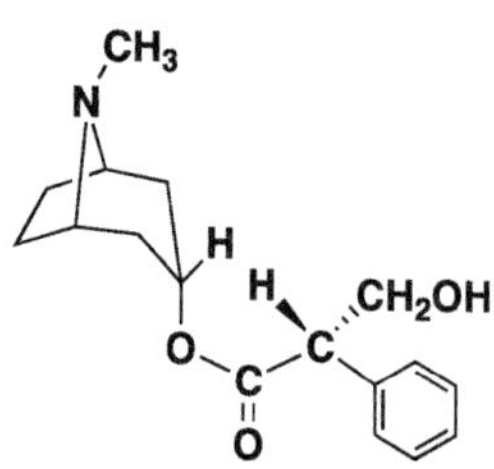

(-)-hyoscyamine

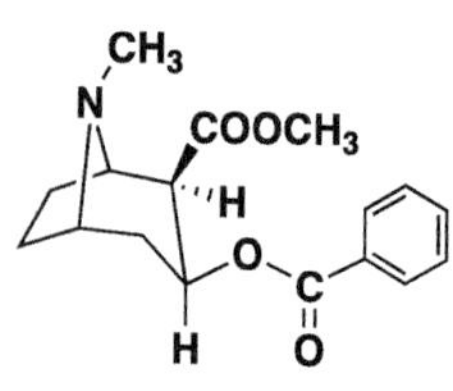

cocaine

This approach though also has some complications, particularly when the alkaloid has more than one nitrogen and more than one biosynthetic amino acid precursor unit. For example, evodiamine and rutaecarpine, from the fruits of *Euodia rutaecarpa* (Rutaceae), can be classified as alkaloids derived from tryptophan. In addition, one of the nitrogen atoms of each alkaloid is derived from an anthranilic acid unit. Therefore, these alkaloids can be classified as alkaloids derived from tryptophan and as alkaloids derived from anthranilic acid. In this volume, these alkaloids are discussed in Chapter 2.19 as alkaloids derived from tryptophan.

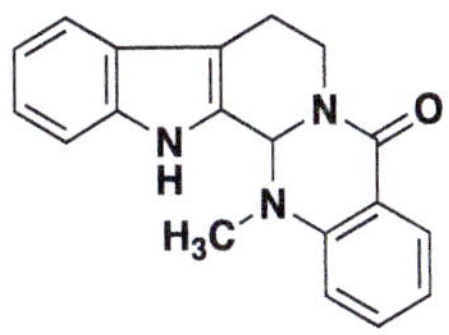

evodiamine

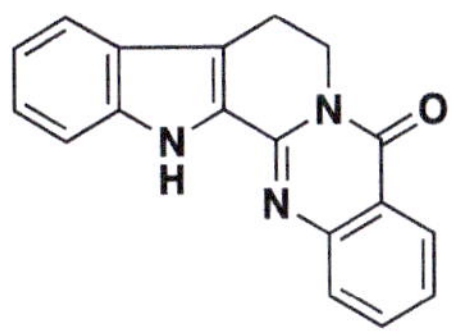

rutaecarpine

Nicotine and anabasine are considered to be alkaloids derived from nicotinic acid (pyridine-3-carboxylic acid). These alkaloids can be also classified as derived from ornithine and lysine, respectively. Therefore, these alkaloids are described in Chapter 10.1 (as alkaloids derived from nicotinic acid), in Chapter 3.1 (alkaloids derived from ornithine and arginine), and in Chapter 4.3 (alkaloids derived from lysine). However, such exceptions are relatively rare if the classification of alkaloids is conducted based on their biosynthetic origin.

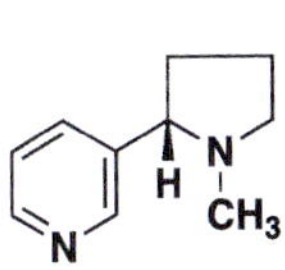

nicotine

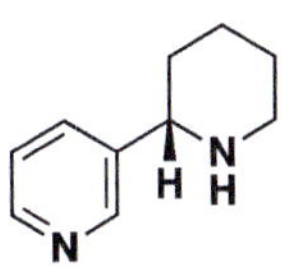

anabasine

From a structural perspective, ephedrine gives the impression that it is derived from phenylalanine, and for some time this alkaloid was regarded this way. However, it was established that the nitrogen of this alkaloid was introduced through the intervention of a C_3N unit. Thus, ephedrine is biosynthesized through the coupling of a C_6C_1 unit and a C_3N unit. As a result, ephedrine is described in Chapter 16 together with other alkaloids derived from a C_6C_1 unit.

LITERATURE CITED

[1] G.A. Cordell, Introduction to Alkaloids. A Biogenetic Approach, Wiley Interscience, New York, NY, 1981. 1055 pages.

3. HUMANS AND ALKALOIDS

Plants of the genus *Magnolia* (Magnoliaceae) are said to be the first angiosperm plants, and they may have appeared on the Earth about 100 million years ago. Magnoflorine is one of the representative alkaloids of this genus (Chapter 1.4), although the simpler alkaloid ephedrine (Chapter 16.1) obtained from *Ephedra* may have appeared on Earth significantly earlier than Magnoliaceae plants, perhaps 160 million years ago. Genes for the formation of the alkaloids of microbial origin must have been present at an even earlier stage in evolutionary history, possibly over 3 billion years ago. However, when they first produced secondary metabolites is not known; clearly though the alkaloids as secondary metabolites must have a very old history on Earth.

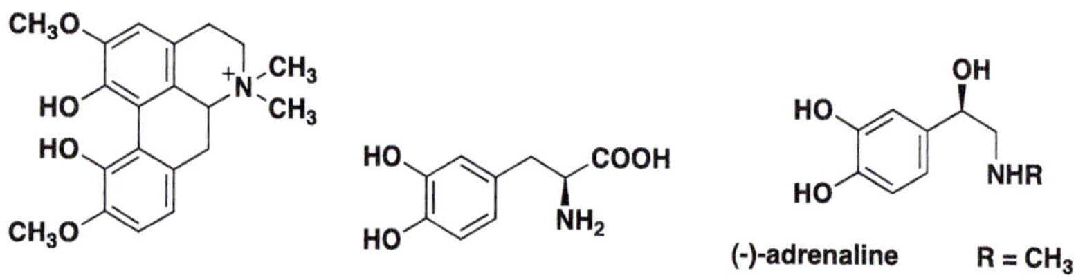

magnoflorine　　L-DOPA　　(-)-adrenaline R = CH_3; (-)-noradrenaline R = H

Humankind also produces alkaloids. For example, we are continuously producing alkaloids such as L-DOPA (3,4-dihydroxyphenylalanine), adrenaline (epinephrine), serotonin, etc. in our bodies (Chapter 1.2). These alkaloids are formed from amino acids through various oxidation and decarboxylation steps. Occasionally, they are referred to as proto alkaloids or biogenic amines.

Humankind has been using the local flora and fauna for food, clothing, shelter, etc. for millions of years. In addition, plants have also been used since antiquity as medicine, as poisons with various purposes, as dyestuffs, as flavor enhancers, etc. From records known at present, there is evidence of the medicinal use of plants in the papyri of ancient times in Egypt about 4000 years ago. In these records, there are already descriptions of the utilization of alkaloid-containing crude drugs such as opium, *Hyoscyamas niger*, containing (−)-hyoscyamine, and *Colchicum*, containing colchicine.

In the meantime, there is also a history of the utilization of alkaloids without any written records; traditions and practices passed down over the millennia. One of these is the use of plants to provide dyestuffs for woven cloth. For example, indigo is regarded as one of the oldest plant dyes. Indigo possesses a bisindole alkaloid skeleton, and is biosynthesized from tryptophan (Chapter 2.4).

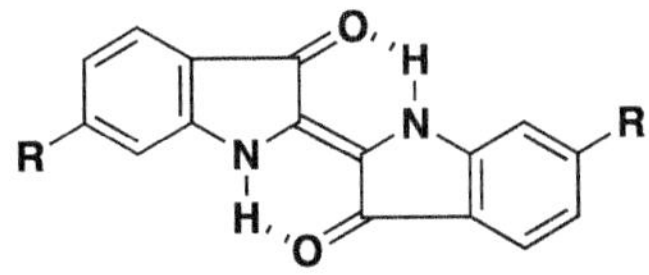

indigo **R = H**
6,6'-dibromoindigo **R = Br**

When the total synthesis of indigo was achieved in 1882 by Baeyer and Drewson [1], it served as one of the origins of the modern synthetic dye industry. On the other hand, there is a famous dye collected from mollusks of the genus *Murex* and originating in the Phoenician city of Tyre in about 1500 BC. It is known as shell purple or Tyrian purple, and the structure of the pigment was shown to be 6,6′-dibromo indigo (Chapter 2.4). Because only a very small amount of dye is obtained from a shell, this purple pigment was regarded as the noblest color and its use limited to the noblest in society. It is said that the sails of Cleopatra's ship were dyed using shell purple.

Many indigenous groups in the world have developed particular teas used as stimulant beverages. Among those are coffee, green tea, black tea, guarana, and mate. It is interesting to note that the xanthine alkaloid caffeine is a principal component of all these drinks. In this book, caffeine is described in Chapter 11.1.

One of the oldest recorded incidents of the deliberate use of a toxic plant concerns the simple alkaloid coniine. Based on the writings of Plato in the *Phaedo*, it is known that the Greek philosopher Socrates (470–399 BC) died by taking an extract of *C. maculatum*. The main toxic component of *C. maculatum* is coniine (Chapter 15.1), whose basic skeleton is derived from a polyketide precursor.

An arrow poison from a blowgun is often used by native peoples of South America for hunting. Three kinds of poisons are known (as explained in Chapter 1.5 and Chapter 2.15). The toxic constituents from these poisons are alkaloids, including *d*-tubocurarine (*d*-Tc), a bisbenzylisoquinoline alkaloid, and C-curarine, a bisindole alkaloid in the *Strychnos* series. Decamethonium (Chapter 1.5), a synthetic compound based on the chemical structure of *d*-Tc, was formerly used as a muscle relaxant in surgery.

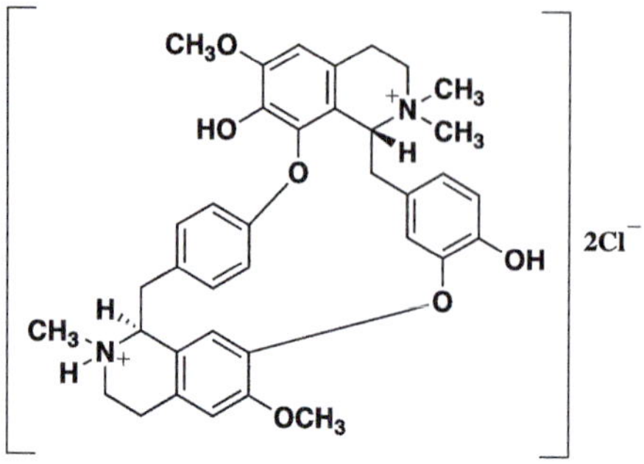

d-tubocuraine Cl (*d*-Tc Cl)

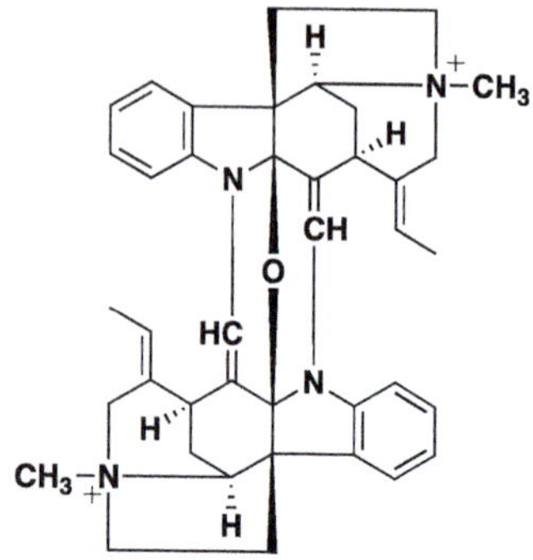

C-curaine

In the meantime, and turning attention to the Far East, there is the long history that the tubers of plants in the genus *Aconitum* (Ranunculaceae) have been used as arrow poisons for harpoons in Japan and neighboring countries. In Japan, the Ainu indigenous people of Hokkaido in northern Japan applied *Aconitum* toxin as an arrow poison. On the other hand, the tubers of *Aconitum* plants have also been used for heart medications under the name "bu-shi" or "uzu" for a long period of time. The main toxic alkaloid of *Aconitum* plants, aconitine, is described in Chapter 14.4.

LITERATURE CITED

[1] A. Baeyer, V. Drewson, Ber. 15 (1882) 2856–2864.

4. DEVELOPMENT OF PHARMACOGNOSY AND NATURAL PRODUCT CHEMISTRY, AND ALKALOIDS

As mentioned, probably since primeval times, humankind has been using plants, animals, and minerals as medicines. Through simple manipulations, humankind began to focus on and dry the most useful portion of the plant (for example, the leaf, fruit, bark, root, milky exudate, animal organ, etc.) as a medicine to gather and stock and/or to improve the drug efficacy. These products are referred to as crude drugs, and the knowledge and study of the identification, preparation, and quality assessment of these crude drugs was the original definition of that area of science called pharmacognosy. It is the oldest of the pharmaceutical sciences.

Crude drugs are used all over the world by the majority of people as their primary health care modality. In some instances the use of these crude drugs was written down and passed on as Ayurvedic, Unani, and other systems of medicine. The use of a systematized mixture of crude drugs is the basis of traditional Chinese medicine. This medical treatment system was subsequently introduced into both Korea and Japan. The system, especially the recipe of the crude drugs was changed, and now the traditional Chinese medicine system as applied in Japan is called Kampo Medicine, rather than traditional Chinese medicine, and uses many of the same plant materials.

As science evolved in the seventeenth century, analysis of plant materials began. Subsequently, it became possible to analyze the chemical substances of medicinal plants and this led to early phases in the development of organic chemistry. Perhaps, not surprisingly, the most famous (indeed notorious) medicinal plant, the opium poppy, was investigated first.

The isolation of a crude preparation of morphine (after Morpheus, the Goddess of sleep), from opium was reported in 1805 by the pharmacist apprentice F.W.A. Sertürner (1783–1841) of Paderborn, Germany. Opium is derived from the milky exudate of the unripe fruits of poppy (*Papaverum somniferum*). The first report of the isolation of morphine appears in the *Journal der Pharmacie* published in 1805, and under the report, isolation of pure "Mohnsäure (Opiumsäure)." At that time, it was assumed that all organic constituents derived from plants are acidic compounds and that is why such a name was put on this compound. Afterward, the name of the biologically active compound isolated from the opium was changed to morphine. Namely, Sertürner also reported the result of re-examination of morphine in the *Annalen der Physik* published in 1817, and in this paper, the name of "Morphium (morphine)" is used for the identical substance. Sometimes the year of the first report of the isolation of morphine is miss-regarded to be 1806. According to Dr I. Arabas of Poland, this is because the 14th volume of the *Journal der Pharmacie* was published in 1805–1806 and the year appeared as 1806 on the cover of the bound volume [1].

Earlier, in 1803, J.F. Derosne (1774–1855), a French pharmacist, had announced the preparation of opium alkaloids, and named it as "sal de Derosne." The preparation became very famous in Europe, it was said that he was the finder of morphine.

Also, A. Seguin (1767–1835), who reported a paper on the opium in 1803, was once considered to be a discoverer of the morphine. [2] However, this also was a misunderstanding.

Therefore, the first report of the isolation of morphine was done by Sertürner, and it was published in 1805. Seventy-six years after the death of

Sertürner, his stone coffin was opened, and beside his mummified body was a container of 1 kg pure morphine. [3] Morphine is described in Chapter 1.10.

In India, *Rauwolfia serpentina* (Apocynaceae) had been used for snake bite, etc., and subsequently reserpine was isolated from this plant. Reserpine is one of the typical monoterpene indole alkaloids, and it has been used as an antipsychotic and antihypertensive drug (Chapter 2.10).

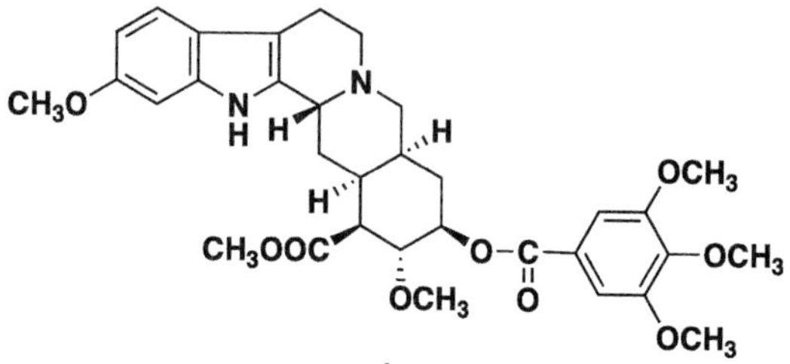

reserpine

These studies on the chemical constituents of crude drugs are a part of contemporary natural product chemistry, and are one aspect of modern pharmacognosy. Among the naturally occurring compounds, alkaloids occupy a very important position.

Although the first report of the isolation of an alkaloid was the isolation of morphine, Leeuvenhoek (1632–1723) observed a crystal in a sperm sample with his microscope and reported it in 1678 [4]. This chemical compound was named spermine after about 210 years, and in 1926 the chemical structure of spermine was clarified, about 250 years after the report of the existence of the crystal (Chapter 3.5). Spermine is an alkaloid biosynthetically derived from ornithine.

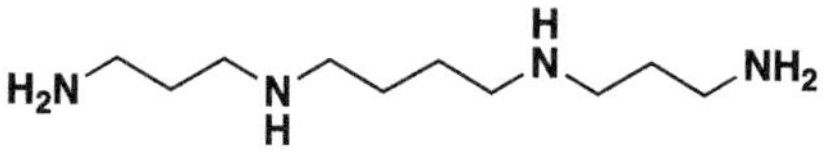

spermine

Following the successful isolation of morphine (1805), attention turned to a number of other prominent traditional medicines or toxic plants in order to derive their active principles. Many of these active compounds were alkaloids. They include the important alkaloids emetine (1816), strychnine (1818), quinine (1820), colchicine (1820), caffeine (1820), nicotine (1828), atropine (1833), cocaine (1860), physostigmine (1864), and ephedrine (1887). Therefore, it is important to understand and appreciate the significance of alkaloids as the source of medicines, directly and indirectly through structure modification.

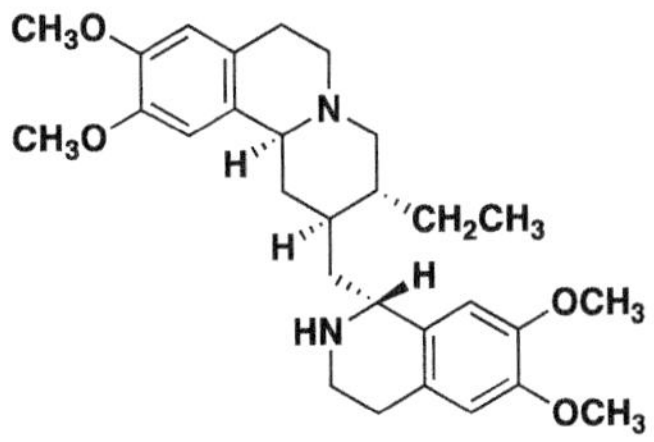

emetine

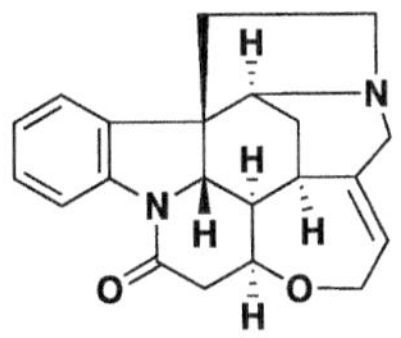

strychnine

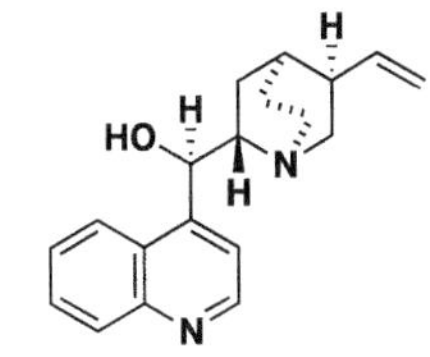

quinine

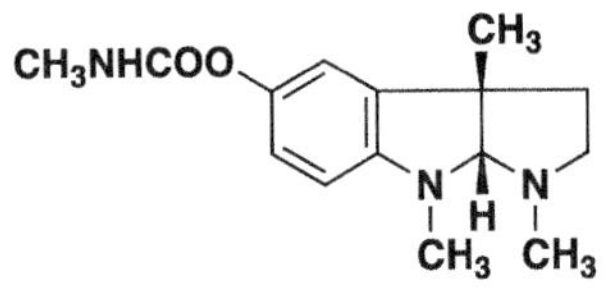

physostigmine

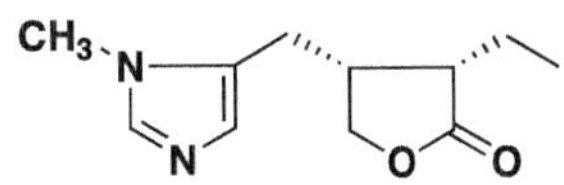

pilocarpine

When microbial, biologically active, systems were examined for the active constituents, in many instances alkaloids of completely new and novel types were found to be the active principles. The discovery of penicillin from the blue mold preparation by A. Fleming in 1929, and the discovery of streptomycin from *Actinomyces* by S.A. Waksman et al. in 1944 completely changed the history of the medicine. These antibiotics, and many others containing one or more nitrogen atoms, some of which are described herein, are also regarded as alkaloids [5].

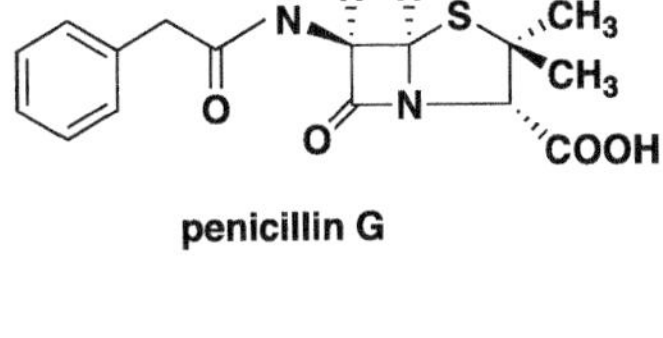

penicillin G

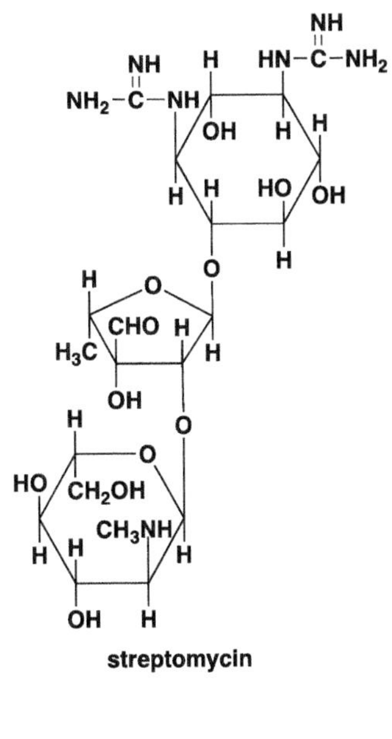

streptomycin

LITERATURE CITED

[1] I. Arabas, Jpn. J. Hist. Pharm. 32 (1997R) 18.
[2] I. Arabas, Investigations on Opium Alkaloids and Their Impact on the 19th Century Polish Research on Drugs, p.23, IHN PAN, Warszawa, 1995 (in Polish).
[3] M.H. Zenk, M. Kawabata, Nat. Med. 50 (1996R) 86.
[4] O. Rosenheim, Biochem. J. 18 (1924R) 1253.
[5] S.A. Waksman, Science 110 (1949) 27.

5. BRIEF HISTORY OF JAPANESE PHARMACEUTICAL SCIENCES AND ALKALOIDS

There is a description of a medicine in an old Japanese book written in 712 AD named "Kojiki." Namely, a god named Ohkuninushino-mikoto used the pollen of *Typha latifolia* (Thyphaceae) for the treatment of injuries.

There are numerous tales of various folklore medicines in Japan, and many of them are still used. The Chinese plant medicines are derived from ancient Chinese practices (Toh era in Japanese; Tang era in Chinese) and were brought by so-called "Kentoh-shi" (the Japanese envoy to Tang Dynasty China) and their introduction affected Japanese medicine very much. Especially, it was said that a high priest Ganjin of Tang brought a large amount of various crude drugs when he at last reached Japan in 753. When the Emperor Shohmu died in 756, the Empress Kohmyo donated various kinds of treasures of the Emperor Shohmu to the great image of Buddha, Daibutsu, of Todaiji Temple in Nara. Those treasures have been preserved in the imperial store-house, Shosoin, at Todaiji Temple. Among the treasures, 60 kinds of crude drugs were also included, and some of these crude drugs are said to be those brought by Ganjin. Some of the crude drugs are still preserved, and the list of the crude drugs (Shuju-Yaku-Cho) is also present. Using modern analytical methods, it was established that some of the crude drugs have retained their active components for over 1260 years. [1,2]

The traditional Chinese medicine system and various books regarding this medical system were also introduced into Japan with those crude drugs. Consequently, the main purpose of Japanese pharmacognosy was to find which medicinal plant explained in the Chinese book was related which plant in Japan. This situation has actually continued for more than 1000 years.

In the Edo era (1603–1867), Seishu Hanaoka (1760–1835) succeeded in conducting the world's first anesthesia operation using an anesthetic comprised of *Aconitum* tuber and *Datura metel* (Solanaceae), etc. (the exact formula remains unknown). Among them, *D. metel* is known to afford atropine. In the meantime, P.F. von Siebold was involved in the so-called Siebold-Incident regarding *S. japonica* (Solanaceae), which also gives atropine. Details of these stories will be given in Chapter 3.2.

At the beginning of Meiji era (1868–1912), science was changed dramatically by the sudden change of the Japanese social system. Namely, Japan at last opened the country to the world. During that period, great scientists appeared who made globally significant achievements. For example, Shibasaburo Kitasato (1853–1931) succeeded in cultivating *Clostridium tetanus* under the supervision of Robert Koch (1843–1910), and on his return to Japan he founded the Kitasato Institute in Tokyo. Kiyoshi Shiga (1871–1957) who discovered *Shigella dysenteriae*, Sahachiro Hata (1873–1938) who found Salvarsan under the supervision of Paul Ehrlich (1854–1915), and Hideyo Noguchi (1876–1928), who succeeded in preparing a pure culture of *Treponema pallidum*, were disciples of S. Kitasato.

On the other hand, Nagayoshi Nagai (1845–1929) et al. isolated ephedrine, which will be discussed in Chapter 16.1. Also, Jokichi Takamine (1854–1922) succeeded in the isolation of adrenaline (epinephrine), and Umetaro Suzuki (1874–1943) discovered oryzanin.

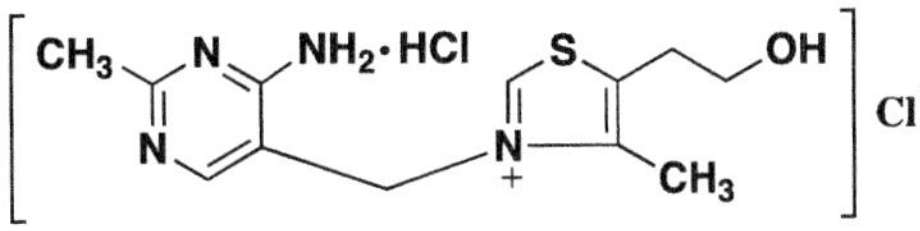

thiamine (vitamin B_1) HCl

It was found that oryzanin corresponded to the beriberi vitamin (vitamin B_1) reported by C. Funk (1884–1967), and actually the report of oryzanin had appeared earlier than that of beriberi vitamin, whereas only the name of vitamin B_1 and the achievement of Funk are recognized at present (Chapter 11.2).

In the meantime, the correct chemical structure of vitamin B_1 was proposed by the Japanese scientist Katashi Makino (1907–1990) for the first time. [3] Makino also proposed the accurate chemical structure of adenosine triphosphate [4,5], which is regarded as one of the most important nucleic acid derivatives.

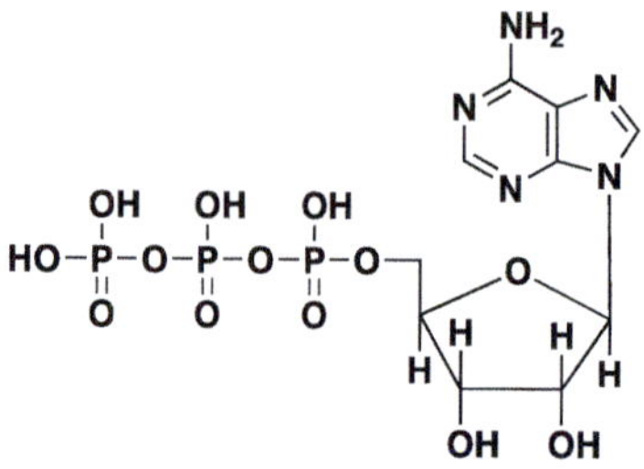

adenosine triphosphate (ATP)

LITERATURE CITED

[1] S. Shibata, Int. J. Pharmacogn. 32 (1994) 75.
[2] S. Funayama, Farumashia 28 (1992R) 1131.
[3] K. Makino, T. Imai, Hoppe-Seiler's Zeitschr. für Physiol. Chem. 239 (I) (1936).
[4] K. Makino, Biochem. *Zeitschr.* 278 (1935) 161 [Chem. Abstr., 29, 8020[1] (1935)].
[5] K. Maruyama, Gendai-kagaku 196 (1987R) 40.

6. CNS STIMULATION AND ALKALOIDS

There are a number of drugs which stimulate the CNS (CNS stimulants), and also there are drugs which lower the function of the CNS (CNS depressants).

Many alkaloids which act as CNS depressants will be discussed in subsequent chapters, and their structures will be shown later. On the other hand, several ethnologically interesting CNS stimulants, such as psychic energizers and hallucinogens in the area of psychotropic drugs, are known. In this section, such alkaloids will be discussed.

Methamphetamine, derived from ephedrine, and cocaine [1], and caffeine are known as CNS stimulants, though the strength of each activity is different. On the other hand, strychnine isolated from the seed of *Strychnos nux-vomica* (Loganiaceae) stimulates the reflex function of the spinal cord. That is to say, it is a stimulant of the spinal cord in the CNS.

In the "Plants of the Gods" (1979) [2] by R.E. Shultes and A. Hofmann (Hofmann was the discoverer of LSD), 14 kinds of hallucinogens are listed (Table 1). Among them, except for *Cannabis* (marihuana), all of the active hallucinogenic constituents described are alkaloids.

For example, the main active component of (1) is muscimol, formed by decarboxylation of ibotenic acid. On the other hand, the plants (2), (5), and (9) are all Solanaceae plants and as the active components of these plants,

Table 1 Kinds of main hallucinogens described in "Plants of the Gods" [1]

1. Mainstay of the heavens	*Amanita* (fly agaric)
2. The hexing herbs	*Atropa* (deadly nightshade), *Hyoscyamus* (henbane), *Mandragora* (mandrake)
3. The nectar of delight	*Cannabis* (marihuana; hashish)
4. St. Anthony's fire	*Claviceps* (ergot)
5. Holy flower of the north star	*Datura* (dhatura; thorn apple; toloache; torna loco)
6. Guide to the ancestors	*Tabernanthe* (iboga)
7. Beans of the hekula spirit	*Anadenanthera* (yopo)
8. Vine of the soul	*Banisteriopsis* (ayahuasca)
9. Trees of the evil eagle	*Brugmansia* (floripondio)
10. The tracks of the little deer	*Lophophora* (peyote)
11. Little flowers of the gods	*Conocybe, Panaeolus, Psilocybe, Stropharia* (teonanacatl)
12. Cactus of the four winds	*Trichocereus* (san pedro)
13. Vines of the serpent	*Ipomoea* (badoh negro; morning glory) *Turbina* (ololiuqui)
14. Semen of the sun	*Virola* (epena')

(−)-scopolamine, (−)-hyoscyamine, and atropine (*dl*-hyoscyamine), all classified as tropane alkaloids, are known.

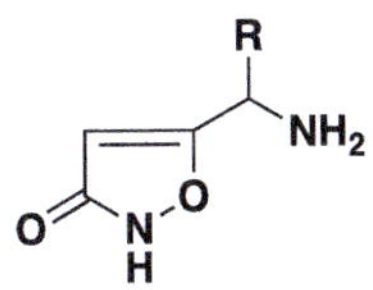

ibotenic acid **R = COOH**
muiscimol **R = H**

From (4), ergot alkaloids, classified as indole alkaloids are isolated. The noted semisynthetic hallucinogen LSD was prepared from lysergic acid derived from these alkaloids. Ibogaine, psilocine, 5-methoxy-*N,N*-dimethyltryptamine and *N,N*-dimethyltryptamine are also hallucinogens which are classified as indole alkaloids.

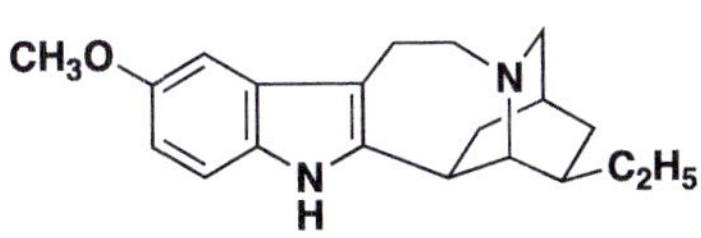

ibogaine

On the other hand, as the active components of (7), indole and β-carboline alkaloids, such as *N,N*-dimethyltryptamine, 5-hydroxy-*N,N*- dimethyltryptamine (bufotenine), and 2-methyl- and 1,2-dimethyl-6-methoxytetrahydro-β-carboline are known. The active components of (8) are also β-carboline-type alkaloids, such as harmine.

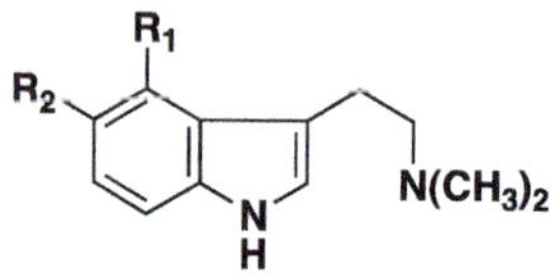

psilocine	$R_1 = OH$, $R_2 = H$
5-methoxy-*N,N*-dimethyltryptamine	$R_1 = H$, $R_2 = OCH_3$
N,N-dimethyltryptamine	$R_1 = R_2 = H$
bufotenine	$R_1 = H$, $R_2 = OH$

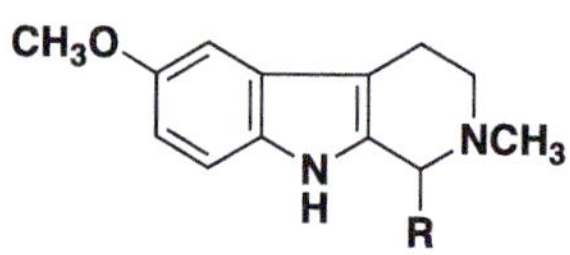

2-methyl-6-methoxytetrahydro-β-carboline	$R = H$
1,2-dimethyl-6-methoxytetrahydro-β-carboline	$R = CH_3$

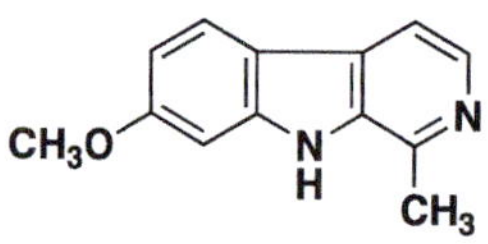

harmine

The simple alkaloid mescaline is the main active component of *L. williamsii* of (10) and of *Trichocereus pachanoi* of (12), and the yield of mescaline is said to be 2% of the dried material. Mescaline is a phenethylamine (=phenylethylamine) type alkaloid, like ephedrine.

Schultes and Hofmann regarded opium, which includes morphine, as a euphoriant, and consequently they distinguished it from the hallucinogens. Thus, opium and its related alkaloids are not included in Table 1. Of course, morphine is one of the most representative alkaloids.

LITERATURE CITED

[1] R.E. Schultes, A. Hofmann, Plants of the Gods, McGraw-Hill Book Company, 1979.
[2] J.L. Phillips, R.D. Wynne, Cocaine: The Mystique and the Reality, Avon Books, 1980.

CHAPTER 1

Alkaloids Derived from Phenylalanine and Tyrosine

Nelumbo nucifera (Nelumbonaceae)

The thousands of alkaloids derived from phenylalanine and tyrosine possess a wide range of important biological activities, and several of them are pharmaceutical agents, present in various traditional medicines in various systems, or serve as biological tools. In this chapter, those alkaloids are discussed, from the simplest alkaloids to those that represent more complex chemical structures. The coclaurine-type alkaloids are one of the simplest of these alkaloids, and are described in Section 1.4. Among them, reticuline, in its antipodal forms, is an important biosynthetic precursor of various alkaloids, including such alkaloids as berberine and morphine. Some of the alkaloids of this type are derived through highly complicated, and incompletely understood, biosynthetic pathways. For example, it is very difficult to elucidate the original amino acid derivations of colchicine (Section 1.11) and lycorine (Section 1.13) from a superficial examination of their chemical structures.

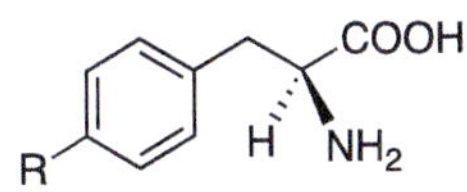

L-phenylalanine **R = H**
L-tyrosine **R = OH**

Both phenylalanine and tyrosine are derived from chorismic acid, which is itself derived from shikimic acid-3-phosphate through the shikimic acid pathway. In this sequence, chorismic acid is first transformed into prephenic acid by chorismate mutase. If prephenic acid is converted into phenylpyruvic acid by

Alkaloids
http://dx.doi.org/10.1016/B978-0-12-417302-6.00001-5

the action of prephenate dehydratase and a transaminase, phenylalanine is formed. On the other hand, if it is transformed into 4-hydroxyphenylpyruvic acid by the action of prephenate dehydrogenase, followed by a transaminase reaction, then tyrosine is formed.

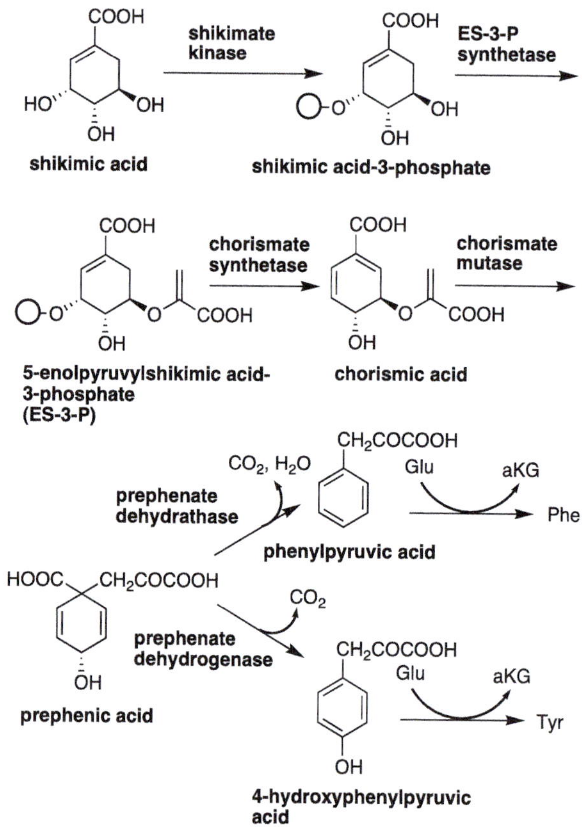

Biosynthetic Pathway of Phenylalanine and Tyrosine

1.1 PHENYLETHYLAMINES (PHENETHYLAMINES)

Peyote (*Lophophora williamsii*, syn. *Anhalonium williamisii*) is a cactus and member of the family Cactaceae, and grows wild in the deserts of Mexico and the southern United States [1]. The cactus is also cultivated in Japan as a decorative plant and known as "Ubatama."

This cactus is an important source of the so-called phenylethylamine (phenethylamine) alkaloids with a C_6C_2N skeleton. The main component of the alkaloid mixture is mescaline. The name mescaline is derived from the name of the cactus, which is also known as "mescal buttons." Mescaline is known to possess hallucinatory effects and a number of undesirable side effects. With respect to the biosynthesis of mescaline, it was shown that L-tyrosine is oxidized to give L-DOPA (L-3,4-dihydroxyphenylalanine), which is transformed to the biosynthetic precursor dopamine (3,4-dihydroxyphenethylamine) [2], which is selectively *O*-methylated to afford 3-*O*-methyldopamine, a key biosynthetic precursor of the alkaloid. The direct biosynthetic precursor of mescaline was determined to be 3,5-dimethoxy-4-hydroxyphenethylamine, because 3,4,5-trihydroxy phenethylamine and 3,4-dimethoxy-5-hydroxyphenethylamine were not incorporated into the biosynthetic pathway to mescaline [2].

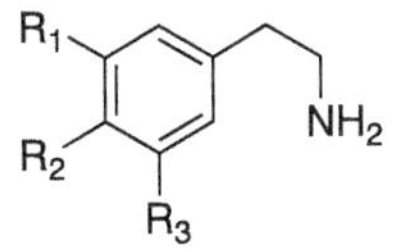

phenethylamine	**$R_1 = R_2 = R_3 = H$**
3,4,5-trihydroxy-phenethylamine	**$R_1 = R_2 = R_3 = OH$**
3,4-dimethoxy-5-hydroxyphenethylamine	**$R_1 = R_2 = OCH_3$, $R_3 = OH$**
3,5-dimethoxy-4-hydroxyphenethylamine	**$R_1 = R_3 = OCH_3$, $R_2 = OH$**
mescaline	**$R_1 = R_2 = R_3 = OCH_3$**

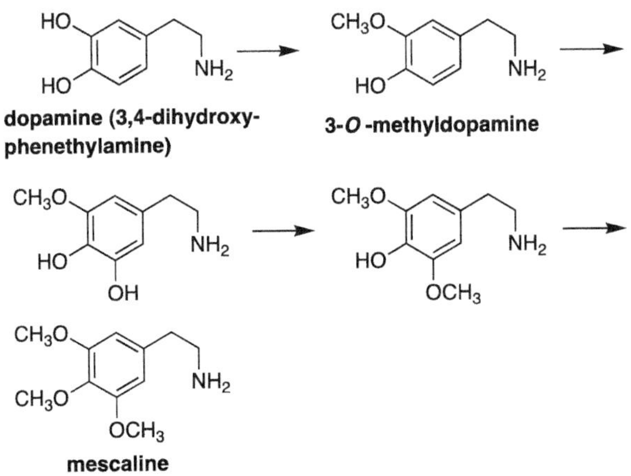

Biosynthetic Route for Mescaline

Hordenine and *N*-methyltyramine are isolates from the young roots of *Hordeum vulgare* var. *hexastichon* (Poaceae), and are simple phenylethylamine-type alkaloids. The biosynthetic precursor of these alkaloids is considered to be tyramine, derived from tyrosine. *dl*-[2-^{14}C]-Tyrosine was fed to *H. vulgare* var. *hexastichon* 4 days after germination, and hordenine and *N*-methyltyramine were isolated after 11 days from the roots. Both alkaloids possessed ^{14}C label at the α-carbon. It was also found that *dl*-[2-^{14}C]-tyrosine was more effectively incorporated into *N*-methyltyramine than into hordenine, and no tyramine was detected in the extract. So, the incorporated tyrosine was converted into tyramine and methylated immediately to give *N*-methyltyramine. Subsequent steps form hordenine by the methylation of *N*-methyltyramine [3].

On the other hand, it was clarified by the incorporation of labeled methionine that the methyl groups incorporated during the biosynthesis of *N*-methyltyramine and hordenine were derived from methionine. The yield of *N*-methyltyramine in these experiments is less than half that of hordenine; however, the incorporation of ^{14}C into *N*-methyltyramine is 1.5 times that of hordenine. This indicates that *N*-methyltyramine is not formed by the demethylation of hordenine, but that *N*-methyltyramine was formed first, and hordenine is then formed by the methylation of the former alkaloid. If *N*-methyltyramine was formed by the demethylation of hordenine, the incorporation rate of ^{14}C of *N*-methyltyramine, which possesses only one *N*-methyl moiety, and is present in lower yield than hordenine, should not be superior to that of hordenine [4].

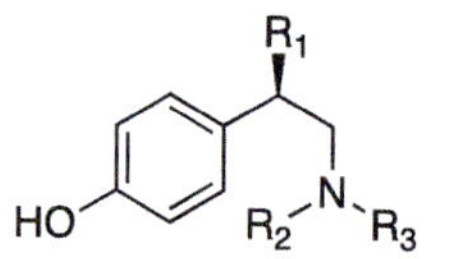

hordenine	**R_1 = H, R_2 = R_3 = CH_3**
tyramine	**R_1 = R_2 = R_3 = H**
***N*-methyltyramine**	**R_1 = R_2 = H, R_3 = CH_3**
synephrine	**R_1 = OH, R_2 = H, R_3 = CH_3**

The dried tubers of *Aconitum* plants (Ranunculaceae) are known as "U-zu" or "Bu-shi" in Japan, and are used in Kampo medicine (formerly regarded as a form of traditional Chinese medicine). The crude drug is known to contain very poisonous aconitine-type alkaloids (Section 14.4). On the other hand, the methanol extract of the tubers of *Anhalonium carmichaeli* showed hypertensive activity against rats by intravenous injection. The active component was identified as the phenylethylamine-type alkaloid coryneine. Coryneine also showed cardiotonic activity against the isolated right atrium of the guinea pig [5].

The dried rinds of the fruits of *Citrus unshiu* and *C. aurantium* var. *daidai* are known as "Chin-pi" or "Toh-hi" in Japan, respectively, and the immature fruits of these plants are called "Ki-jitsu." These are also used in Kampo medicine, and from them, the phenylethylamine-type alkaloids *N*-methyltyramine and synephrine were isolated [6,7].

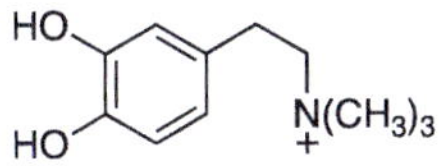

coryneine

LITERATURE CITED

[1] E.F. Anderson, Peyote – The Divine Cactus, The University of Arizona Press, 1980.
[2] A.G. Paul, K.L. Khanna, H. Rosenberg, M. Takido, Chem. Commun. 838 (1969).
[3] E. Leete, L. Marion, Can. J. Chem. 31 (1953) 126.
[4] E. Leete, L. Marion, Can. J. Chem. 32 (1954) 646.
[5] C. Konno, M. Shirasaka, H. Hikino, Planta Med. 35 (1979) 150.
[6] T. Namba, Y. Tuda (Eds.), Shoyakugaku-Gairon, Nankodo Co., Ltd., Tokyo, 1993R, p. 272 (Outline of Pharmacognosy, a Textbook).
[7] F. Kusu, X.-D. Li, K. Takamura, Chem. Pharm. Bull. 40 (1992) 3284.

1.2 L-DOPA AND DOPAMINE

Noradrenaline (NAdr = norepinephrine, NE) is formed by the stereospecific oxidation of the β-carbon of dopamine, which itself is formed by the decarboxylation of L-DOPA (L-3,4-dihydroxyphenylalanine). Epinephrine (Adr) is formed by the *N*-methylation of norepinephrine. Thus, a compound that blocks the enzyme dopa decarboxylase (responsible for the decarboxylation of dopa) causes a decline in the level of catecholamines, such as dopamine and norepinephrine, and hypotensive activity is expected. The absolute stereostructures of norepinephrine and epinephrine were determined by the transformation of each of these alkaloids into *R*-mandelic acid [1].

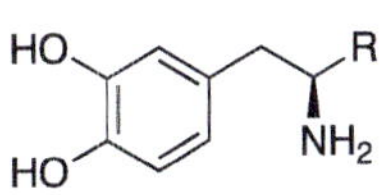

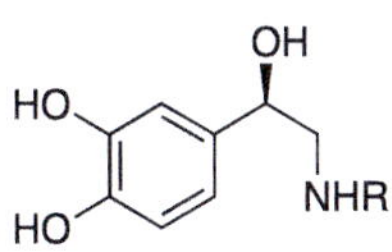

L-DOPA	**R = COOH**	**epinephrine (adrenaline)**	**R = CH_3**
dopamine	**R = H**	**norepinephrine (noradrenaline)**	**R = H**

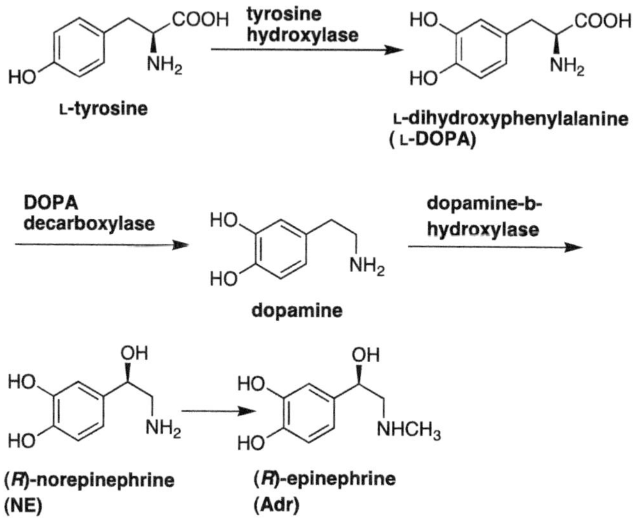

Biosynthetic Route to Norepinephrine (Noradrenaline) and Epinephrine (Adrenaline)

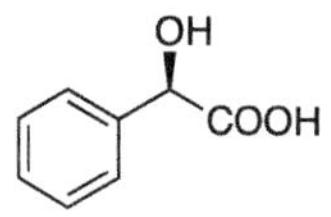

***R*(-)-mandelic acid**

Parkinsonism occurs as a metabolic disorder of the extrapyramidal tract involving dopamine. Namely, to keep the extrapyramidal tract functioning, a balance of the amount of acetylcholine (ACh) and dopamine is needed; in Parkinsonism, dopamine is lacking. Thus, for the treatment of Parkinsonism, medication with anti-acetylcholine agents or with L-DOPA, the precursor of dopamine, in order to increase the amount of dopamine, is effective.

As expected, L-DOPA is also constituent of higher plants. For example, it was isolated from the seedlings and pods of *Vicia faba* (Fabaceae), and its plane structure was determined [2,3]. DL-, D-, and L-DOPA were chemically synthesized [4].

Dopamine was also obtained by the transformation of aminotryptamine and homoveratrylamine [5–7] and isolated from *Hermidium alipes* (Nyctaginaceae) [8]. The pharmacology and the curative effects of dopamine have been reviewed [9].

The black pigments isolated from the peel of the seeds of watermelon and sunflower are melanins formed by the oxidation, cyclization, and polymerization of dopa with the aid of oxidative enzymes. Melanins in the skin and hair of humans and animals, and in the ink of cuttlefish, are formed by the polymerization of alkaloids with indole skeletal, as discussed in the next chapter [10].

LITERATURE CITED

[1] P. Pratesi, A. La Manna, A. Campiglio, V. Ghislandi, J. Chem. Soc. 4062 (1959).
[2] T. Torguati, Arch. Farm. Sper. 15 (1913) 308 [Chem. Abstr., 7, 2774 (1913)].
[3] M. Guggenheim, Z. Hoppe-Seyler's, Physiol. Chem. 88 (1913) 276.
[4] S. Yamada, T. Fujii, T. Shioiri, Chem. Pharm. Bull. 10 (1962) 693.
[5] E. Waser, H. Sommer, Helv. Chim. Acta 6 (1923) 54.
[6] C. Schöpf, H. Bayerle, Liebigs Ann. Chem. 513 (1934) 190.
[7] G. Hahn, K. Stiehl, Chem. Ber. 69 (1936) 2627.
[8] D.W. Buelow, O. Gisvold, J. Am. Pharm. Assoc. 33 (1944) 270 [Chem. Abstr., 38, 53615 (1944)].
[9] J.F. Dasta, M.G. Kirby, Pharmacotherapy 6 (1986R) 304.
[10] R.A. Nicolaus, M. Piattelli, E. Fattorusso, Tetrahedron 20 (1964) 1163.

1.3 THYROID GLAND AND THYROXINE

The thyroid is an endocrine gland that is located in the upper part of the trachea and the front of the throat, and its shape is a flat H or horseshoe. In humans, this gland weighs about 20–23 g. Compounds that are effective against the symptoms due to the loss of the thyroid gland are known as thyroid hormones. Among them are L-thyroxine (L-3,5,3′,5′-tetra-iodothyronine $C_{15}H_{11}NO_4I_4$, mw 777) and L-liothyronine (L-3,5,3′-tri-iodothyronine, $C_{15}H_{12}NO_4I_3$, mw 651) [1]. Both of these alkaloids possess iodine in the molecule and are stored in the follicle as thyroglobulins (protein).

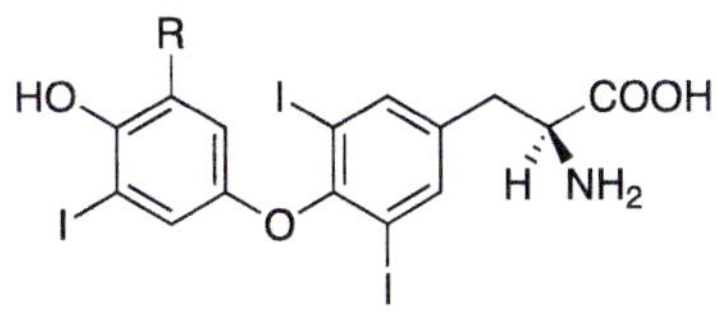

L-thyroxine **R = I**
L-liothyronine **R = H**

L-Thyroxine was isolated in 1919 by Kendall from the thyroid gland and its molecular formula was reported as $C_{11}H_{10}NO_3I_3$. On the other hand, Harington obtained it from the dried thyroid gland of +++++++ in a yield of 0.125%, and reported the molecular formula as $C_{15}H_{11}NO_4I_4$ [2].

The D,L-form of this alkaloid was synthesized [2,3]. The natural form of thyroxine is L, and the D -form is only obtained by chemical synthesis. The biological activity of D-thyroxine is very weak, and, as expected, the hormonal activity of D,L-thyroxine is about half that of the L-thyroxine.

L-Liothyronine was found subsequently to L-thyroxine, although its hormonal activity is said to be five times stronger than that of L-thyroxine [1]. L-Liothyronine was also obtained by the iodination of 3,5-diiodothyronine [4].

It is thought that L-thyroxine is formed by the electrophilic iodination of L-tyrosine followed by dimerization and loss of the side chain with the mediation of PLP (pyridoxal phosphoric acid).

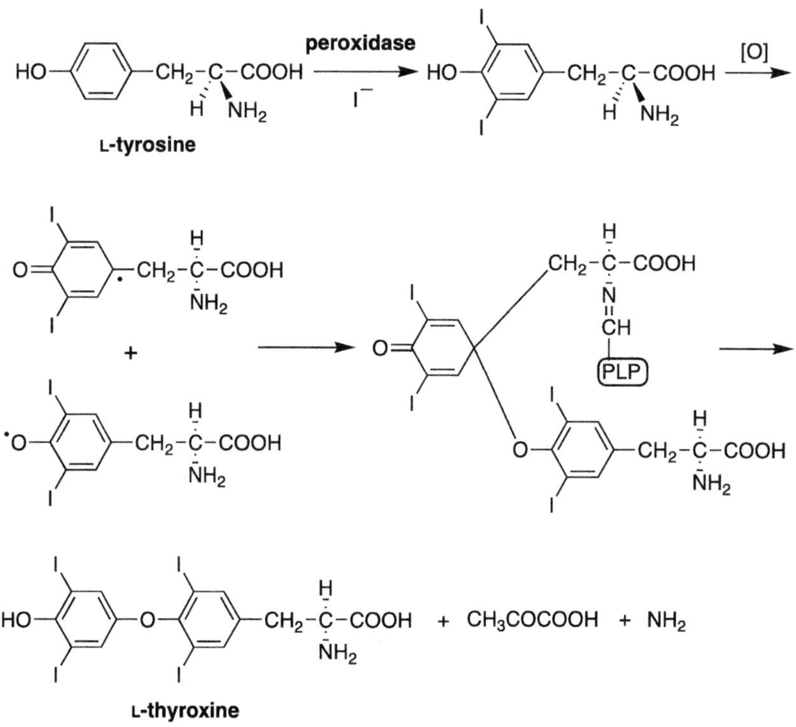

Biogenesis of L-Thyroxine

LITERATURE CITED

[1] J. Gross, R. Pitt-Rivers, Biochem. J. 53 (1953R) 645.
[2] C.R. Harington, Biochem. J. 20 (1926) 293.
[3] C.R. Harington, G. Barger, Biochem. J. 21 (1927) 169.
[4] J. Roche, S. Lissitzky, R. Michel, Biochem. Biophys. Acta 11 (1953) 215.

1.4 COCLAURINE AND COCLAURINE-TYPE ALKALOIDS

Cocculus laurifolius (Menispermaceae) is an evergreen tree that grows wild in the southern area of Kyushu Island in Japan. The dried roots of this plant are known as "Kohshu-Uyaku" in Japan, and are used in traditional Chinese medicine as an anthelmintic and a diuretic. From this crude drug, coclaurine [1] and coclanoline [2] were isolated, and are classified as benzylisoquinoline alkaloids. Benzylisoquinoline alkaloids are one of the most typical alkaloids derived from phenylalanine/tyrosine. Coclaurine showed convulsive toxicity and the minimum lethal dose for rabbits is 70 mg/kg (i.v.) [3]. From this crude drug, aporphine-type alkaloids, such as laurifoline [4] and magnoflorine [5], which are discussed in the following section, were also isolated.

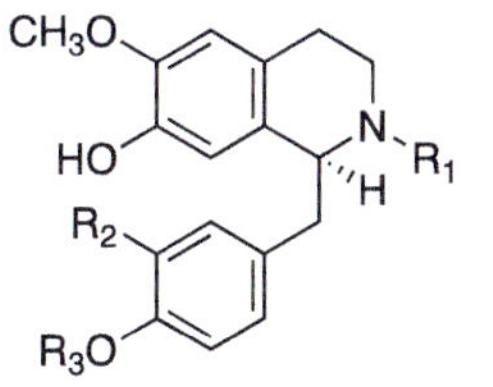

coclaurine $R_1 = R_2 = R_3 = H$
coclanoline $R_1 = R_3 = CH_3, R_2 = OH$

It is reported that the extract of the tubers of *Aconitum* plants showed cardiotonic activity, and, as already discussed, coryneine, which is classified as a phenylethylamine-type alkaloid, showed cardiotonic activity against the isolated guinea pig heart (Section 1.1). From the Kampo medicine "Bushi" prepared from the tubers of *Aconitum* sp., higenamine (DL-demethylcoclaurine) was isolated [6]. DL-demethylcoclaurine was already synthesized at that time [7].

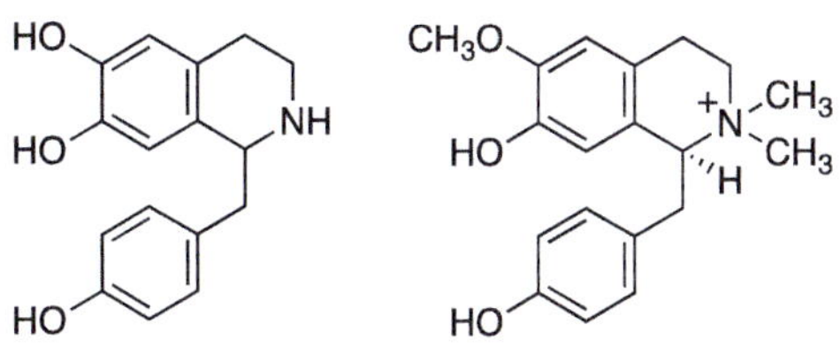

higenamine **magnocurarine**

Dried embryos of the matured fruits of *Nelumbo nucifera* (Nymphaeaceae) are known as "Renjishin" in Japan, and from this crude drug,

R(+)-demethylcoclaurine, which possesses the same plane chemical structure as that of higenamine, was isolated as a smooth muscle and uterine relaxant constituent [8]. According to Kosuge and Yokota [6], it is known that catecholamines with this type of structure also show adrenaline β-type stimulant activity, and that the activity is stronger in the *S* series than in the *R* series. The strong cardiotonic activity of higenamine isolated from the roots of *Aconitum* sp. is explained by the fact that this alkaloid is in the racemic *dl*-form, and thus half of it is *S*(−)-demethylcoclaurine.

The dried stem bark material of *Magnolia obovata* (Magnoliaceae) is known as "Koh-boku." It is an important crude drug in Kampo medicine, used for the treatment of stomach ache and rush of the blood to the head, etc. From this crude drug, magnocurarine, classified as a benzylisoquinoline alkaloid, was isolated [9], in addition to sesquiterpenoids, such as α-eudesmol, and phenylpropanoids, such as magnolol.

Papaver somniferum (Papaveraceae) is a very important plant in the world, and is famous as the source material for "Ahen (Opium)," which is discussed subsequently (Section 1.10). From opium, papaverine and related alkaloids belonging to the benzylisoquinoline alkaloids were isolated, in additon to morphine and codeine, which belong to the morphine alkaloid series. Papaverine possesses smooth muscle relaxant activity and is used clinically.

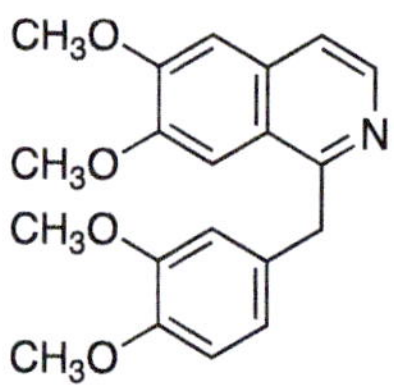

papaverine

Alkaloids possessing the 1-benzylisoquinoline skeleton are biosynthesized from two molecules of tyrosine, which are differentiated at the beginning of the biosynthetic pathway. Namely, the isoquinoline moiety, except for one carbon, originates from 3′,4′-dihydroxyphenylethylamine (dopamine), which is formed from tyramine following the decarboxylation of tyrosine. The other part of the alkaloid is derived from 4′-hydroxyphenylacetaldehyde, which is formed from tyrosine via 4′-hydroxyphenyl pyruvic acid. These two compounds are coupled stereoselectively to give (*S*)-norcoclaurine. (*S*)-Norcoclaurine is transformed into (*S*)-coclaurine by 6-*O*-methylation, and further *N*-methylation, hydroxylation at the

3′-position, followed by 4′-*O*-methylation of (*S*)-coclaurine affords (*S*)-reticuline [10–12]. (*S*)-Reticuline is the common biosynthetic precursor of the aporphine, morphine, protoberberine, and benzophenanthiridine alkaloids, as discussed later.

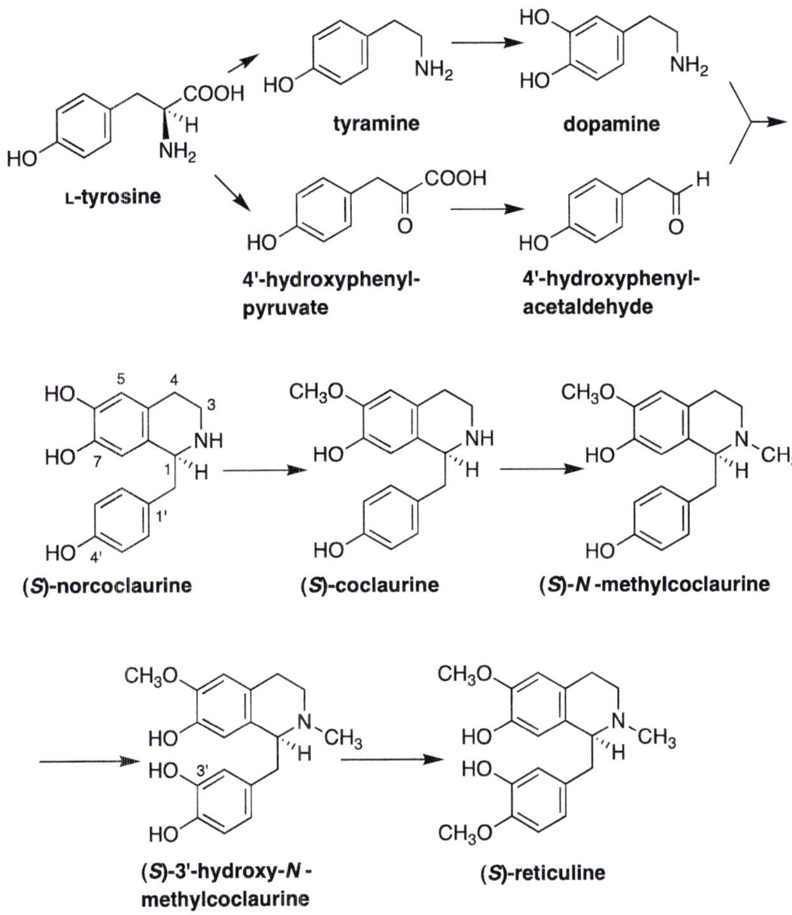

Biosynthetic Route of (*S*)-Reticuline

LITERATURE CITED

[1] M. Tomita, F. Kusuda, Yakugaku Zasshi 72 (1952) 280.
[2] F. Kusuda, Chem. Pharm. Bull. 1 (1953) 189.
[3] S. Tsuruta, Nihon Yakubutsugaku Zasshi 2 (1926) 269.
[4] M. Tomita, F. Kusuda, Chem. Pharm. Bull. 1 (1953) 5.

[5] T. Nakano, M. Uchiyama, Chem. Pharm. Bull. 4 (1956) 407.
[6] T. Kosuge, M. Yokota, M. Nagasawa, Yakugaku Zasshi 98 (1978) 1370.
[7] H. Yamaguchi, Yakugaku Zasshi 78 (1958) 692.
[8] H. Koshiyama, H. Ohkuma, H. Kawaguchi, H.-Y. Hsu, Y.-P. Chen, Chem. Pharm. Bull. 18 (1970) 2564.
[9] T. Nakano, Chem. Pharm. Bull. 3 (1955) 234.
[10] R. Stadler, T.M. Kutchan, S. Loeffler, N. Nagakura, B. Cassels, M.H. Zenk, Tetrahedron Lett. 28 (1987) 1251.
[11] R. Stadler, T.M. Kutchan, M.H. Zenk, Phytochemistry 28 (1989) 1083.
[12] R. Stadler, M.H. Zenk, Liebigs Ann. Chem. 555 (1990).

1.5 TUBOCURARE AND *d*-TUBOCURARINE

Many indigenous groups in the South American rain forests of the Amzon basin use blowguns with blowpipe darts poisoned at the tip to paralyze hunted game. The poisonous material is known as curare (which means "poison" in several local languages). There are three kinds of curare which are named by the difference in the containers used to carry them; these are (1) tubo curare (tube of bamboo curare), (2) pot curare, and (3) calabash curare, and are described below.

1. Tubocurare is used in the Amazon basin, and is kept in a container known as a tubo or tube. The poison comprises extracts of the bark of *Chondodendron tomentosum* and *C. platyphyllum* (Menispermaceae), which grow wild in Brazil. The toxic principle is the biscoclaurine-type alkaloid *d*-tubocurarine (*d*-Tc), which is described later in this section.

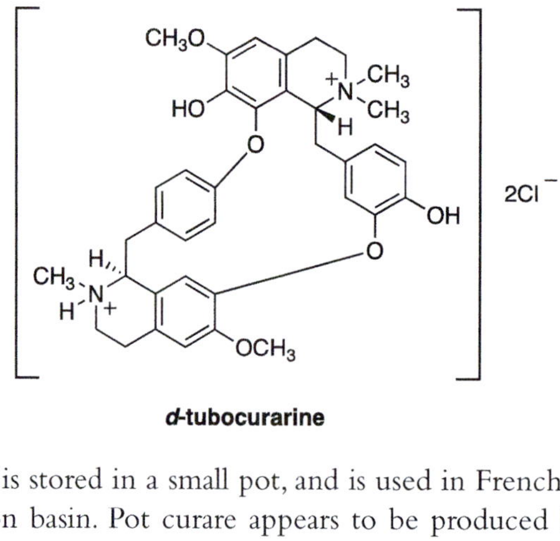

***d*-tubocurarine**

2. Pot curare is stored in a small pot, and is used in French Guiana and in the Amazon basin. Pot curare appears to be produced by extracting a

mixture of *Strychnos* (Loganiaceae) plants (such as *S. castelnaei*) and the previously mentioned *Chondodendron* plants, such as *C. tomentosum*.

3. Calabash curare is produced from *Strychnos toxifera* and related plants of the same genus [1]. Calabash curare is named after its container, and is used by the indigenous peoples of the Rio-Negro river, and in the upper reaches of the Orinoco river.

As described previously, the origins of the three types of curare are the extracts of *Chondodendron* (Menispermaceae) and/or *Strychnos* (Loganiaceae) plants. Among them, the toxic components of *Strychnos* plants are C-curarine and C-toxiferine I, etc. These constituents are alkaloids derived from tryptophan and are described in the next (Section 2.15). The "C-" is the initial of calabash. On the other hand, the toxic principle of tubocurare is *d*-tubocurarine, an alkaloid derived from phenylalanine. The alkaloid was first isolated as a hydrochloride from tubocurare and is preserved in the Museum of the British Society of Pharmacy [1]. The chemical structure of *d*-tubocurarine was first proposed as a bisbenzylisoquinoline with two quaternary ammonium moieties. This was revised in 1970 to be the structure with one tertiary and one quaternary amine after X-ray crystallographic analysis [2].

d-Tubocurarine shows competitive antagonist activity against acetylcholine, which is released from the end of the motor nerve as an excitation transmitting compound; blocking the transmission relaxes the muscle. The muscle-relaxing activity of *d*-tubocurarine is useful during surgery and also to prevent the convulsions caused by strychnine poisoning. Curare is also used in the field of psychiatry to prevent convulsions during shock therapy, and for convulsive diseases, such as tetanus and hydrophobia. On the other hand, *d*-tubocurarine has also served as a pharmacophoric template for synthetic anticonvulsants, such as decamethonium and suxamethonium, and hypotensive drugs, such as hexamethonium.

$$\underset{X^-}{(CH_3)_3N^+}-(CH_2)_n-\underset{X^-}{N^+(CH_3)_3}$$

decamethonium $n = 10$
hexamethonium $n = 6$

$$\underset{X^-}{(CH_3)_3N^+}-CH_2CH_2OCOCH_2CH_2COOCH_2CH_2-\underset{X^-}{N^+(CH_3)_3}$$

suxamethonium

Pharmacological studies of curare were initiated in the middle of the nineteenth century by the French physiologist Claude Bernard (1813–1878). The dominant pharmacological effect of curare is the paralysis of the whole skeletal muscle in both warm- and cold-blooded animals. The sequence of the paralytic effects of the skeletal muscle is well defined. At first, the eyes, ears, and toes are affected, and next the muscles of the arms and legs. Paralysis of the muscles of the neck then occurs, and finally the respiratory muscles are affected, resulting in suffocation in warm-blooded animals. The muscle relaxant activity of curare is strengthened by diethylether. Because curare can release stored histamine in the tissues, it may cause hypotension and excessive secretion in the trachea.

Curare is not absorbed promptly from the digestive organs, and when it is absorbed, it is metabolized in the liver. Thus, curare can be used as an arrow poison to kill game for food. Curare is used as an injection in both animal experiments and clinical use.

There are more than 350 bisbenzylisoquinoline alkaloids now known that typically possess one or more ether linkage(s), and one or more phenyl–phenyl linkages between the two benzylisoquinoline molecules. Although it is considered that these alkaloids are formed by various forms of the oxidative coupling between the two benzylisoquinoline moieties, their biosynthesis is not well studied.

LITERATURE CITED

[1] H. King, J. Chem. Soc. 1381 (1935).
[2] A.J. Everett, L.A. Lowe, S. Wilkinson, Chem. Commun. 1020 (1970).
[3] K.W. Bentley, The Isoquinoline Alkaloids, Harwood Academic Publishers, Amsterdam, 1998. 59.

1.6 APORPHINE-TYPE ALKALOIDS

Aporphine-type alkaloids are formed by the intramoleculer oxidative coupling of the benzylisoquinoline alkaloid, *S*-reticuline. Thus, from the *ortho*, *ortho'*-, and *ortho*, *para'*-intramolecular coupling of the biradicals formed from *S*-reticuline, bulbocapnine- and isoboldine-type aporphine alkaloids are formed, respectively. When *para*, *ortho'*-intramolecular coupling of the biradicals formed from *S*-reticuline occurs, the morphine skeleton, described in Section 1.10, is formed. On the other hand, nor-protosinomenine is the biosynthetic precursor of boldine, glaucine, and dicentrine in *Dicentra eximia* (Fumariaceae). In this case, the biradical formed from norprotosinomenine is changed into a dienone by *para, para'*-coupling, followed by a rearrangement reaction to produce these alkaloids [1].

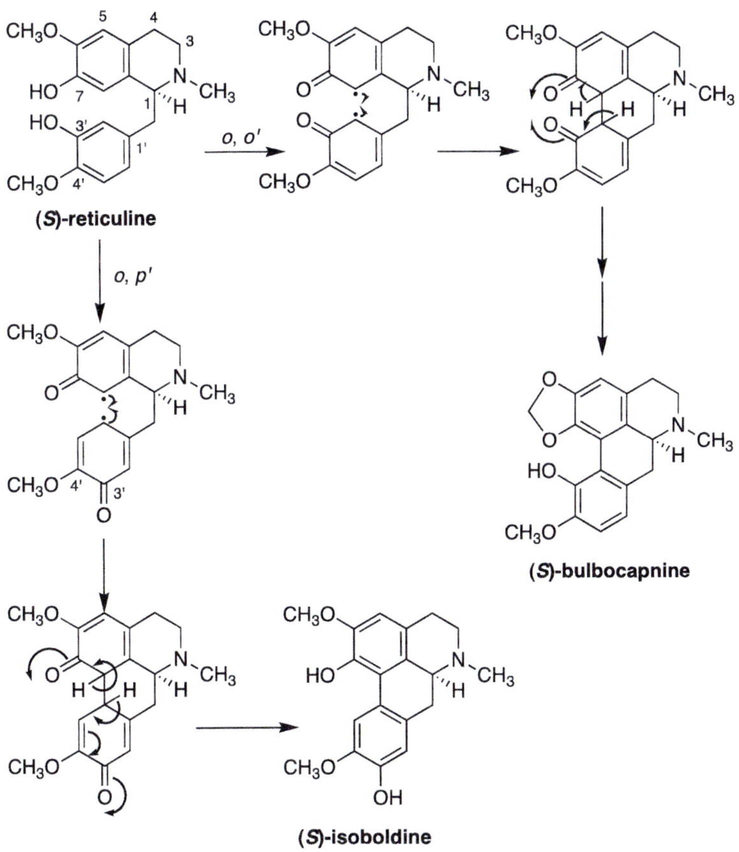

Biosynthetic Route from (*S*)-Reticuline to (*S*)-Bulbocapnine and (*S*)-Isoboldine

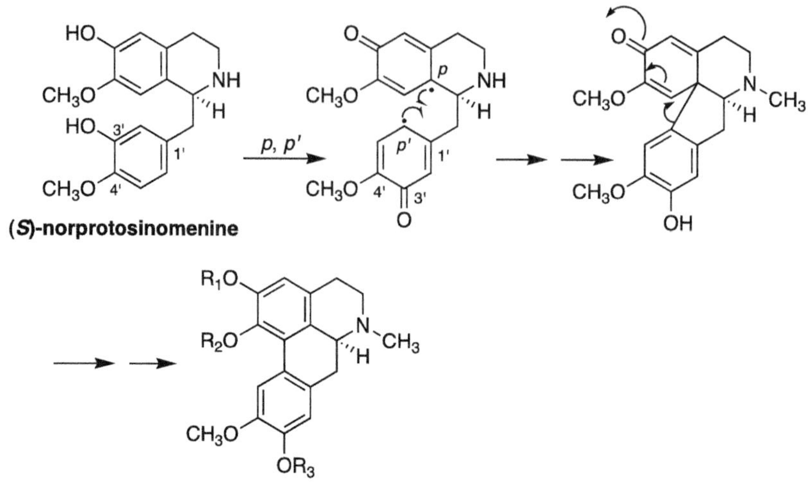

(S)-boldine $R_1 = R_3 = H, R_2 = CH_3$
(S)-glaucine $R_1 = R_2 = R_3 = CH_3$
(S)-dicentrine $R_1, R_2 = \text{-}CH_2\text{-}, R_3 = CH_3$

Biogenesis of (S)-Norprotosinomenine to (S)-Boldine, (S)-Glaucine and (S)-Dicentrine

There are more than 700 aporphine alkaloids isolated, and they have been reviewed by Guinaudeau et al. [2–6]. Typical examples of this type of alkaloid are magnoflorine, isolated from the stems and leaves of *Epimedium macranthum* var. *violaceum* (Berberidaceae), domesticine from the fruits of *Nandina domestica* (Berberidaceae), nuciferine from the seeds of *N. nucifera* (Nymphaeaceae), and bulbocapnine isolated from the tubers of *Corydalis decumbens* (Fumariaceae).

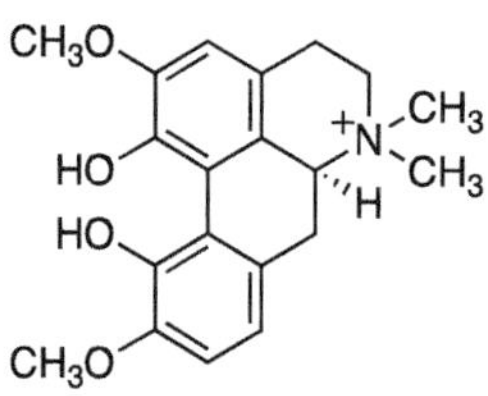

(S)-magnoflorine

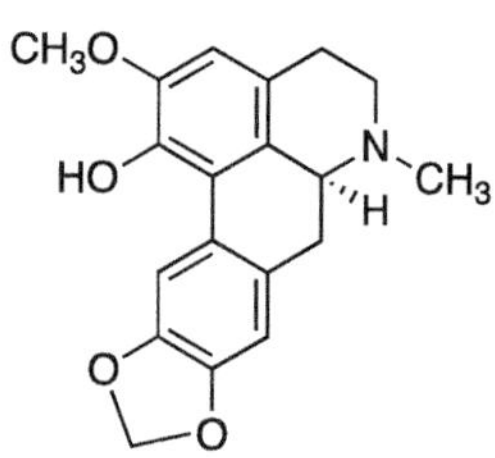

(S)-domesticine

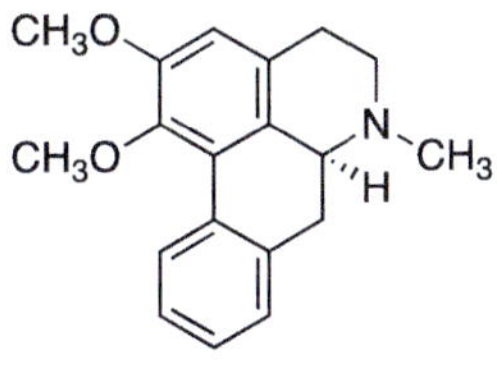

(*S*)-nuciferine

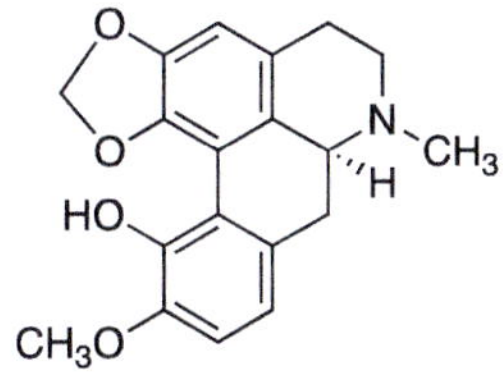

(*S*)-bulbocapnine

LITERATURE CITED

[1] A.R. Battersby, J.L. McHugh, J. Staunton, M. Todd, Chem. Commun. 985 (1971).
[2] H. Guinaudeau, M. Leboeuf, A. Cave, J. Nat. Prod. 38 (1975R) 275.
[3] H. Guinaudeau, M. Leboeuf, A. Cave, J. Nat. Prod. 42 (1979R) 325.
[4] H. Guinaudeau, M. Leboeuf, A. Cave, J. Nat. Prod. 46 (1983R) 761.
[5] H. Guinaudeau, M. Leboeuf, A. Cave, J. Nat. Prod. 51 (1988R) 389.
[6] H. Guinaudeau, M. Leboeuf, A. Cave, J. Nat. Prod. 57 (1994R) 1033.

1.7 *Aristolochia* SPP. AND ARISTOLOCHIC ACID

Aristolochia debilis (Aristolochiaceae) is a perennial vine that grows wild in the plains and on the banks of the rivers in Japan. The dried roots are called "Sei-mokkoh" or "Do-sei-mokkoh," and the dried fruits are called "Ba-to-rei," and are used in Kampo medicine. The dried roots are used as an anti-pyretic, and the dried fruits as a sedative, antitussive, and to clear the throat. In Korea and China, *Anhalonium contorta* and *Anhalonium kaempferi* are used instead of *A. debilis*.

The dried roots of this plant contain, other than magnoflorine described in Section 1.6, aristolochic acid-I, which possesses a very rare nitro moiety [1]. From other plants in the genus *Aristolochia*, other than aristolochic acid-I, at least 14 related alkaloids, each of them possessing a nitro moiety, such as aristolochic acids II–IV, were isolated [2]. These alkaloids are often isolated with their corresponding lactams (aristolactams). As described subsequently, the aristolactams can be easily formed from the aristolochic acids, and these lactams might be formed during the extraction process. Aristolochic acid-I was synthesized by the application of a photochemical reaction [3].

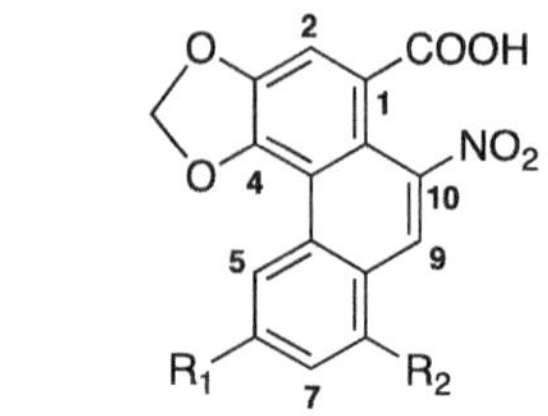

aristolochic acid I	**$R_1 = H, R_2 = OCH_3$**
aristolochic acid II	**$R_1 = R_2 = H$**
aristolochic acid III	**$R_1 = OCH_3, R_2 = H$**
aristolochic acid IV	**$R_1 = R_2 = OCH_3$**

When aristolochic acid-I is reduced under usual conditions (for example: $LiAlH_4$/THF, reflux for 2 h), aristololactam-I is formed. However, if this alkaloid is dissolved in 1% aq. ammonia and $NaBH_4$ is added, aristolochic acid-I loses its nitro moiety, and aristolic acid is obtained in more than 90% yield [4].

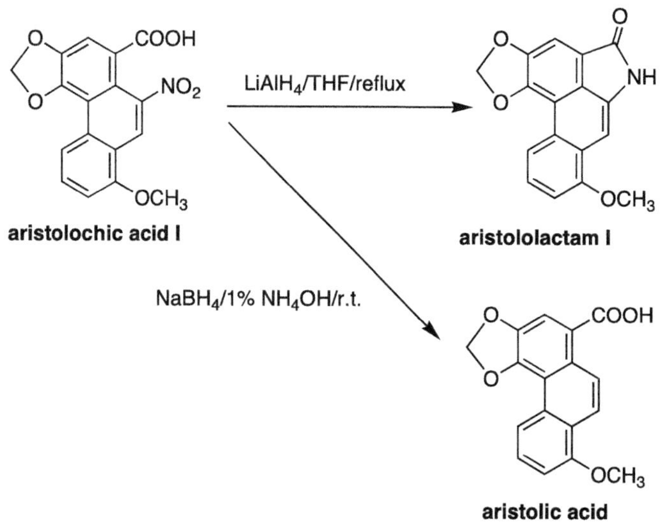

Hydrogenation of Aristolochic Acid I

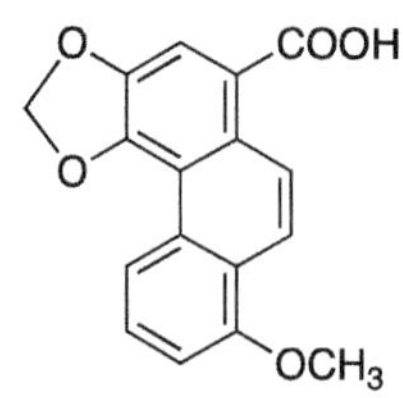

aristolic acid

The acute toxicities of aristolochic acid-I (LD_{50}, i.v.) are as follows: male mice (38.4 mg/kg), female mice (70.1 mg/kg), male rats (82.5 mg/kg), and female rats (74.0 mg/kg). On the other hand, the LD_{50} values of this alkaloid p.o. are the following: male mice (55.9 mg/kg), female mice (106.1 mg/kg), male rats (203.4 mg/kg), and female rats (183.9 mg/kg), respectively [5].

It is considered that these alkaloids are biosynthesized from the aporphine-type alkaloids (Section 1.6). Experimentally, various precursors of aporphine-type alkaloids labeled with ^{14}C were fed to *Anhalonium sifo* and the formation of aristolochic acid-I was examined [6]. According to the results, 60% of the radioactivity was lost by the decarboxylation of the labeled aristolochic acid-I obtained by feeding (3-^{14}C)DL-tyrosine (Figure a). On the other hand, no radioactivity was lost by the decarboxylation of the labeled aristolochic acid-I obtained by feeding (2-^{14}C)DL-3′,4′-dihydroxyphenylalanine (Figure b). Furthermore, in the case of the nitrophenanthrene formed after decarboxylation of the labeled aristolochic acid-I obtained after feeding (2-^{14}C)3′,4′-dihydroxyphenylethylamine, all of the radioactivity was lost (Figure a).

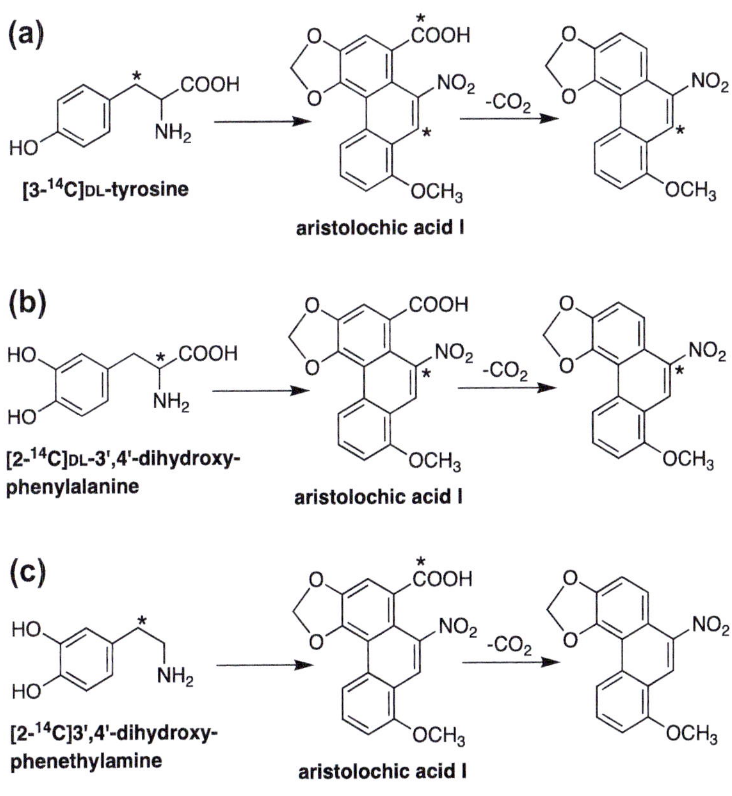

Incorporation of ^{14}C Labeled Compounds into Aristolochic Acid I

From these observations, it was shown that aristolochic acid-I is formed from two molecules of tyrosine, and that, during the biosynthesis of this alkaloid, 3′,4′-dihydroxyphenylalanine became the origin of the benzene ring with a methoxyl moiety and two carbons and a nitrogen, namely, the C_6–C_2–N unit of this alkaloid. On the other hand, 3,4-dihydroxyphenylethylamine, derived from tyrosine, was the origin of the aromatic ring with a methylenedioxy moiety and a carbon of the attached carboxylic acid of aristolochic acid-I. Because C-2 of 3′,4′-dihydroxyphenylethylamine is the origin of the carbonyl carbon, it is lost during the decarboxylation of labeled aristolochic acid-I to give the unlabeled nitrophenanthrene.

LITERATURE CITED

[1] M. Tomita, S. Kura, Yakugaku Zasshi 77 (1957) 812.
[2] D.B. Mix, H. Guinaudeau, M. Shamma, J. Nat. Prod. 45 (1982R) 657.
[3] S.M. Kupchan, H.C. Wormser, J. Org. Chem. 30 (1965) 3792.
[4] S. Mukhopadhyay, S. Funayama, G.A. Cordell, H.H.S. Fong, J. Nat. Prod. 46 (1983) 507.
[5] S. Budavari (Ed.), The Merck Index, twelfth ed, Merck and Co., Inc., Whitehouse Station, NJ, 1996R, p. 133.
[6] I.D. Spenser, H.P. Tiwari, Chem. Commun. 55 (1966).

1.8 *Phellodendron amurense* AND BERBERINE

Phellodendron amurense is a deciduous tree that grows wild in Japan, Korea, and the northern areas of China, and often grows to be 25 m high and 1 m in diameter.

The dried bark of this tree is known as "Ohbaku" and is used in Kampo medicine as a tonic for the stomach, as a medicine for intestinal disorders, and as an antiinflammatory and antipyretic agent. The crude drug has been used for a long time and is mentioned in the "Shinno-Honzo-Kyo," which was written in ancient China (the original is said to be written in around 480 AD, but no longer exists). An extract of this crude drug is also used as a constituent in local medicines for stomach ache in Japan, and these medicines are produced in various places under names such as "Daranisuke," "Hyakuso," and "Nerikuma."

The active constituent of *Papaver amurense* is berberine, and it is usually isolated as berberine chloride. Pills containing berberine chloride are marketed as a tonic for stomach and intestinal disorders in Japan.

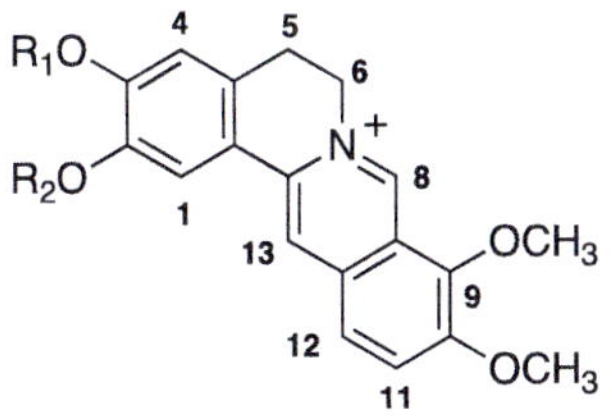

berberine	R_1, R_2 = -CH_2-
palmatine	$R_1 = R_2 = CH_3$
jateorrhizine	$R_1 = CH_3, R_2 = H$
columbamine	$R_1 = H, R_2 = CH_3$

Berberine is also obtained from *Hydrastis canadensis* (Berberidaceae), *Coptis japonica* (Ranunculaceae), and several other species [1]. Its chemical structure was clarified at the beginning of the twentieth century [2], and the alkaloid was synthesized by several routes [3]. Regarding the many biological activities of berberine, other than antimicrobial activity against *Staphylococcus aureus*, dysentery *bacillus*, cholera *vibrio*, and gonococcus organisms, hypertensive, sedative, anti-inflammatory [4], and cytotoxic [5,6] activities are reported.

Alkaloids of *P. amurense* with the same skeleton as berberine, for example palmatine and jatrorrhizine, were also isolated. Magnoflorine, described previously (Section 1.6), is also obtained from this plant. Palmatine and jatrorrhizine are, together with columbamine, the main constituents of *Jateorhiza columba* (Menispermaceae). *Jateorhiza columba* is a large vine that grows wild in the forests of the east coast of Africa, and it possesses many underground tubers. The cut and dried tubers of this plant are known as Columbo-roots and are used as a stomach tonic having a bitter taste.

The biosynthetic precursor of berberine is (*S*)-reticuline, the biosynthesis of which was reviewed in Section 1.4. First, (*S*)-reticuline is converted into (*S*)-scoulerine, and the berberine bridge enzyme responsible for this step has been purified from the cultivated cells of *Berberis beaniana* (Berberidaceae) [7]. Through enzymatic reaction, 1 mol of oxygen is digested, and when the *N*-methyl moiety of (*S*)-reticuline is labeled with ^{3}H (★ in the Figure), H_2O labeled with ^{3}H is obtained.

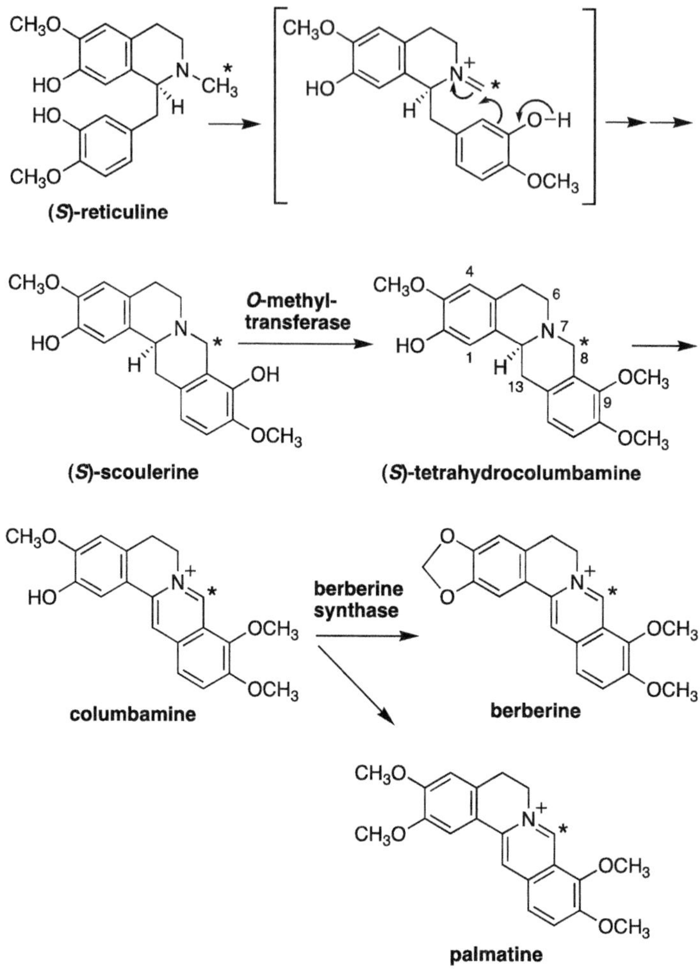

Biosynthetic Route of Berberine and Related Alkaloids

The hydroxyl moiety at the C-9 position of (*S*)-scoulerine is methylated by the action of an *O*-methyltransferase to give (*S*)-tetrahydrocolumbamine [8]. Oxygen (1.5 moles) is needed when the tetrahydroberberine-type alkaloid is transformed to columbamine, a protoberberine-type alkaloid, and it was clarified that a flavine enzyme which produces 1 mol each of H_2O_2 and H_2O is involved in this reaction. Also, it was shown that the 7,14-dehydroberberinium ion was involved as a biosynthetic intermediate [9].

Berberine synthase converts columbamine into berberine, which possesses a methylenedioxy moiety between the C-2 and C-3 positions. The enzyme possesses a Fe^{2+} ion and is blocked by cyanide ion [10]. On the other hand, an enzyme that produces palmatine by methylating columbamine, was isolated from the cultured cells of *Berberis wilsoniae* var. *subcaulioalata* and *B. aggregata* [11]. It was also reported by the same group that the methylation process occurs at the last stage, namely after columbamine was formed from (*S*)-tetrahydrocolumbamine.

LITERATURE CITED

[1] R. Wasicky, M. Joachimotz, Arch. Pharm. 255 (1917R) 497 [Chem. Abstr., 14, 999 (1920)].
[2] W.H. Perkin, R. Robinson, J. Chem. Soc. 97 (1910) 305.
[3] T. Kametani, I. Noguchi, K. Saito, S. Kaneda, J. Chem. Soc. (C) 2036 (1969).
[4] H. Fujimura, T. Sawada, M. Goto, Yakugaku Zasshi 90 (1970) 782.
[5] Y. Kondo, H. Suzuki, Shoyakugaku Zasshi 45 (1991) 35.
[6] S. Funayama, N. Haga, K. Kurumada, S. Nozoe, M. Hayashi, K. Komiyama, Nat. Med. 50 (1996) 427.
[7] P. Steffens, N. Nagakura, M.H. Zenk, Phytochemistry 24 (1985) 2577.
[8] S. Muemmler, M. Rueffer, N. Nagakura, M.H. Zenk, Plant Cell Rep. 3 (1985) 26.
[9] M. Amann, N. Nagakura, M.H. Zenk, Tetrahedron Lett. 25 (1984) 953.
[10] M. Rueffer, M.H. Zenk, Tetrahedron Lett. 26 (1985) 201.
[11] M. Rueffer, M.H. Zenk, Tetrahedron Lett. 27 (1986) 923.

1.9 *Chelidonium majus* AND CHELIDONINE

Chelidonium majus (Papaveraceae) is a biennial plant that is widespread in the temperate zones of Asia and is also found widely in Japan in sunny locations. The dried, above-ground part of this plant in the flowering season is known as "Hakkutsu-sai" in Japan, and was formerly used as an analgesic, cough medicine, diuretic, and for detoxication. In traditional medicine, the pressed juice of the fresh leaves of *Chelidonium majus* is used for the treatment of edemas and warts, and for the bites of insects and snakes.

The whole plant material contains alkaloids, and one of the main alkaloids is (+)-chelidonine [1,2]. The chemical structure of this alkaloid was deduced about 100 years after the first report of its isolation [3–5]. The absolute chemical structure was determined by X-ray crystallography of the 4-bromobenzoate derivative [6].

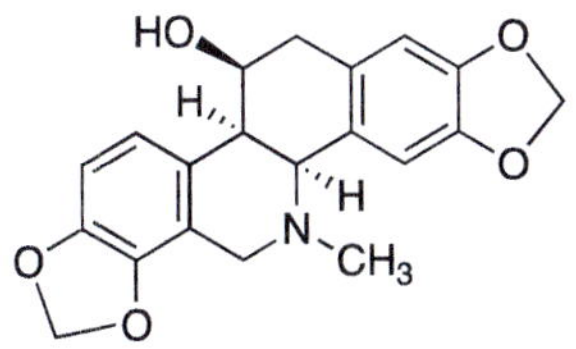

chelidonine

Experiments on the biosynthesis of this alkaloid using ^{14}C-labeled precursors showed that chelidonine was formed through *N*-methylstylopine derived from *S*-reticuline, an important biosynthetic precursor of the isoquinoline-type alkaloids [7,8], and (*S*)-stylopine, the biosynthetic precursor of (*S*)-*N*-methylstylopine, was isolated from this plant. Although chelidonine possesses an isoquinoline moiety, as shown in the Figure, the isoquinoline moiety is not formed directly through the incorporation from (*S*)-reticuline, but is newly formed, and is derived from the berberine bridge carbon after rearrangement. These biosynthetic routes were clarified by isolating each of the enzymes in the pathway [9,10]. Alkaloids possessing the chelidonine skeleton are known as benzophenanthridine alkaloids.

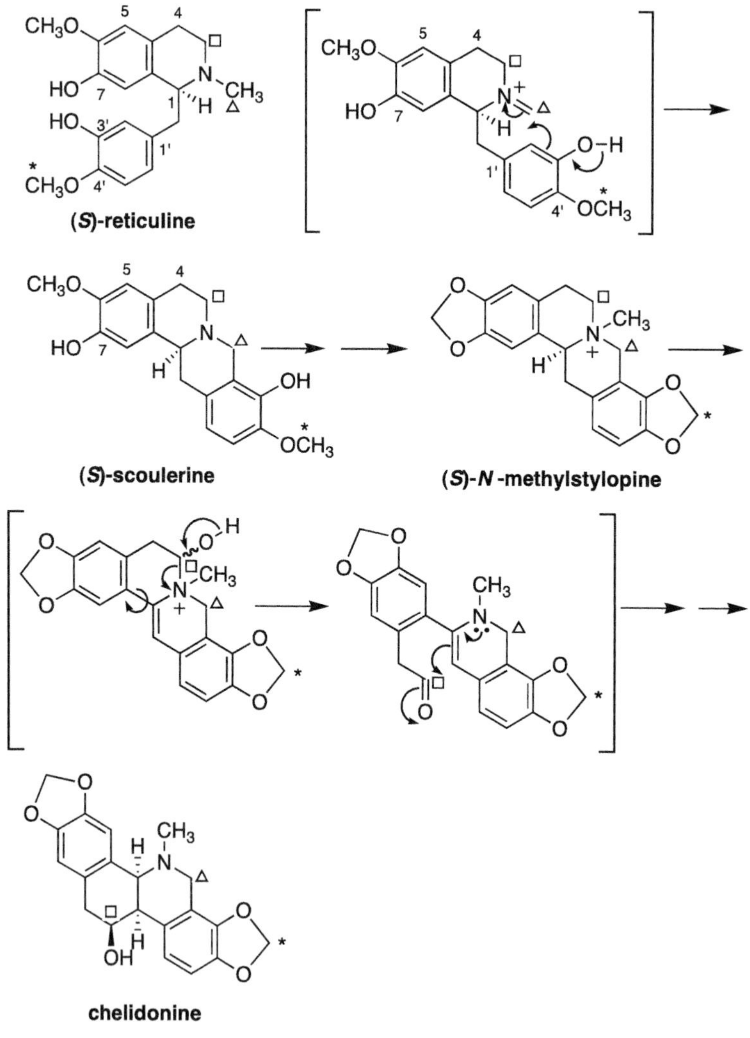

Biosynthesis of Chelidonine

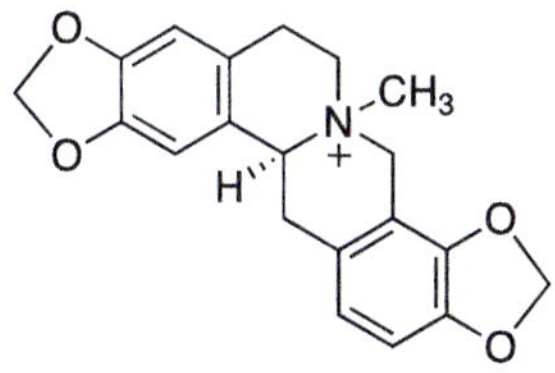

(S)-N-methylstylopine

(S)-stylopine

LITERATURE CITED

[1] J.M. Probst, Ann. Pharm. 29 (1839) 113.
[2] G.L.W. Reuling, Ann. Pharm. 29 (1839) 131.
[3] F. Bruchhausen, H.W. Bersch, Chem. Ber. 63 (1930) 2520.
[4] E. Späth, F. Kuffner, Chem. Ber. 64 (1931) 370.
[5] H.W. Bersch, Arch. Pharm. 291 (1958) 491.
[6] N. Takao, N. Bessho, M. Kamigauchi, K. Iwasa, K. Tomita, T. Fujiwara, S. Fujii, Tetrahedron Lett. 20 (1979) 495.
[7] A.R. Battersby, R.J. Francis, M. Hirst, E.A. Ruveda, J. Staunton, J. Chem. Soc. Perkin Trans. 1140 (1975).
[8] A.R. Battersby, J. Staunton, H.R. Wiltshire, R.J. Francis, R. Southgate, J. Chem. Soc. Perkin Trans. 1147 (1975).
[9] J. Gollwitzer, R. Lenz, N. Hampp, M.H. Zenk, Tetrahedron Lett. 34 (1993) 5703.
[10] R. Lenz, M.H. Zenk, Eur. J. Biochem. 233 (1995) 132.

1.10 OPIUM AND MORPHINE

Papaver somniferum (Papaveraceae) is a biennial originating in the southeastern area of Europe, and is cultivated as the material for opium ("A-hen" in Japanese) and for morphine. *Papaver somniferum* is a typical narcotic plant, and is cultivated under strict control in a limited number of countries, including Pakistan, Bulgaria, Turkey, Australia, and Japan.

Papaver somniferum flowers around May in Japan with red, white, partially colored, and double-petaled flowers. After flowering, it produces a large capsule, which, after ripening, contains many seeds. Opium is the dried latex (it is first white and becomes black subsequently) exuded by scratching the fruit slightly while it is immature. Opium is a coal-black block and the shape is different according to the producing districts. In the Japanese Pharmacopoeia, it is mixed with starch or lactose to make the content of morphine to be in the range of 9.5–10.5%, this is called opium powder, and is used as the material for medicines containing morphine.

Opium contains 10–25% alkaloids, and the main constituent is morphine. Other than morphine, more than 25 alkaloids are known, and among them are other morphinan-type alkaloids (codeine and thebaine), benzylisoquinoline-type alkaloids (papaverine and noscapine), and protopine-type alkaloids (protopine). The biosynthetic precursor of all of these alkaloids is phenylalanine.

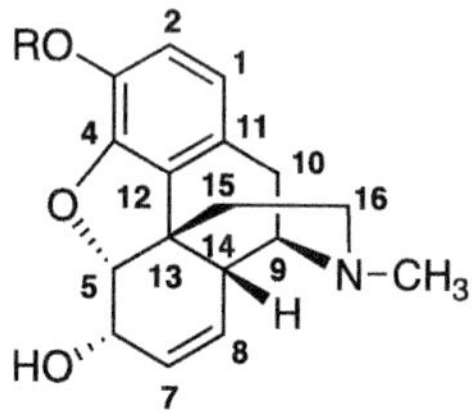

codeine **R = CH$_3$**
morphine **R = H**

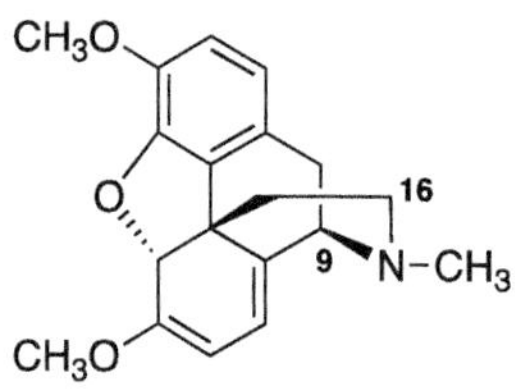

thebaine

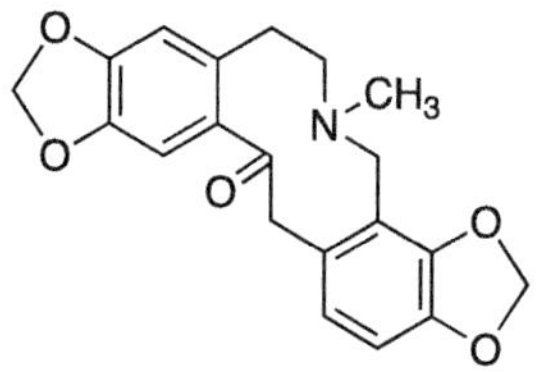

protopine

A number of medicines derived from opium are clinically very important. Morphine hydrochloride, purified from opium powder, is used as an analgesic and an anesthetic, and codeine phosphate and noscapine hydrochloride are used in cough medicines. Papaverine hydrochloride is used as a smooth muscle relaxant.

Among the species of *Papaver*, only *P. somniferum* described above and *Papaver setigerum* and their varieties, are legally controlled. *Papaver setigerum* is similar to *P. somniferum*, but its overall size, as well as the capsule, are smaller than those of the latter. For these reasons, this plant is rarely cultivated. The cultivation of poppies in ornamental gardens, such as *Papaver rhoeas*, *Papaver orientale*, *Papaver bracteatum*, and *Papaver nudicaule*, which do not contain narcotic alkaloids, is quite common and not controlled legally.

As described in the previous section, morphinan alkaloids are produced by the *para, ortho'*-coupling of a biradical derived from (*S*)-reticuline. During biosynthesis, (*S*)-reticuline is transformed into (*R*)-reticuline via a 1,2-dehydroderivative, which is then used in the formation of the morphinan alkaloids.

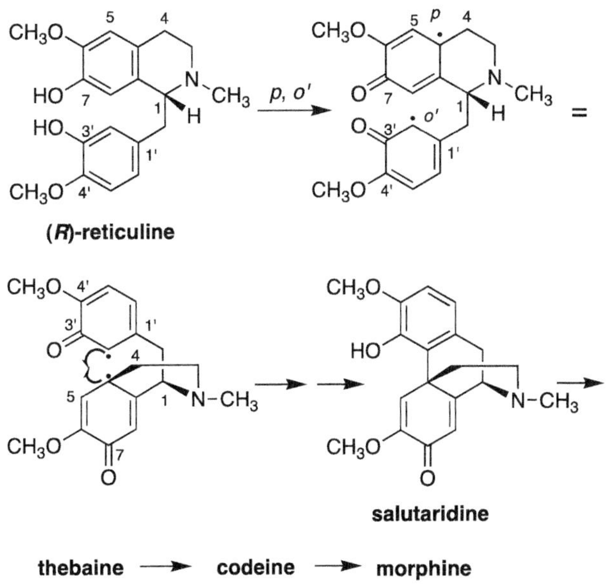

Biosynthetic Route of Thebaine, Codeine and Morphine

When (*S*)-[*N*-$^{13}CH_3$]reticuline is fed to the sprouts of *P. somniferum*, the *N*-methyl moiety of thebaine is labeled. Also, when (*R,S*)-[1-^{13}C]coclaurine, the precursor of reticuline, is fed to 5-week old sprouts of *P. somniferum*, morphine labeled at the C-9 position is obtained [3].

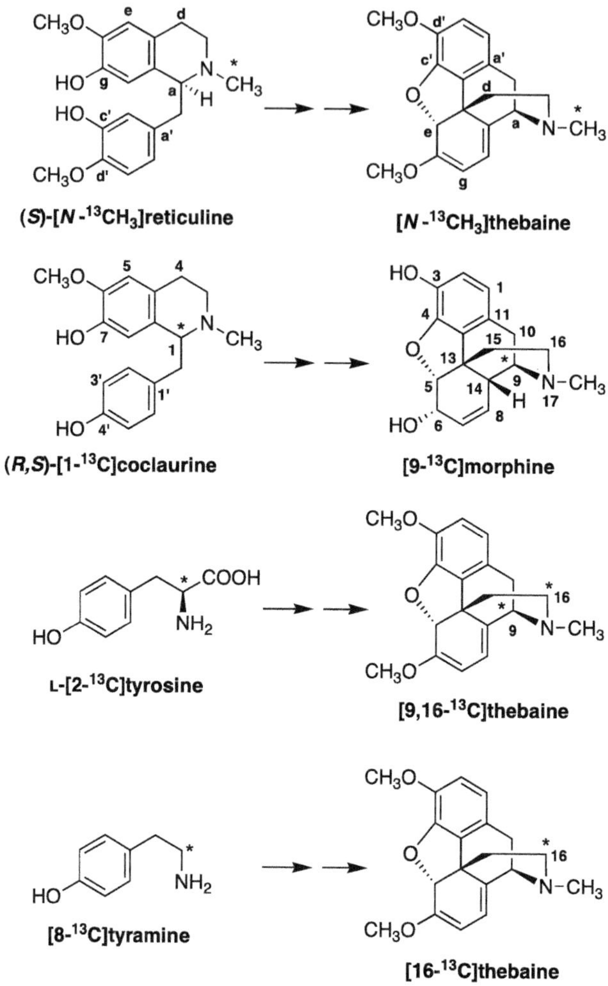

Incorporation Pattern of ^{13}C Labelled Precursors into Thebaine and Morphine

The same group also reported that (*S*)-[6-$O^{14}CH_3$]coclaurine was incorporated into thebaine, but that (*R*)-[6-$O^{14}CH_3$]coclaurine was not incorporated, suggesting the importance of (*S*)-reticuline. In addition, it was clarified that when L-[2–^{13}C]tyrosine was incorporated, the C-9 and C-16 positions of thebaine were labeled. On the other hand, only the C-16 position of thebaine was labeled when [8-^{13}C]tyramine was fed. Two biosynthetic routes from thebaine to morphine, namely, via neopinone, codeinone, and codeine, and via oripavine and morphonone, were clarified by isolating each of the enzymes in the pathway [4,5].

Other examples of alkaloids possessing the morphinan skeleton include sinomenine [6–8] isolated from the roots of *Sinomenium acutum* (Menispermaceae), and metaphanine [9–14] isolated from the stems of *Stephania japonica* (Menispermaceae). Sinomenine possesses the mirror image skeleton to that of morphine, and is derived from (*S*)-reticuline. On the other hand, metaphanine possesses the hasubanan skeleton. Total syntheses of metaphanine have been reported [15,16].

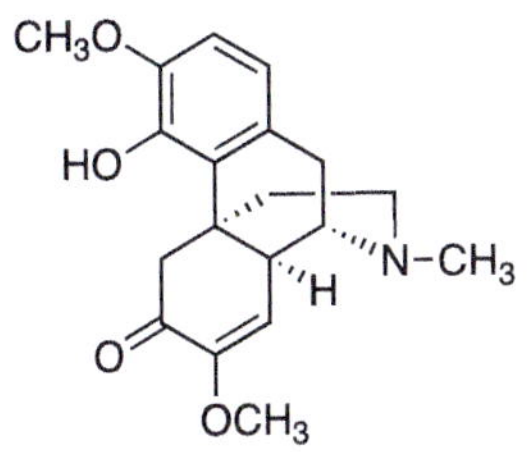

sinomenine

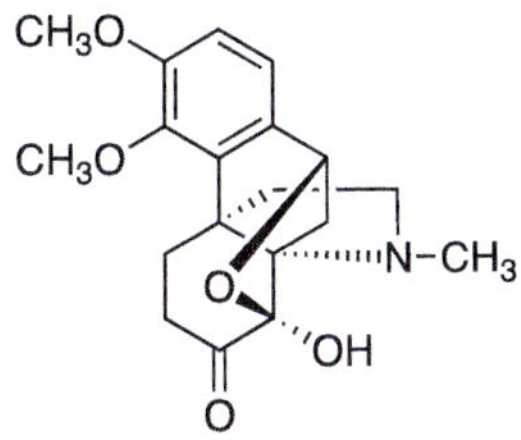

metaphanine

LITERATURE CITED

[1] E. Brochman-Hanssen, T. Furuya, J. Pharm. Sci. 53 (1964) 575.
[2] E. Brochman-Hanssen, T. Furuya, Planta Med. 12 (1964) 328.
[3] S. Loeffler, R. Stadler, N. Nagakura, M.H. Zenk, J. Chem. Soc. Chem. Commun. 1160 (1987).
[4] J. Gollwitzer, R. Lenz, N. Hampp, M.H. Zenk, Tetrahedron Lett. 34 (1993) 5703.
[5] R. Lenz, M.H. Zenk, Eur. J. Biochem. 233 (1995) 132.
[6] H. Kondo, E. Ochiai, Liebigs Ann. der Chem. 470 (1929) 224.
[7] Y. Sasaki, Yakugaku Zasshi 80 (1960) 270.
[8] H. Kondo, T. Sanada, Yakugaku Zasshi 44 (1924) 8.
[9] H. Kondo, T. Sanada, Yakugaku Zasshi 44 (1924) 1034.
[10] H. Kondo, T. Sanada, Yakugaku Zasshi 47 (1927) 177.
[11] H. Kondo, T. Sanada, Yakugaku Zasshi 47 (1927) 930.
[12] H. Kondo, T. Watanabe, Yakugaku Zasshi 58 (1938) 268.
[13] M. Tomita, T. Ibuka, Y. Inubushi, K. Takeda, Tetrahedron Lett. 5 (1964) 3605.
[14] M. Tomita, T. Ibuka, Y. Inubushi, K. Takeda, Chem. Pharm. Bull 13 (1965) 695.
[15] T. Ibuka, K. Tanaka, Y. Inubushi, Tetrahedron Lett. 13 (1972) 1393.
[16] T. Ibuka, K. Tanaka, Y. Inubushi, Chem. Pharm. Bull. 22 (1974) 907.

1.11 *Colchicum* AND COLCHICINE

Colchicum autumnale (Colchicaceae) is a perennial plant that grows wild in Europe and North Africa. *Colchicum autumnale*, the autumn crocus, is also cultivated as a decorative plant. Colchicine, a modified isoquinoline alkaloid isolated from the corms of this plant [1], is used for the treatment of rheumatism and gout. It prevents mitosis of cells, and is also used to form a polyploid in gardening, and to form a watermelon without seeds.

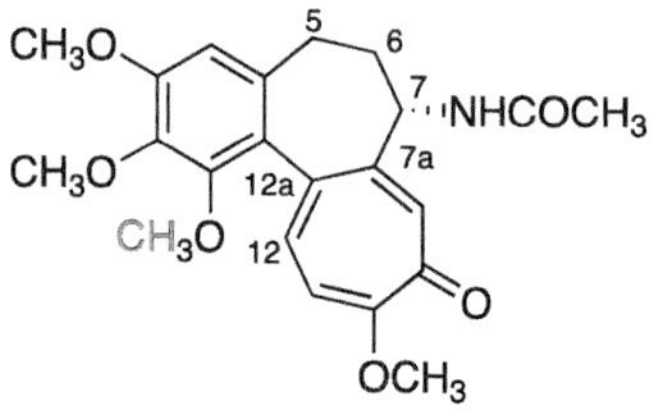

colchicine

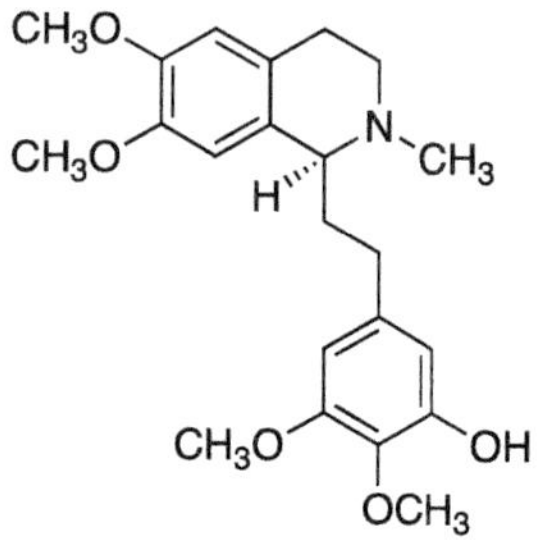

autumnaline

The biosynthetic route to this alkaloid was clarified by the use of labeled precursors. Namely, it was found that tyrosine with ^{14}C at the 3-position was incorporated into colchicine [2], and it was also found that dopamine was formed from labeled tyrosine, and cinnamic acid was formed from phenylalanine [3]. The coupling of these compounds formed autumnaline.

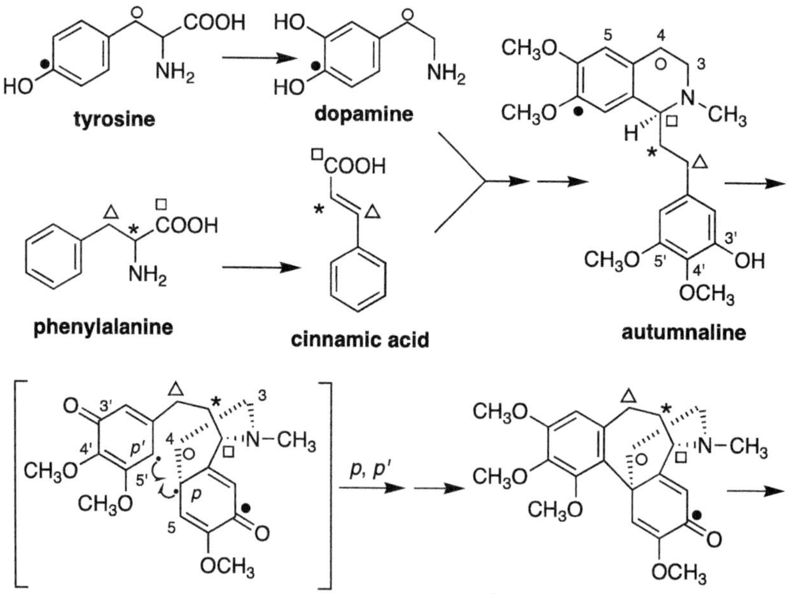

tyrosine **dopamine** **phenylalanine** **cinnamic acid** **autumnaline**

***O*-methylandrocymbine**

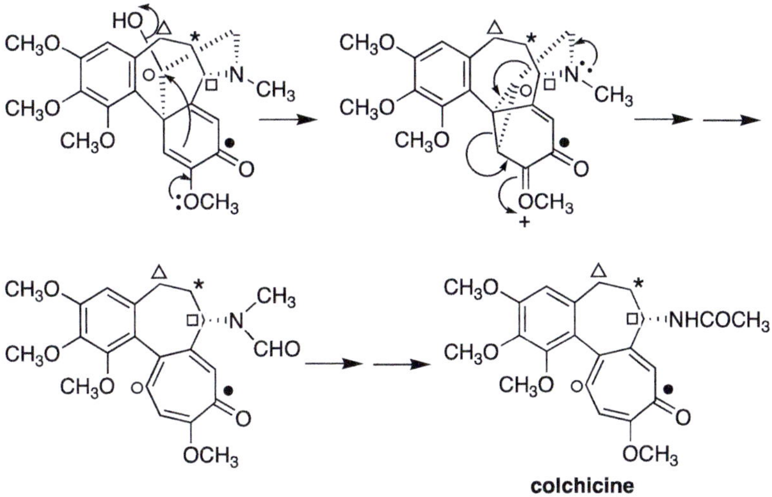

Biosynthetic Route of Colchicine

Autumnaline is a phenethylisoquinoline alkaloid, and possesses an additional C_1 unit compared with the benzylisoquinoline alkaloids. Autumnaline, postulated to be the biosynthetic precursor of colchicine, was isolated from *Colchicum cornigerum* (Colchicaceae) [4]. *O*-methylandrocymbine is formed through the intramolecular *para,para'*-oxidative phenolic coupling of autumnaline. Ring expansion of the six-membered ring to the seven-membered ring then occurs to give the colchicine nucleus [5–9]. The movement of the benzyl carbon of the tyrosine molecule during this process should be noted. It was established that [1-^{14}C]acetic acid was incorporated into the *N*-acetyl moiety at the 7 position, and that [Me-^{14}C] methionine was incorporated into the *O*-methyl moiety [10].

Colchicum alkaloids and its congeners have been reviewed [11].

LITERATURE CITED

[1] M.J.S. Dewer, Nature 155 (1945) 141.
[2] E. Leete, Tetrahedron Lett. 6 (1965) 333.
[3] A.R. Battersby, R. Binks, J.J. Reynolds, D.A. Yeowell, J. Chem. Soc. 4257 (1964).
[4] A.R. Battersby, R. Ramage, A.F. Cameron, C. Hannaway, F. Santavy, J. Chem. Soc. (C) 3514 (1971).
[5] A.R. Battersby, R.B. Herbert, E. McDonald, R. Ramage, J.H. Clements, Chem. Commun. 603 (1966).
[6] A.C. Barker, A.R. Battersby, E. McDonald, R. Ramage, J.H. Clements, Chem. Commun. 390 (1967).
[7] A.R. Battersby, R.B. Herbert, L. Pijewska, F. Santavy, P. Sedmera, J. Chem. Soc. Perkin Trans. 1736 (1972).

[8] A.R. Battersby, R.B. Herbert, E. McDonald, R. Ramage, J.H. Clements, J. Chem. Soc. Perkin Trans. 1741 (1972).
[9] A.R. Battersby, P.W. Sheldrake, J.A. Milner, Tetrahedron Lett. 15 (1974) 3315.
[10] R.D. Hill, A.M. Unrau, Can. J. Chem. 43 (1965) 709.
[11] O. Boye, A. Brossi, Alkaloids 41 (1992) 125.

1.12 *Erythrina indica* AND ERYTHRINA ALKALOIDS

Erythrina indica (=*E. variegata* var. *orientalis*) (Fabaceae) is a deciduous tree, and is the symbolic flower of Okinawa prefecture in Japan. It grows wild in eastern Africa, India, southeastern Asia, the Pacific islands, and New Guinea, and is typically found at the seashore. The seeds of this plant are light and float on the seawater, so *Erythrina indica* in the southern area of Taiwan and on the seashore of Rhukyu and Ogasawara islands may have drifted to these places from other areas. The genus name "*Erythrina*" comes from the color of the flowers of this plant and originates from "erythros (red)" in Greek.

A group of alkaloids with a unique skeleton known as the *Erythrina* alkaloids have been isolated from the seeds of *Erythrina* plants; examples include erythraline and erysodienone obtained from the seeds of *E. indica* [1,2]. The relative and absolute stereochemistry of these alkaloids was determined by X-ray crystallography and chemical degradation studies [3]. The chemical synthesis of the Erythrina alkaloid nucleus was achieved biomimetically through phenol oxidative coupling [4].

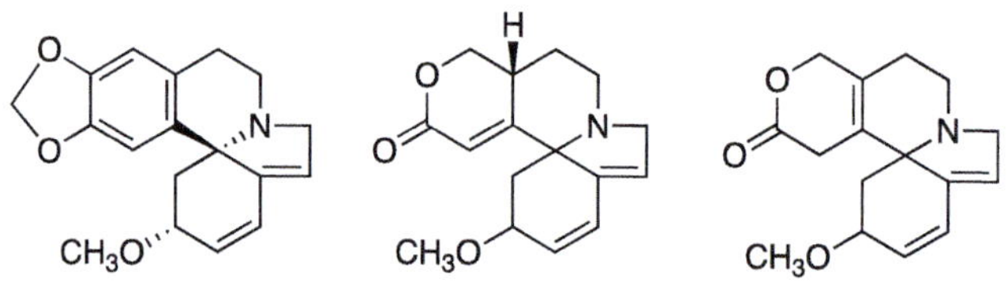

erythraline **α-erythroidine** **β-erythroidine**

Although the alkaloids of this group show biological activities that are similar to those of curare, they are not used clinically.

At first, it was thought that the Erythrina alkaloids were biosynthesized through a C_6–C_2–N–C_2–C_6 intermediate derived from DOPA. More recently it was clarified that these alkaloids are derived from a benzylisoquinoline alkaloid precursor. It was previously established that tyrosine was involved in the biosynthesis of α- and β-erythroidine from the incorporation of labeled precursors into *E. berterana* [5]. Further studies using *E. crista galli* showed that erythraline was formed through erythodienone derived from the intramolecular *para,para'*-oxidative coupling of *S*-(+)-*N*-norprotosinomenine [6,7].

In this case, it was also found that only *S*-(+)-*N*-norprotosinomenine was incorporated through the feeding experiments of DL-norprotosinomenine7, or (+)- and (−)-norprotosinomenine separately [8].

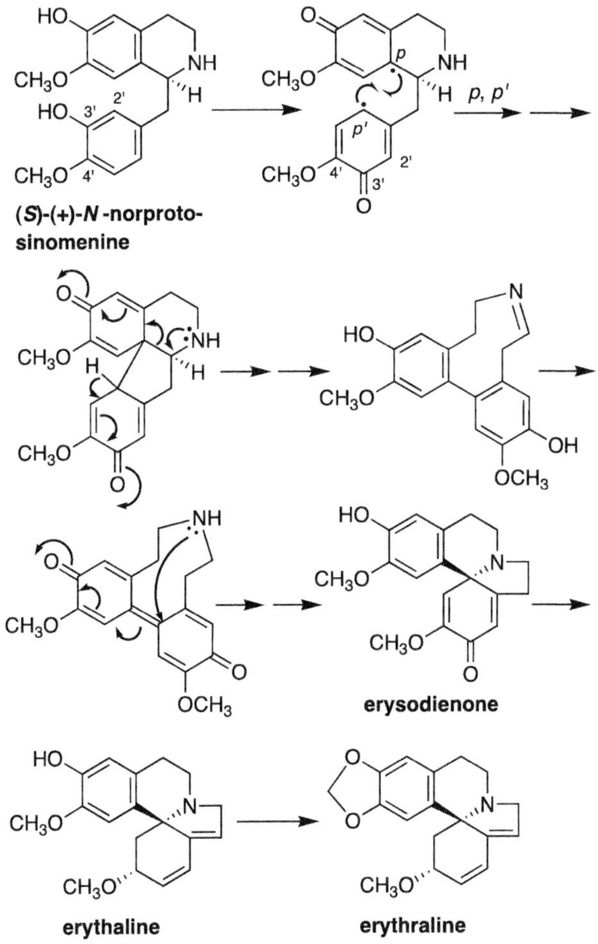

Biogenetic Route to Erythraline and Related Alkaloids

On the other hand, the existence of a symmetrical intermediate in the pathway was established. When the [7-$O^{14}CH_3$]*dl*-*N*-norprotosinomenine was fed to *Cocculus laurifolius* (Menispermaceae), the two methoxyl moieties of the isococculidine and cocculidine isolated were labeled [8]. It was also found that cocculidine was incorporated into cocculine efficiently; thus it was suggested that de-*O*-methylation occurred as the final stage in the biosynthesis of cocculine. The isolation of erybidine, which is estimated to be a biosynthetic precursor, from *E. xbidwilli*, also supported the biosynthetic proposals described above [9].

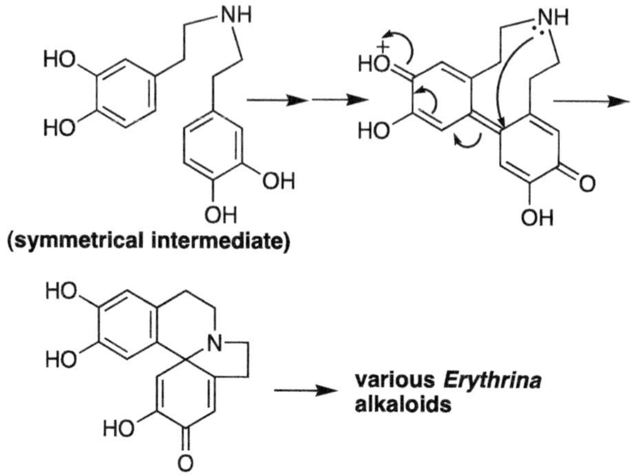

Biogenesis of *Erythrina* Alkaloids

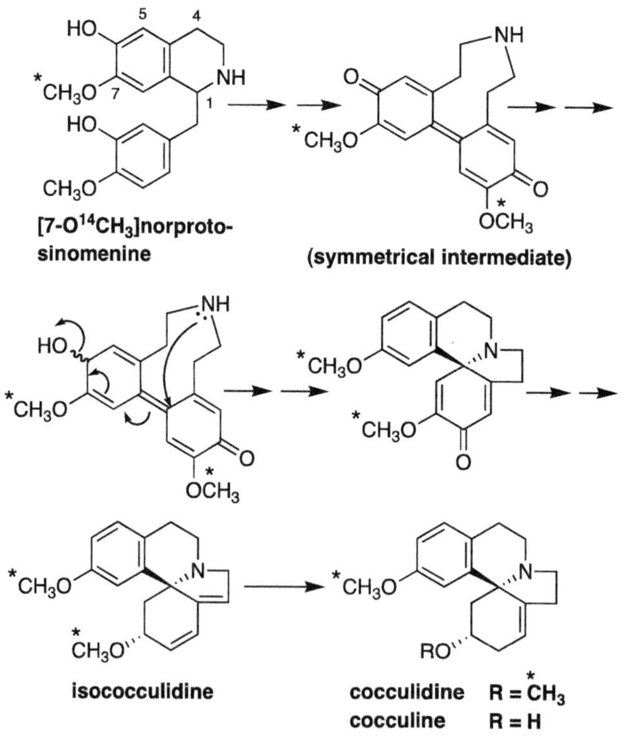

Incorporation Studies of [7-O^{14}CH$_3$]norprotosinomenine into Cocculine and Related Alkaloids

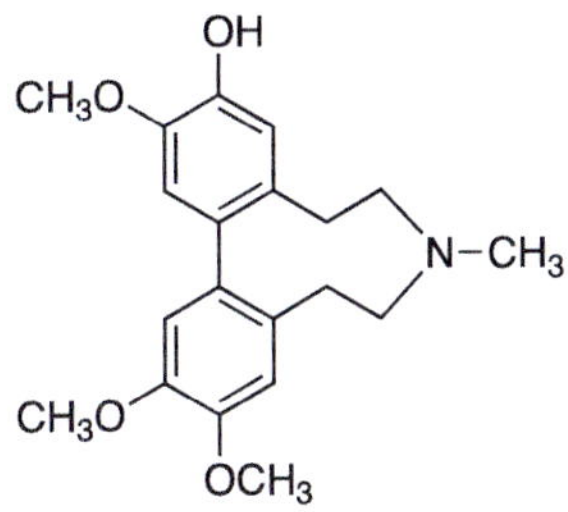

erybidine

LITERATURE CITED

[1] R.B. Boar, D.A. Widdowson, J. Chem. Soc. (B) 1591 (1970).
[2] S. Ghosal, S.K. Dutta, S.K. Bhattacharya, J. Pharm. Sci. 61 (1972) 1274.
[3] W. Klyne, J. Buckingham, Atlas of Stereochemistry, Chapman and Hall Ltd., London, 1974R, p. 146.
[4] T. Kametani, R. Charubala, M. Ihara, M. Koizumi, K. Fukumoto, Chem. Commun. 289 (1971).
[5] E. Leete, A. Ahmad, J. Am. Chem. Soc. 88 (1966) 4722.
[6] D.H.R. Barton, R.B. Boar, D.A. Widdowson, J. Chem. Soc. (C) 1213 (1970).
[7] D.H.R. Barton, C.J. Potter, D.A. Widdowson, J. Chem. Soc. Perkin Trans. I 346 (1974).
[8] D.S. Bhakuni, A.N. Singh, J. Chem. Soc. Perkin Trans. I 618 (1978).
[9] K. Ito, H. Furukawa, H. Tanaka, Chem. Pharm. Bull. 19 (1971) 1509.

1.13 *Lycoris* spp. AND LYCORINE

Some Amaryllidaceae plants possess a group of alkaloids containing a C_6–C_2–N–C_1–C_6 unit. In this unit, the C_6–C_2–N moiety is derived from tyrosine or tyramine, and the C_6–C_1 part is derived from phenylalanine through cinnamic acid, *p*-coumaric acid, and protocatechualdehyde [1]. Tyramine (C_6–C_2–N unit) and protocatechualdehyde (C_6–C_1 unit) are then combined and methylated to form *O*-methylnorbelladine, which is a common biosynthetic precursor of various Amaryllidaceae alkaloids. Through *para,ortho′*-, *ortho,para′*- and *para,para′*-phenol coupling of norbelladine, the lycorine, galanthamine, and crinine type alkaloids are formed, respectively [2].

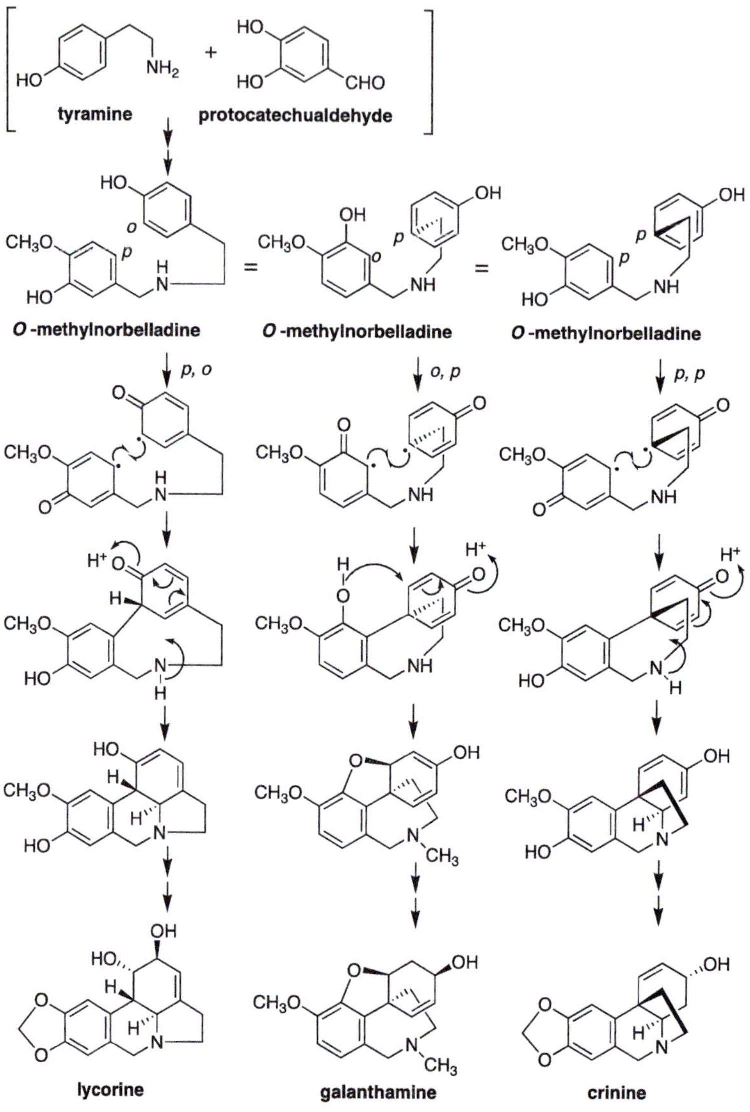

Biosynthesis of Lycorine, Galanthamine and Crinine

Lycorine was isolated from the bulbs of *Lycoris radiata* (Amaryllidaceae), and its chemical structure, including the absolute stereochemistry, was reported [3,4]. The biological activities of this alkaloid are similar to those of emetine (Section 1.14), and it was formerly used for the treatment of amoebic dysentery. Galanthamine was shown to be the same alkaloid as lycoremine, which had been isolated from the bulbs of the same plant [5,6], and the chemical structure was determined by total synthesis [7]. Galanthamine was shown to be present in *L. squamigera* and until now was isolated from about 60 plants in this family.

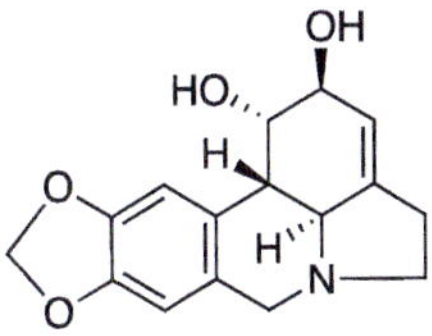

lycorine

galanthamine

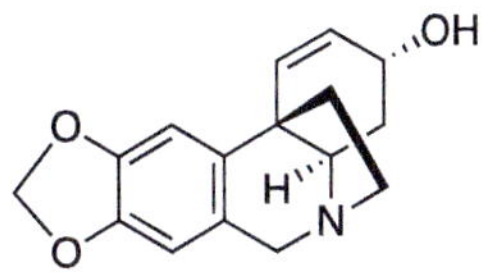

crinine

Among the *Lycoris* plants growing wild in Japan, the content of lycorine is reported to be the highest in *L. aurea* (0.017%) which grows wild in the Shikoku and Kyushu islands, and is cultivated as an ornamental plant [8]. Galanthamine possesses anti-cholinesterase activity, and was formerly applied to the treatment of the sequelae of polio in Japan. The side effects of this alkaloid are said to be weaker than those of drugs such as physostigmine (Section 2.5) used for this purpose. It has been used for the treatment of Alzheimer's disease from 2011. The total synthesis of galanthamine was achieved by several groups [9].

Starch of the bulbs of *L. radiata* was formerly used to stave off hunger during times of famine until the middle of the nineteenth century in Japan. In that instance, these toxic alkaloids should be washed out thoroughly.

Crinine is isolated from the bulbs of *Crinum powellii* and other species [10,11], and is considered to be formed by the *para,para'*-phenol oxidative coupling of *O*-methylnorbelladine.

Pretazettine is formed from crinine by cleavage between N and the adjacent methylene, followed by re-cyclization. Pretazettine is transformed into tazettine. Tazettine was isolated from the bulbs of *Narcissus tazetta* var. *chinensis*, and the total synthesis of this alkaloid was achieved [12–15].

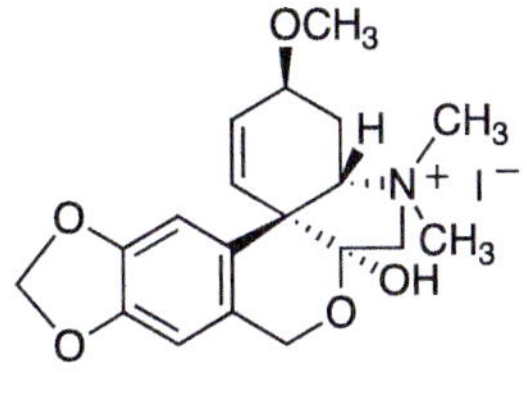

tazettine methiodide

When the alkaloids from the bulbs of *Sprekelia formosissima* (Amaryllidaceae) were isolated without using strong alkali and alumina column chromatography, tazettine was not isolated, and instead pretazettine was isolated as the main constituent [11]. In addition, pretazettine was stable when it was refluxed with 0.2 N HCl for 12 h, pretazettine was transformed into tazettine when pretazettine was treated with 0.1 N NaOH sol. (room temperature (r.t.), 30 min), $CHCl_3$ (r.t., 12 h), or water (70 °C, 1 h). Consequently, it was estimated that pretazettine is the natural form, and that tazettine is an artifact formed through the purification procedure. The biosynthetic route from crinine to pretazettine, and the transformation of pretazettine to tazettine, are indicated in the Figure. The absolute stereochemistry of (+)-tazettine was determined by X-ray crystallography of the methiodide [16].

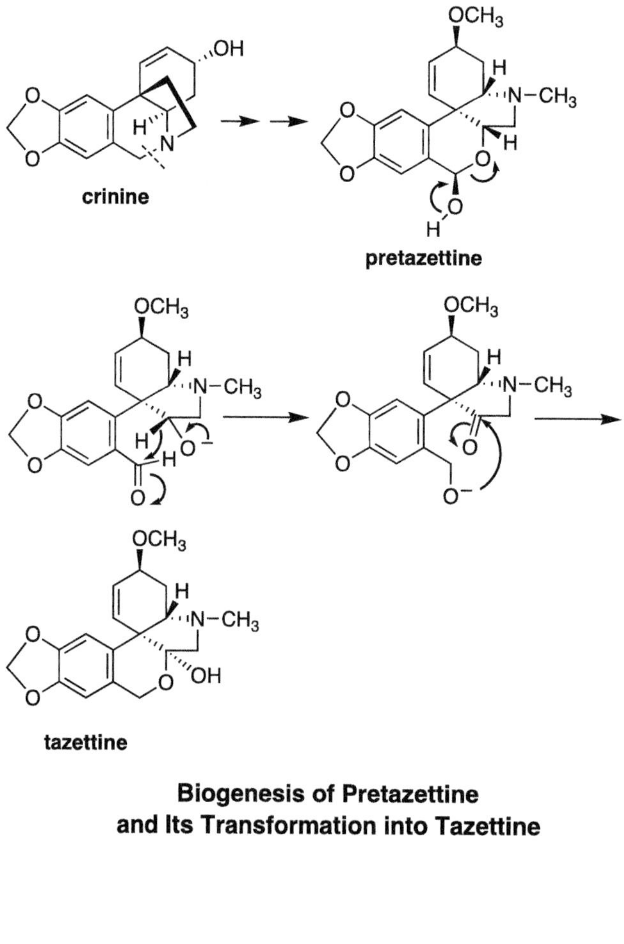

Biogenesis of Pretazettine and Its Transformation into Tazettine

From the bulbs of *L. radiata*, examples of the lycorine-type, galanthamine-type, and crinine-type alkaloids were isolated, in addition to pretazettine [17]. According to Uyeo and Yajima, the amount of total alkaloids obtained from the bulbs of *L. radiata* was almost the same throughout the year, whereas the amount of lycorine varied, and the highest amount (December 3) was 0.0277% and the lowest (May 10) was 0.0026%, about a 10-fold difference.

LITERATURE CITED

[1] A.R. Battersby, R. Binks, S.W. Breuer, H.M. Fales, W.C. Wildman, R.J. Highet, J. Chem. Soc. 1595 (1964).
[2] D.H.R. Barton, G.W. Kirby, J.B. Taylor, G.M. Thomas, J. Chem. Soc. 4545 (1963).
[3] K. Kotera, Y. Hamada, K. Tori, K. Aono, K. Kuriyama, Tetrahedron Lett. 7 (1966) 2009.
[4] C. Fuganti, M. Mazza, Chem. Commun. 936 (1972).
[5] H. Kondo, K. Tomimura, S. Ishido, Yakugaku Zasshi 52 (1932) 433.
[6] S. Uyeo, S. Kobayashi, Chem. Pharm. Bull. 1 (1953) 139.
[7] D.H.R. Barton, G.W. Kirby, J. Chem. Soc. 806 (1962).
[8] S. Uyeo, Handbook of International Congress, IUPAC, 1957. p.209.
[9] T. Kametani, C. Seino, K. Yamaki, S. Shibuya, K. Fukumoto, K. Kigasawa, F. Satoh, M. Hiiragi, T. Hayasaka, J. Chem. Soc. (C) 1043 (1971).
[10] H.M. Fales, W.C. Wildman, J. Am. Chem. Soc. 82 (1960) 3368.
[11] W.C. Wildman, D.T. Bailey, J. Org. Chem. 33 (1968) 3749.
[12] E. Späth, L. Kahovec, Chem. Ber. 67 (1934) 1501.
[13] W.C. Wildman, C.J. Kaufman, J. Am. Chem. Soc. 76 (1954) 5815.
[14] T. Ikeda, W.I. Taylor, Y. Tsuda, S. Uyeo, H. Yajima, J. Chem. Soc. 4749 (1956).
[15] Y. Tsuda, A. Ukai, K. Isobe, Tetrahedron Lett. 13 (1972) 3153.
[16] T. Sato, H. Koyama, J. Chem. Soc. (B) 1070 (1971).
[17] S. Uyeo, H. Yajima, J. Chem. Soc. 3392 (1955).

1.14 *Cephaelis ipecacuanha* AND EMETINE

Cepahelis species (Rubiaceae) are shrubs that grow wild in areas of Brazil, Colombia, and Malaysia. The roots of this plant (Ipecacuanhae Radix) are used as an emetic and to clear the throat of phlegm. There are three kinds of Ipecacuanhae Radix, and they are known as Rio-, Carthagena- and Johore-Ipecacuanhae Radix.

Among them, Rio-Ipecacuanhae Radix is prepared from *C. ipecacuanha,* which grows wild in the Amazon area and is exported from Rio de Janeiro, Brazil. On the other hand, Carthagena-Ipecacuanhae Radix is prepared from the roots of *C. acuminata,* which grows wild in Colombia, South America, and is exported from Carthagena, Colombia. Furthermore,

Johore-Ipecacuanhae Radix is prepared from the roots of *C. ipecacuanha*, which was transplanted from Brazil to Ceylon (now Sri Lanka) and Johore in the Malay peninsula in the late nineteenth century. Nowadays, the cultivation of this plant in these areas has declined.

Emetine (70% of the total alkaloids) and cephaeline were isolated as the main constituents of Ipecacuanhae Radix (*C. ipecacuanha*), whereas Ipecacuanhae Radix originating from *C. acuminata* contains almost the same amount of cephaeline as emetine.

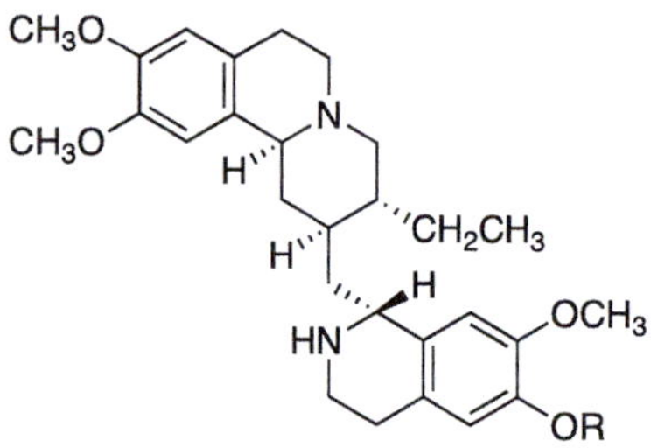

emetine **R = CH$_3$**
cephaeline **R = H**

Emetine shows an emetic effect by stimulating the gastric mucous membranes. It causes an increase of secretion in the trachea in a small dose and is used to clear throat of phlegm. Cephaeline shows stronger emetic activity and higher toxicity than emetine. Because emetine kills protozoa, even at a concentration of 0.5–1.0 mg/mL, it is also used as hydrochloride as a specific medicine for amoebic dysentery. Amoebic dysentery is caused by the protozoan *Endamoeba histolytica* and is widespread in the tropics. It is distinct from the dysentery caused by the *Shigella bacillus*, which occurs in both tropical and temperate zones.

Emetine and cephaeline possess two phenylethylamine units derived from phenylalanine, and a central structure consisting of nine carbons of a monoterpene unit, which is also found in the ajmaline (Section 2.10), strychnine (Section 2.14), and quinine (Section 2.17) alkaloids. This partial structure is derived from geraniol through secologanin by the loss of one carbon, as was established in the case of cephaeline [1].

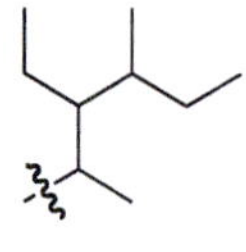

C$_9$ unit

Emetine was isolated at the beginning of the nineteenth century, and its chemical structure was determined in the middle of the twentieth century [2]. The absolute stereochemistry was clarified [3], and the total synthesis of this alkaloid was achieved [4,5]. On the other hand, cephaeline was isolated early in the twentieth century [6], and its chemical structure, including absolute stereochemistry, was reported at almost the same time as that of emetine [7].

LITERATURE CITED

[1] A.K. Garg, J.R. Gear, Tetrahedron Lett. 10 (1969) 4377.
[2] R. Robinson, Nature 162 (1948) 524.
[3] A.R. Battersby, S. Garratt, J. Chem. Soc. 3512 (1959).
[4] E.E. van Tamelen, C. Placeway, G.P. Schiemenz, I.G. Wright, J. Am. Chem. Soc. 91 (1969) 7359.
[5] T. Kametani, Y. Suzuki, H. Terasawa, M. Ihara, J. Chem. Soc. Perkin Trans. I 1211 (1979).
[6] F.H. Carr, F.L. Pyman, J. Chem. Soc. 1591 (1914).
[7] E.E. van Tamelen, P.E. Aldrich, J.B. Hester Jr., J. Am. Chem. Soc. 79 (1957) 4817.

CHAPTER 2

Alkaloids Derived from Tryptophan

Catharanthus roseus (Apocynaceae)

The essential amino acid tryptophan is the precursor to thousands of alkaloids. Many of the alkaloids derived from tryptophan are known to be important medicines, dyes, and phytohormones. Several alkaloids of this group also possess hallucinogenic activity. Most of the alkaloids described in this chapter contain an indole moiety in the structure, so these alkaloids possessing indole skeletons are also called indole alkaloids.

First, the alkaloids possessing an almost unchanged tryptophan moiety are shown. These alkaloids are sometimes called protoalkaloids. Next, the more complicated alkaloids with an attached moiety such as a C_{10} unit derived from two isoprene units, etc., are described.

The biosynthesis of many indole alkaloids can be discussed in terms of the tryptophan-derived moiety and the non–tryptophan-derived moiety. Examples of non-tryptophan moieties include isoprene and secoiridoid units, polyketide moieties, and other amino acids.

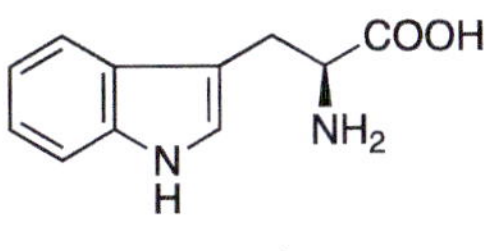

L-tryptophan

Among the alkaloids in this chapter derived from tryptophan, some do not retain the original nucleus of tryptophan, such as quinine, obtained from *Cinchona* plants (Rubiaceae), camptothecin, originally from *Camptotheca acuminata* (Cornaceae), and pyrrolnitrin of microbial origin.

Alkaloids
http://dx.doi.org/10.1016/B978-0-12-417302-6.00002-7

2.1 SEROTONIN

Serotonin is also known as 5-hydroxytryptamine (5-HT), and is widely distributed in animals and plants. In the higher orders of animals, a greater amount of serotonin is found in the hypothalamus, limbic system, glandula pinealis, and thrombocyte. In cells, a high content is found in the endoplasmic reticulum of the synapse. Serotonin is one of the most important neurotransmitting substances, and it is known as a bioamine, together with histamine, etc.

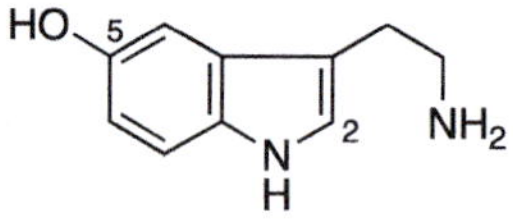

serotonin (5-HT)

It was well known that defibered blood contracted the blood vessels and showed hypertensive activity. The active principle was found by Rapport et al., in 1948 in the blood of cows, and they named the compound serotonin [1,2]. Serotonin shows writhing activity of the gut, in addition to contraction of the capillary vessels.

Serotonin is formed directly from tryptophan via hydroxylation and decarboxylation, and is metabolized to 5-hydroxy-3-indoleacetic acid by monoamine oxidase (MAO) and excreted.

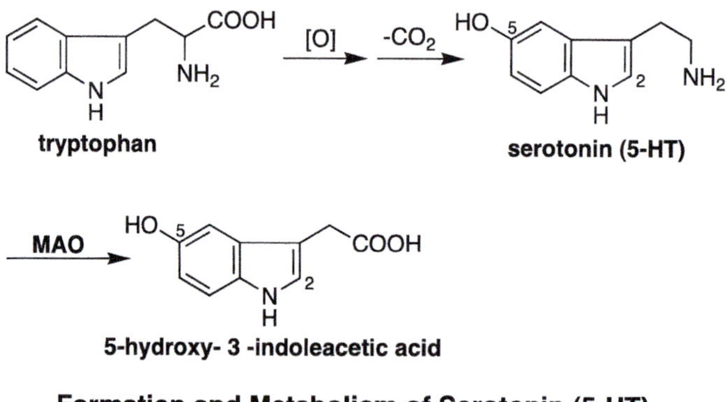

Formation and Metabolism of Serotonin (5-HT)

LITERATURE CITED

[1] M.M. Rapport, A.A. Green, I.H. Page, Science 108 (1948) 329.
[2] M.M. Rapport, A.A. Green, I.H. Page, J. Biol. Chem. 176 (1948) 1243.

2.2 AUXIN AND INDOLE-3-ACETIC ACID

It has been known from ancient times that the urine of mammals, including humans, possessed growth-promoting activity for plants. Kögl et al. isolated the growth-promoting active compound from human urine and named it auxin [1]. It was shown subsequently that auxin comprises four compounds, and these were named auxin-a, auxin-a-lactone, auxin-b, and heteroauxin; now only the existence of heteroauxin is recognized.

The chemical structure of heteroauxin was established to be indole-3-acetic acid (indole-β-acetic acid (IAA)). It is considered that IAA is formed by the enzymatic oxidation of indole-3-acetaldehyde derived from tryptophan via indole-3-pyruvic acid or tryptamine. Between these two routes, the main route is thought to be through indole-3-pyruvic acid.

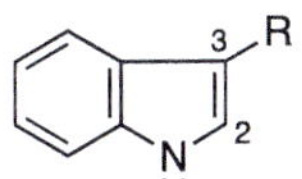

indole-3-acetic acid (IAA)	**R = CH_2COOH**
indole-3-ethanol (IEt)	**R = CH_2CH_2OH**
indole-3-aldehyde	**R = CHO**
indole-3-carboxylic acid	**R = COOH**
indole-3-acetonitrile (IAN)	**R = CH_2CN**

The chemical synthesis of IAA was first reported by Johnson and Crosby [2]. Many IAA-related compounds have been synthesized, and α-naphthaleneacetic acid possesses almost the same activity as IAA. Because IAA can be produced inexpensively, it is used in agricultural practice. On the other hand, (2,4-dichlorophenoxy) acetic acid (2,4-D, 2,4-Cl_2-C_6H_3O-CH_2COOH) possesses strong root elongation activity. Since overstimulation of such activity blights plants, it is used as a herbicide.

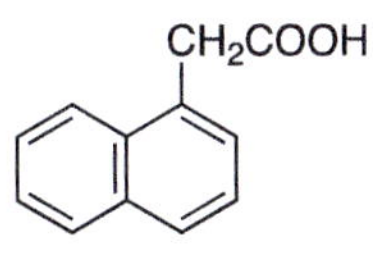

α-naphthaleneacetic acid

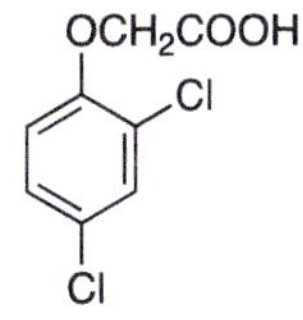

(2,4-dichlorophenoxy) acetic acid (2,4-D)

Internal auxins of plants, other than IAA, are also known, and include indole-3-ethanol (IEt), indole-3-aldehyde, indole-3-carboxylic acid, and indole-3-acetonitrile (IAN) [3].

Simple alkaloids, such as IAA described in this section, serotonin described in the previous (Section 2.1), and psilocin described in the next (Section 2.3), are sometimes known as protoalkaloids. Protoalkaloids possess a comparatively simple chemical structure, are formed in very few steps from amino acids, and may serve as biosynthetic precursors of true alkaloids.

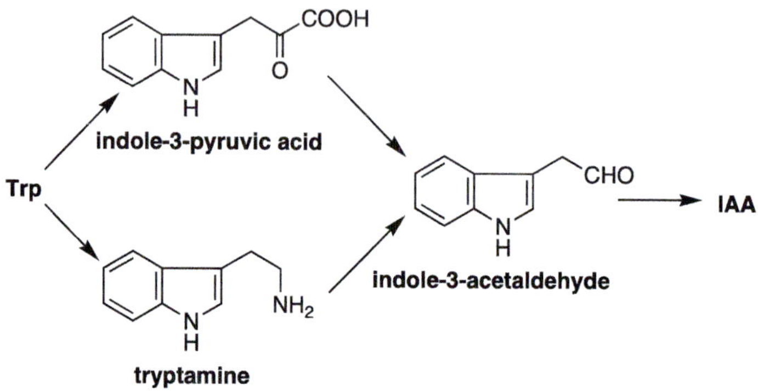

Biosynthetic Route to Indole-3-acetic acid (IAA)

LITERATURE CITED

[1] S. Shibata (Ed.), Seirikassei Tennenbutsu Kagaku (Natural Products Chemistry with Biological Activities), I-Shi-Yaku Shuppan Co. Ltd, 1978R, p. 1.
[2] H.E. Johnson, D.G. Crosby, J. Org. Chem. 28 (1963) 1246.
[3] P.J. Davies (Ed.), Plant Hormones and Their Role in Plant Growth and Development, Kluwer Academic Publishers, Dordrecht, 1990R.

2.3 TEONANACATL AND PSILOCYBIN

Mushrooms classified to the genus *Psilocybe* are distributed in the central parts of North America, Central America, the northern part of South America, and Europe. The use of these mushrooms as hallucinogens appears to be limited to Mexico and Guatemala. In these regions, the mushrooms are known as "teonanacatl," and are used in religious ceremonies. According to Schultes and Hofmann [1], there are seven kinds of *Psilocybe* mushroom, and of these, *Psilocybe mexicana* is the most famous.

Many stone dolls in the curious shape of a mushroom from the Mayan civilization were excavated in Guatemala, southern areas of Mexico, and El Salvador. At first, the meaning of these dolls was not known. Subsequently, it was found that the shape of these dolls indicated reverence for the

Psilocybe mushroom, and this demonstrated the long history of the use of this mushroom in these cultures.

As hallucinogenic substances of *Psilocybe* origin, psilocybin was isolated as the main constituent, and psilocin was obtained as a minor constituent. Psilocybin is a phosphoric acid ester of psilocin, and both of these alkaloids are 4-substituted tryptamine derivatives. The chemical structures of these alkaloids are therefore similar to that of serotonin (5-HT), as mentioned in the previous (Section 2.2).

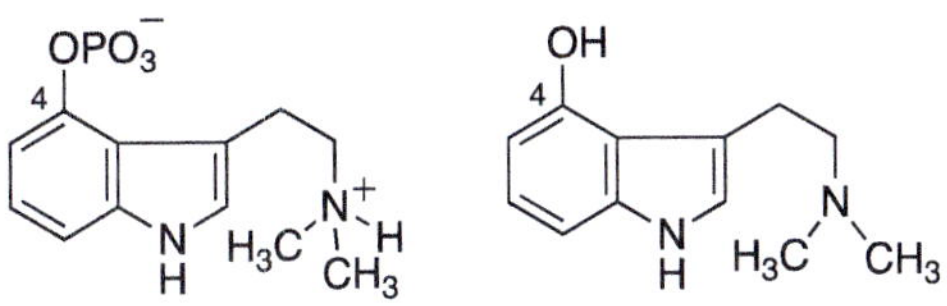

LITERATURE CITED

[1] R.E. Schultes, A. Hofmann, Plants of the Gods (Origins of Hallucinogenic Use), McGraw-Hill Book, Co, New York, 1979R. 144–145.

2.4 INDIGO AND THE ANCIENT PURPLE

Indigo is probably one of the oldest known natural dyes. Indigo is produced from *Indigofera* spp. (Fabaceaae), as well as some other plants, and is also known as indigotin. The material used for indigo in Japan is *Polygonum tinctorium*, and Tokushima is the place famous for its production.

Indigo originally existed as indican in the plant; it is an indoxyl glucoside, and it affords indoxyl by the action of a glucosidase. Indoxyl yields indigo, having a dimeric structure, by oxidation with air. Indigo is a dark blue pigment and is insoluble in water, alcohol, and ether, but leucoindigo is soluble in alkaline solution, and indigo is applied for dyeing in this form. The chemical structure of indigo was verified by total synthesis [1]. *Trans*- and *cis*-forms are possible for the structure of indigo, and it was confirmed that only the *trans*-form existed by X-ray crystallography, and the existence of intramolecular hydrogen bonding, as shown, was also confirmed [2]. Indigo was one of the first dyes to be produced industrially [3,4], and it was in 1897 when BASF of Germany succeeded in its industrial production.

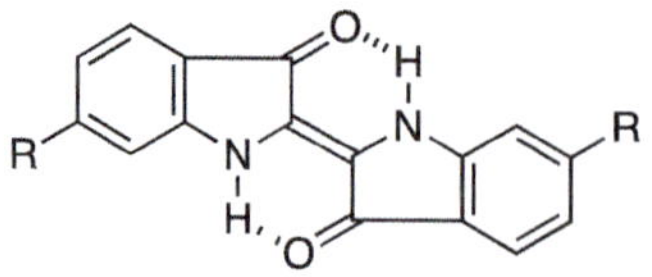

indigo R = H
6,6'-dibromoindigo R = Br

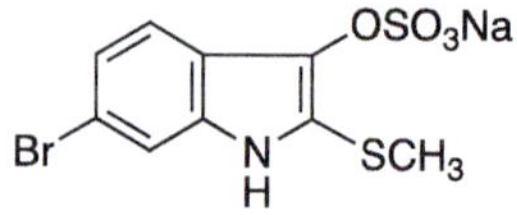

sodium tyrindoxyl sulfate

On the other hand, the almost colorless fluid secreted by *Murex brandaris*, a shellfish of the Mediterranean Sea region, becomes reddish purple on contact with air. The fluid was used by the Phoenicians as a dye from about 1500 BC, and the purple dye is known as Tyrian purple, ancient purple, or shellfish purple, etc. Friedländer, in 1915, isolated about 1.5 g of the pigment from 12,000 specimens of *M. brandaris*, and clarified its chemical structure as 6,6′-dibromoindigo [5]. Subsequently, the origin of this pigment was identified as sodium tyrindoxyl sulfate [6].

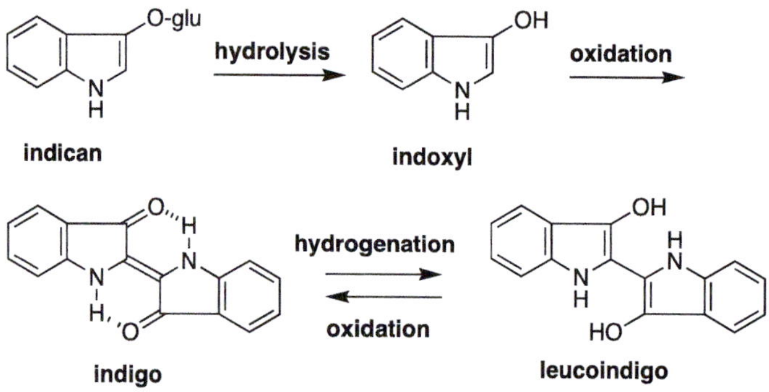

Formation of Indigo from Indican

The same pigment containing bromine atoms in the molecule was also isolated from the secretion of the marine sea shell *Purpura aperta* used for dyeing by the native peoples of Mexico and Costa Rica, and also from European *P. lapillus* [7].

LITERATURE CITED

[1] P. Friedländer, Naturwissenschaften 3 (1915) 573.
[2] T. Posner, Chem. Ber. 59 (1926) 1799.
[3] K. Heumann, Chem. Ber. 23 (1890) 3043.
[4] K. Heumann, 34, Chem. Ber. 23 (1890).
[5] P. Friedländer, Rev. Gen. Mat. Color 14 (1) (1910), [*Chem. Abstr., 4,* 1378(1910)].
[6] K. Shiomi, Y. Nagashima, Kaiyo Dobutsu No Doku (Poisonous Substances from Marine Animals), Seizando Pub. Co, 1997R, p. 143.
[7] P. Friedländer, Chem. Ber. 55 (1922) 1655.

2.5 CALABAR BEANS AND PHYSOSTIGMINE

Physostigma venenosum (Fabaceae) is a perennial vine that grows in the Calabar district in southeastern Nigeria. The lignified stems of *P. venenosum* grow to be about 4 cm in diameter and 15 m in length. The capsules contain one to three dark brown seeds in the shape of a kidney, and these seeds are known as Calabar beans.

Calabar beans are noted for their strong toxic activity, and were used as a truth serum in trials conducted by indigenous peoples. Namely, the accused person was given Calabar beans or their aqueous extract, and if the accused person died, the person was concluded to be guilty. The toxic substance of Calabar beans at first causes feelings of nausea and vomiting, and finally causes paralysis of breathing. It was said that if the person was innocent they would drink the poison quickly to demonstrate that he (or she) was innocent. By doing so, the person would vomit violently all of the stomach contents and he (or she) would not die. On the other hand, if the person was guilty, he (or she) would drink the poison little by little, and by doing so paralysis of breathing occurred, and the person would die.

The main poisonous substance of Calabar beans was isolated in 1864, and was named physostigmine [1]. The plane structure of this alkaloid was determined in 1925, and the absolute stereochemistry of (−)-physostigmine was clarified by degradation studies in 1969 [2,3]. Subsequently, the total synthesis of DL-physostigmine was achieved [4]. Calabar beans were known as "Eséré" by the indigenous people, and so physostigmine is also known as eserine.

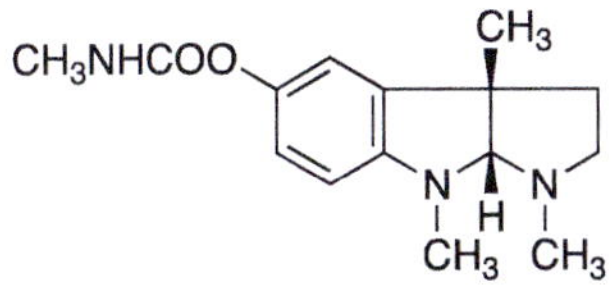

physostigmine (eserine)

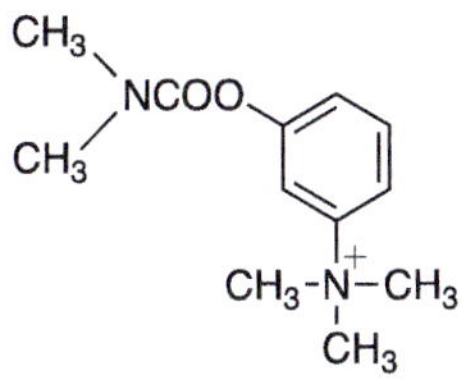

neostigmine

Physostigmine causes excitement of the parasympathetic nervous system and contraction of skeletal muscles. These activities are attributed to blocking of the destruction of acetylcholine (ACh) by the hindrance of choline esterase (ChE), an activity that is reversible. This explains why strong miosis and the decrease in intraocular pressure are among the biological effects, and consequently the alkaloid is used for the treatment of glaucoma and as an antagonist of mydriatic agents such as atropine. On the other hand, atropine is applied to detoxify the poisonous activity of physostigmine. Physostigmine prevents the decomposition of acetylcholine mainly at the end of motor nerves. Thus, physostigmine causes contraction of the skeletal muscle, and is also an antagonist to the muscle relaxation activity of D-tubocurarine (Section 1.5).

Synthetic parasympathetic nerve stimulants, such as neostigmine, are designed based on the chemical structure of physostigmine [5].

LITERATURE CITED

[1] J. Jobst, O. Hesse, Liebigs Ann. der Chem. 129 (1864) 115.
[2] E. Stedman, G. Barger, J. Chem. Soc. 127 (1925) 247.
[3] R.B. Longmore, B. Robinson, Chem. Ind. (Lond.) 622 (1969).
[4] J.B.P.A. Wijnberg, W.N. Speckamp, Tetrahedron 34 (1978) 2399.
[5] J.A. Aeschlimann, M. Reinert, J. Pharmacol. 43 (1931) 413 [*Chem. Abstr.*, 26, 2519(1932)].

2.6 HARMALA AND HARMINE

Peganum harmala (Zygophyllaceae) is a shrub that grows wild in western Asia, northern India, and Mongolia, and is also known as "Syrian Rue." The seeds of this plant are similar to those of the smaller seeds of the morning glory, and are known as "Harmala." From harmala, β-carboline alkaloids, such as harmine and harmaline, were isolated [1]. The structures of these alkaloids were determined about 50 years after their initial isolation [2,3], and their chemical synthesis is well established [4]. These two alkaloids were also isolated from *Banisteriopsis caapi* (Malpighiaceae), and harmine was isolated from the stem bark of *B. inebrians* [5,6]. These plants are used as monoamine oxidase inhibitors in ayahuasca, in the western Amazon and Orinoco basins, in conjunction with *Psychotria virida* (Rubiaceae), which provides the main psychoactive agent *N,N*-dimethylytryptamine. Biosynthesis of these β-carboline derivatives has been reported [7].

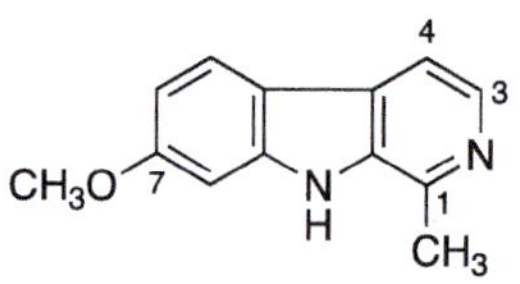

harmine
harmaline **3,4-H$_2$**

In "Rig-veda," the oldest document composed in India, in the 13–12 century BC, a drink called "Soma" is described. It is said that persons who consumed this drink felt as if they were flying high in the sky like a bird, and could hear the songs of the poets. The constitution of "Soma" has been long discussed. Some suggest that it is a plant of *Ephedra* sp. (Ephedraceae), and others suggest that it is *Cannabis sativa* (Moraceae), or the mushroom *Amanita muscaria.* According to Schultes and Hofmann, the harmala described herein may be "Soma" [8]. On the other hand, a dimeric β-carboline derivative was isolated from the marine invertebrate *Didemnum* sp. (a sea squirt) [9]. This type of sea squirt is beautifully colored, and among them, a green specimen was collected at the southern area of the Great Barrier Reef, off the northeastern coast of Australia. The pigment of this sea squirt was isolated, and it was clarified that the chemical structure of this pigment is an unusual *N–N* dimer of the β-carboline alkaloid norharmane. This alkaloid was prepared previously from norharmane by photochemical reaction [10], and this was the first isolation from natural sources.

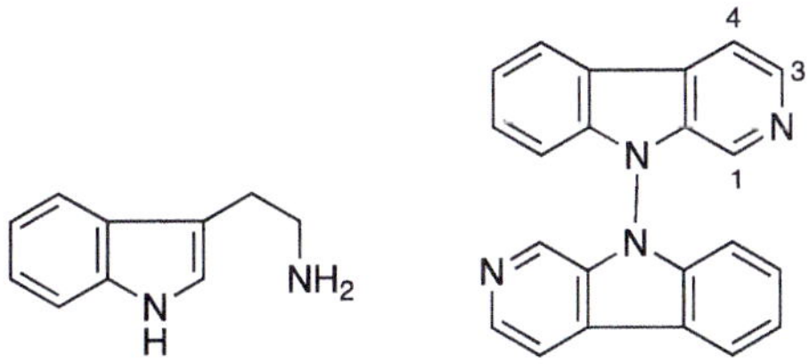

tryptamine **β -carboline dimer**

LITERATURE CITED

[1] F. Goebel, Liebigs Ann. der Chem. 38 (1841) 363.
[2] R.H.F. Manske, J. Chem. Soc. 1 (1927).
[3] R.H.F. Manske, R. Robinson, J. Chem. Soc. 240 (1927).
[4] E. Späth, E. Lederer, Chem. Ber. 63B (1930) 120.
[5] F.D. O'Connell, E.V. Lynn, J. Am. Pharm. Assoc. 42 (1953) 753.
[6] F.A. Hochstein, A.M. Paradies, J. Am. Chem. Soc. 79 (1957) 5735.
[7] R.B. Herbert, J. Mann, The biosynthesis of the β-carboline alkaloids, harman and eleagnine, J. Chem. Soc. Perkin Trans. 1 (1982) 1523–1525.
[8] R.E. Schultes, A. Hofmann, Plants of the Gods, p. 53, McGraw-Hill, New York, 1979R.
[9] P.S. Kearns, J.C. Coll, J.A. Rideout, J. Nat. Prod. 58 (1995) 1075.
[10] R. Erra-Balsells, A.R. Frasca, Tetrahedron 39 (1983) 33.

2.7 *Picrasma* sp. AND NIGAKINONE

Picrasma ailanthoides (Simaroubaceae) is a deciduous tree that grows wild in the mountains and plains of Japan. The dried wood without the bark is called "Nigaki" in Japan, and is used for the treatment of gastrointestinal disorders such as indigestion, diarrhea, and inflammation of the stomach and intestines.

As bitter principles of this material, the nigakilactone group of compounds, such as the degraded triterpenes, quassinoids, were isolated [1]. On the other hand, from the heart wood of old timber, alkaloids such as nigakinone, methylnigakinone, 1-hydroxymethyl-β-carboline, and methyl β-carboline-1-carboxylate were obtained [2–4]. Among these alkaloids, the former two alkaloids are members of the canthin-6-one group of alkaloids. Utilizing this core structure name, nigakinone is also known as 5-hydroxy-4-methoxycanthin-6-one, and methylnigakinone is 4,5-dimethoxycanthin-6-one.

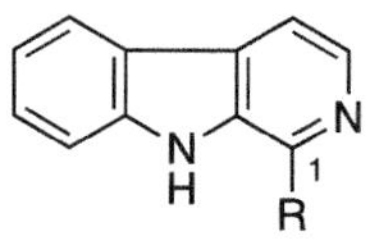

1-hydroxymethyl-β-carboline $R = CH_2OH$
methyl β-carboline-1-carboxylate $R = COOCH_3$

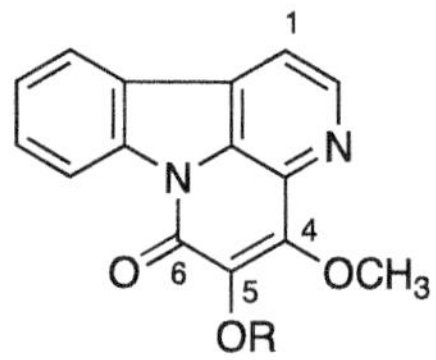

nigakinone $R = H$
methylnigakinone $R = CH_3$

It is considered that the β-carboline skeleton is formed by adding a C_2 unit derived from the polyketide biosynthetic route to tryptamine. On the other hand, the canthin-6-one skeleton is formed by adding a C_4 unit, such as acetoacetate, derived from the polyketide biosynthetic route.

Only two species of Simaroubaceae plants are known in Japan, namely *P. ailanthoides* described above, and *Ailanthus altissima*. The former plant grows wild in Japan, whereas the latter plant is a deciduous ornamental tree native to China that was imported into Japan at the end of the nineteenth century. This tree is known as "Tree of Heaven," meaning that this tree grows very fast to be able to touch heaven quickly. Canthin-6-one alkaloids isolated from Simaroubaceae plants were reviewed by Ohmoto and Zasshi [5].

The biosynthetic route of 1-methoxycanthin-6-one was studied [6], and it was clarified that the methoxyl moiety of this alkaloid was introduced after the canthin-6-one skeleton was formed.

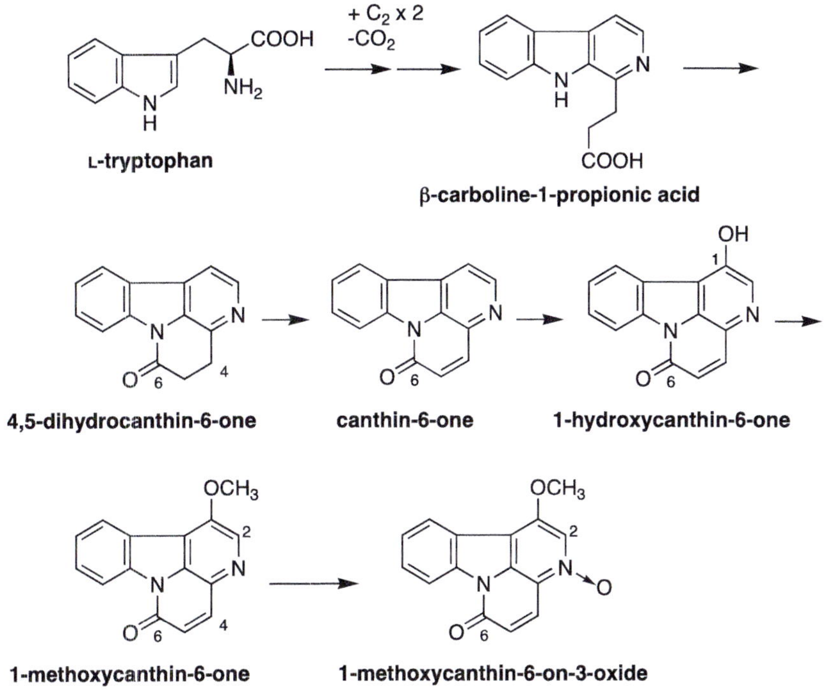

Biosynthetic Pathways of 1-Methoxycanthin-6-one and Related Alkaloids

LITERATURE CITED

[1] T. Murae, T. Tsuyuki, T. Takahashi, Chem. Pharm. Bull. 19 (1971) 1747.
[2] N. Inamoto, S. Masuda, O. Simamura, T. Tsuyuki, Bull. Chem. Soc. Jpn. 34 (1961) 888.
[3] Y. Kimura, M. Takido, S. Koidzumi, Yakugaku Zasshi 87 (1967) 1371.
[4] Y. Kondo, T. Takemoto, Chem. Pharm. Bull. 21 (1973) 837.
[5] T. Ohmoto, Yakugaku Zasshi 115 (1995R) 261.
[6] N. Crespi-Perellino, A. Guicciardi, G. Malyszko, A. Minghetti, J. Nat. Prod. 49 (1986) 814.

2.8 *Clerodendron* AND TRICHOTOMINE

Clerodendron trichotomum (Verbenaceae) is a deciduous shrub that is widely distributed in Japan, Taiwan, Korea, and China. The leaves have an unpleasant odor, but the plant puts forward sweet-scented white flowers in the summer. The fruit are globular, and the color of the ripe fruit is deep blue. The leaves and stems of *C. trichotomum* are used for the treatment of hypertension in China.

From the seeds, trichotomine was isolated as a deep blue pigment and its chemical structure determined [1]. The skeleton of this alkaloid is, as in the case of nigakinone, formed by the addition of a C_4 unit to a tryptamine moiety. However, in this instance, the direction of the ring formation is towards N_b rather than N_a. Trichotomine is considered to be formed by a dimerization reaction, as in the case of indigo (Section 2.4) [1].

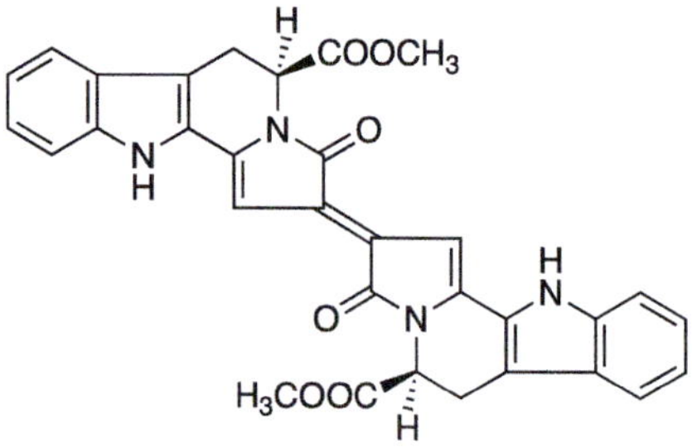

trichotomine

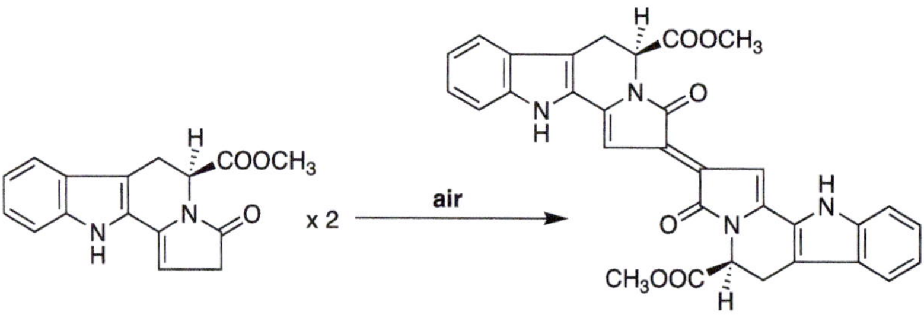

trichotomine

Formation of Trichotomine

Trichotomine possesses hypotensive and sedative activities, and the same alkaloid was also isolated from *Premna microphylla* (Verbenaceae) [2]. Trichotomine also possesses a DNA fragmentation effect, and it was clarified that the activity is not affected by light, or by the presence of metal ions, such as Cu^{2+}, Fe^{2+}, Fe^{3+}, Co^{2+}, and Zn^{2+}. This activity occurs by the formation of superoxide anion in the presence of oxygen, and, in the absence of oxygen, the alkaloid acts with DNA directly [3].

LITERATURE CITED

[1] S. Iwadare, Y. Shizuri, K. Sasaki, Y. Hirata, Tetrahedron Lett. 15 (1974) 1051.
[2] S. Iwadare, Y. Shizuri, K. Sasaki, Y. Hirata, Jpn. Kokai Sho 51 (1976) 41415.
[3] M. Haruna, M. Hata, H. Ohnishi, K. Ito, in: The 115th Annual Meeting of the Pharmaceutical Society of Japan, Vol. 2, 1995, p. 246.

2.9 IBOGA AND IBOGAINE

Tabernanthe iboga (Apocynaceae) is a shrub that grows wild in the Gabonese Republic, the Congo, and Zaire, and it is cultivated in western Africa.

The roots of this plant are known as "Iboga" by the indigenous peoples, and are used as an aphrodisiac and hallucinogen in male adult initiation cermonies. Among the alkaloids contained in *T. iboga*, the main alkaloid is ibogaine. The presence of ibogaine has been known from the beginning of the twentieth century [1], whereas the partial structure [2], total plane structure [3], and absolute structure [4] were clarified much later. ^{13}C nuclear magnetic resonance (NMR) spectral data are available [5]. Ibogaine was also isolated from other Apocynaceae plants [6] and is currently of great interest for use against narcotic and other forms of addiction.

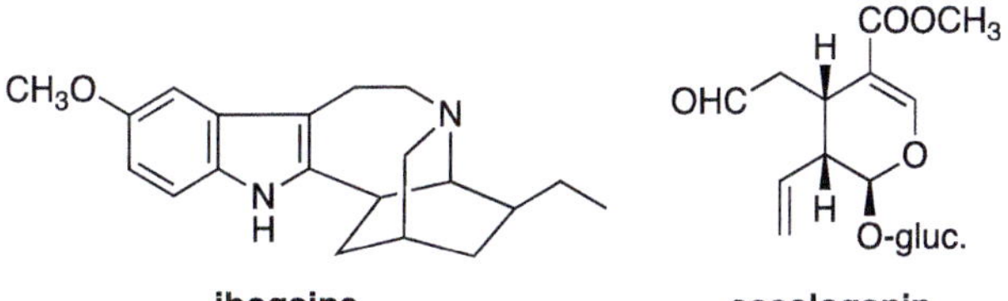

ibogaine **secologanin**

Several thousand complex indole alkaloids are composed of tryptamine and a monoterpenoid secologanin unit, which is subsequently extensively modified. The most frequently occurring carbon skeletons derived from secologanin are those of the corynanthe, iboga, and Aspidosperma types. These alkaloids possess the basic carbon frameworks indicated in the Figure. Among them, some are constructed with nine carbons, and in that case the carbon indicated by a dotted line is eliminated in the biosynthetic process [7,8]. A unique feature of the numbering system of all monoterpenoid indole alkaloids is that, irrespective of the skeleton, the numbering of individual carbons is tracked biogenetically [9].

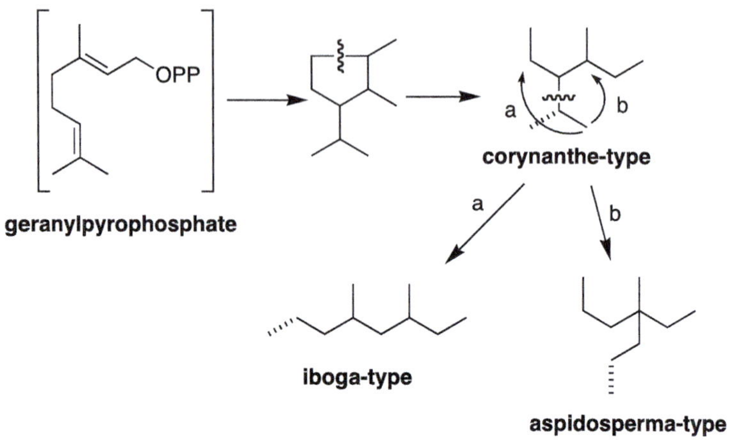

Biogenetic Pathways of the Corynanthe-Aspidosperma and Iboga-type Partial Structures of Monoterpenoid Indole Alkaloids (Carbons indicated by a dotted line may be omitted)

LITERATURE CITED

[1] J. Dybowski, E. Landrin, Compt. Rend. 133 (1901) 748.
[2] E. Schlittler, C.A. Burckhardt, E. Gellért, Helv. Chim. Acta 36 (1953) 1337.
[3] M.F. Bartlett, D.F. Dickel, W.I. Taylor, J. Am. Chem. Soc. 80 (1958) 126.
[4] K. Blàha, Z. Koblicová, J. Trojánek, Tetrahedron Lett. 13 (1972) 2763.
[5] E. Wenkert, D.W. Cochran, H.E. Gottlieb, E.W. Hagaman, R.B. Filho, F.J. de Abreu Matos, M.I.L.M. Madruga, Helv. Chim. Acta 59 (1976) 2437.
[6] N. Ghorbel, M. Damak, A. Ahond, E. Philogène, C. Poupat, P. Potier, H. Jacquemin, J. Nat. Prod. 44 (1981) 717.
[7] G.A. Cordell, Lloydia 37 (1974) 219–298.
[8] S.E. O'Connor, J.J. Maresh, Nat. Prod. Rep. 23 (2006) 532.
[9] W.I. Taylor, J. Le Men, Experientia 21 (1965) 508.

2.10 *Rauvolfia* AND RESERPINE

The roots of *Rauvolfia serpentina* (Apocynaceae) have been used for the treatment of venomous snake bites, for mental diseases, and as an antipyretic. In Japan, the crude drug is known as "Indo-Jaboku."

In 1933, Chopra et al. found that a crystallized alkaloid isolated from the roots of this plant showed hypotensive activity [1]. Subsequently, reserpine, an alkaloid possessing sedative activity, was isolated and characterized in 1952 [2]. The chemical structure, including stereochemistry, was determined subsequently [3,4], and the first total synthesis of reserpine [5] was achieved by Woodward et al.

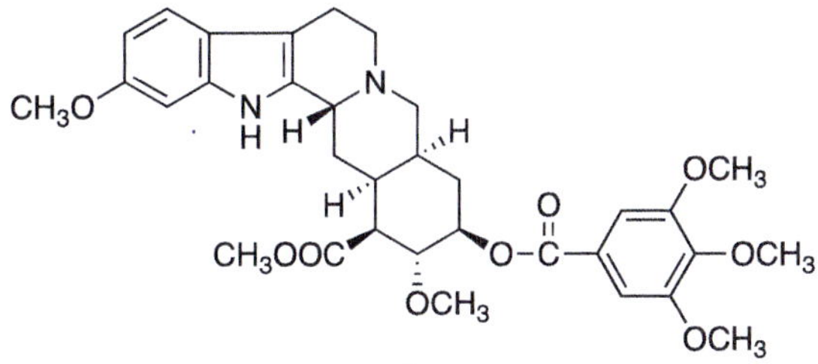

reserpine

Reserpine is an important hypotensive drug that was used clinically and that also possesses sedative activity. Reserpine is a major tranquilizer because it exhausts the noradrenaline (NAdr = norepinephrine, NE) of the adrenergic nerve and suppresses the function of the sympathetic nerves. In particular, the incorporation of dopamine, the precursor of NE, is inhibited. On the other hand, stored NE is emitted by the excitement of the sympathetic nerve, and is destroyed by monoamine oxidase (MAO). Thus the amount of NE is gradually reduced, and, as the NE in the nerve ending is depleted, the sedative and hypotensive activities occur. Reserpine is rarely used clinically in the field of psychiatry now because of the availability of the phenothiazine-type drugs.

In addition to reserpine, ajmaline was isolated from the roots of *R. serpentina* at almost the same time [6,7]. Its chemical structure, including absolute stereochemistry, was established in 1962 [8–10]. The biological activity of ajmaline is quite different from that of reserpine. It possesses anti-arrhythmia activity and is used clinically for this purpose.

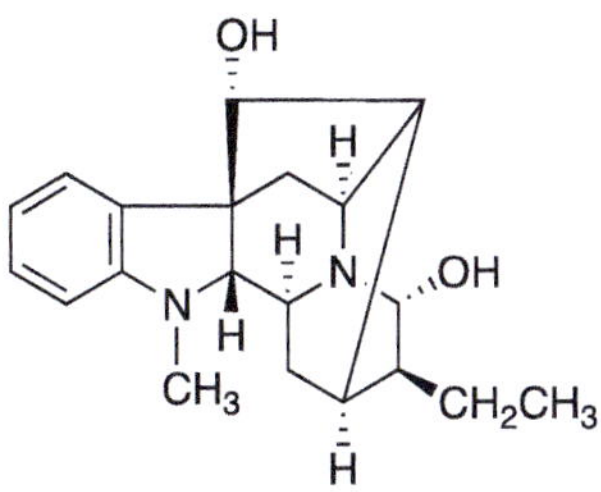

ajmaline

Both reserpine and ajmaline possess chemical structures composed of a tryptamine and a corynanthe unit (Section 2.9). Although the complete C_{10} unit of the corynanthe system is incorporated in reserpine, in the case of ajmaline one of the carbons is lost.

LITERATURE CITED

[1] R.N. Chopra, J.C. Gupta, B. Mukherjee, Indian J. Med. Res. 21 (1933) 261 [*Chem. Abstr.*, 28, 8275 (1934)].
[2] J.M. Mueller, E. Schlitter, H.J. Bein, Experientia 8 (1952) 338 [*Chem. Abstr.*, 47, 8084d (1953)].
[3] N. Neuss, H.E. Boaz, J.W. Forbes, J. Amer. Chem. Soc. 76 (1954) 2463.
[4] P.A. Diassi, F.L. Weisenborn, C.M. Dyrion, O. Wintersteiner, J. Amer. Chem. Soc. 77 (1955) 4687.
[5] R.B. Woodward, F.E. Bader, H. Bickel, A.J. Frey, R.W. Kierstead, Tetrahedron 2 (1958) 1.
[6] S. Siddiqui, R.H. Siddiqui, J. Ind. Chem. Soc. 9 (1932) 539.
[7] S. Siddiqui, R.H. Siddiqui, J. Ind. Chem. Soc. 12 (1935) 37.
[8] A. Chatterjee, S. Bose, J. Ind. Chem. Soc. 31 (1954) 17.
[9] F.A.L. Anet, D. Chakravarti, R. Robinson, E. Schlittler, J. Chem. Soc. 1242 (1954).
[10] M.F. Bartlett, R. Sklar, W.I. Taylor, E. Schlittler, R.L.S. Amai, P. Beak, N.V. Bringi, E. Wenkert, J. Amer. Chem. Soc. 84 (1962) 622.

2.11 YOHIMBE AND YOHIMBINE

Pausinystalia yohimba (*Corynanthe johimbe*) (Rubiaceae) is an evergreen tree that grows wild in the southern areas of Africa, and is known locally as yohimbe. The bark of *P. yohimba* has been used as an aphrodisiac drug for centuries.

The main alkaloid of this plant material was isolated in 1896 by Spiegel [1] and named yohimbine. The alkaloid was the same as one obtained previously from *Aspidosperma quebracho-blanco* (Apocynaceae) [2]. Isoyohimbine and allo-yohimbine, possessing structures similar to that of yohimbine, were also isolated from *P. yohimba* [3,4]. Isoyohimbine, based on the studies of Warnat [3], is known as α-yohimbine.

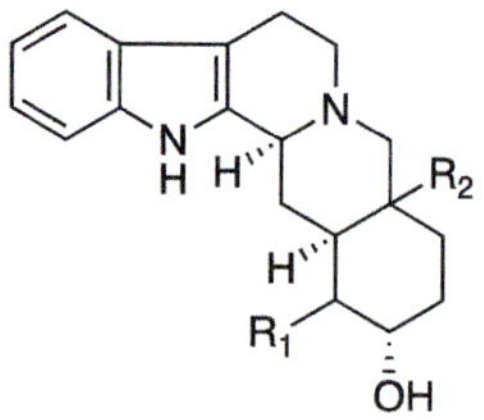

yohimbine	**R_1 = α-COOCH$_3$, R_2 = β-H**
α-yohimbine	**R_1 = β-COOCH$_3$, R_2 = α-H**
***allo*-yohimbine**	**R_1 = α-COOCH$_3$, R_2 = α-H**

The planar structure of yohimbine was reported by Witkop in 1943 [5] and by Clemo and Swan in 1946 [6], and its absolute chemical structure was clarified in 1964 [1]. Total synthesis of yohimbine has been achieved by several groups, including Kametani et al. [7]. On the other hand, the chemical structure, including absolute stereochemistry, of α-yohimbine was not clarified until 1961 [8]. The total synthesis of this alkaloid was completed, and, during these studies, the chemical structure of *allo*-yohimbine was revised [9].

α-Yohimbine was also isolated from *Rauvolfia tetraphylla* (Apocynaceae) and the biosynthetic route of this alkaloid was studied using this plant [10]. The alkaloid is derived from tryptamine and secologanin through strictosidine, followed by extensive rearrangements to form the carbocyclic ring. Reserpine (Section 2.10) is an elaborated member of this group.

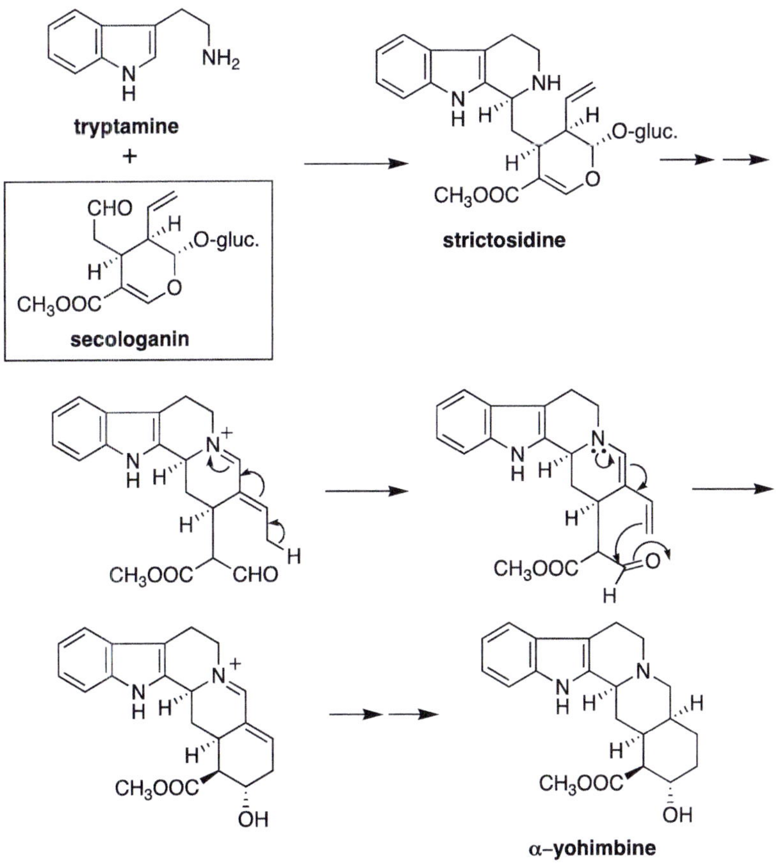

Biosynthetic Route to α-Yohimbine

Yohimbine causes the blocking of the α_2 receptor of the sympathetic nerve, and inhibits the separation of NE from the receptor ending. As a result, extension of the blood vessels in the skin and mucous membranes, and especially of the external genitals, occurs. This alkaloid is also said to possess an activity to stimulate the central nervous system and cause priapism. That is why this alkaloid is used for erectile dysfunction; but the effective dose is only one-third of the toxic dose.

LITERATURE CITED

[1] Y. Ban, O. Yonemitsu, Tetrahedron 20 (1964R) 2877.
[2] O. Hesse, Ber. Dtsch. Chem. Ges. (Chem. Ber.) 13 (1880) 2308.
[3] K. Warnat, Chem. Ber. 59 (1926) 2388.
[4] G. Hahn, W. Brandenberg, Chem. Ber. 60 (1927) 669.
[5] B. Witkop, Liebigs Ann. der Chem. 554 (1943) 83.
[6] G.R. Clemo, G.A. Swan, J. Chem. Soc. 617 (1946).
[7] T. Kametani, Y. Hirai, K. Fukumoto, Chem. Pharm. Bull. 24 (1976) 2500.
[8] M.-M. Janot, R. Goutarel, E.W. Warnhoff, A.L. Hir, Bull. Soc. Chim. Fr. 637 (1961).
[9] L. Töke, K. Honty, L. Szabó, G. Blaskó, Cs. Szántay, J. Org. Chem. 38 (1973) 2496.
[10] N. Nagakura, M. Rüffer, M.H. Zenk, J. Chem. Soc. Perkin Trans. 1 (1979) 2308.

2.12 *Camptotheca* AND CAMPTOTHECIN

Camptothecin is a quinoline alkaloid isolated from the stems of *C. acuminata* (Cornaceae) [1], and 10-hydroxycamptothecin and 10-methoxycamptothecin were also obtained as minor alkaloids [2]. *C. acuminata* is an ornamental plant in China. Campthothecin has also been isolated from both the Apocynaceae and Rubiaceae plant families.

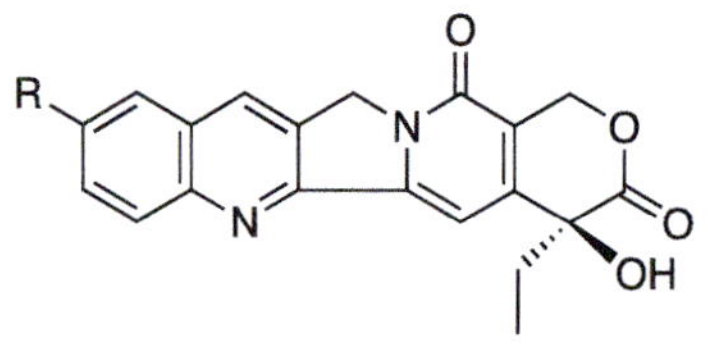

camptothecin	**R = H**
10-hydroxycamptothecin	**R = OH**
10-methoxycamptothecin	**R = OCH_3**

Although camptothecin possesses a quinoline skeleton, this alkaloid is derived from tryptamine. Thus tryptamine, as well as mevalonic acid and secologanin, are incorporated into this alkaloid [3]. It was also reported that [5-^{13}C]strictosamide is incorporated into camptothecin, and that the C-5 position of this alkaloid is labeled [4]. The incorporation rate of strictosamide

(3*S*-isomer) was 2.0%, whereas the incorporation rate of 18,19-dihydrostrictosamide was higher than that of strictosamide and was 4.7%. Furthermore, the incorporation rate of vincosamide, the 3*R* isomer of strictosamide, was very low. This was an early report that the 3*S* isomer is necessary as a precursor for the biosynthesis of monoterpene indole alkaloids.

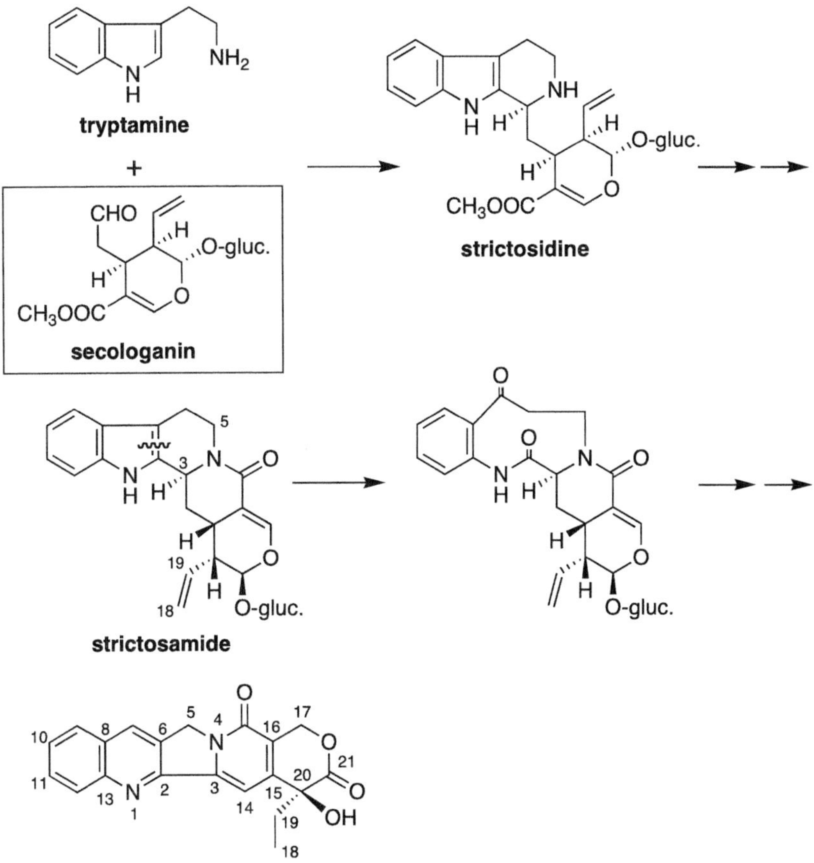

Biogenetic Pathway of Camptothecin

It was shown that camptothecin inhibits DNA and RNA synthesis at a concentration of 5 mmol/l by using HeLa and L1210 cell lines. It shows inhibitory activity against a number of tumor cell lines and possesses a wide spectrum of antitumor activity [5]. Clinically it has side effects, such as disorder of the bone marrow, disorder of the digestive system such as nausea, vomiting, and diarrhea, and a decrease in white blood cells. Consequently it is only occasionally used for bladder cancer in Japan. Two other derivatives, irinotecan and topotecan, have greater bioavailability and were approved for use in the treatment of refractory ovarian cancer [6].

LITERATURE CITED

[1] M.E. Wall, M.C. Wani, C.E. Cook, K.H. Palmer, A.T. McPhail, G.A. Sim, J. Am. Chem. Soc. 88 (1966) 3888.
[2] M.C. Wani, M.E. Wall, J. Org. Chem. 34 (1969) 1364.
[3] C.R. Hutchinson, A.H. Heckendorf, P.E. Daddona, E. Hagaman, E. Wenkert, J. Am. Chem. Soc. 96 (1974) 5609.
[4] G.M. Sheriha, H. Rapoport, Phytochemistry 15 (1976) 505.
[5] J.W. Lown, H.-H. Chen, Biochem. Pharmacol. 29 (1980) 905.
[6] M. Potmesil, H. Pinedo (Eds.), Camptothecins: New Anticancer Agents, CRC Press, Boca Raton, FL, 1995, p. 149.

2.13 *Uncaria* AND RHYNCHOPHYLLINE

The dried thorn hooks of *Uncaria sinensis, U. rhynchophylla*, and *U. kawakamii* (Rubiaceae) are known as "Cho-Toh-Koh" in the Chinese herbal medicine system in Japan, and have been used as a sedative for a long time. An ethanol extract of this crude drug provides a temporary hypotensive response in the rabbit (150 mg/kg), with an increase in breathing rate and a decrease in arterial blood flow [1].

As one of the active principles, rhynchophylline, was isolated from the stems and roots of *U. sinensis* [2], and its chemical structure [3] and stereochemistry [4] were determined by chemical degradation and partial synthesis. A stereoselective total synthesis of this alkaloid was achieved [5].

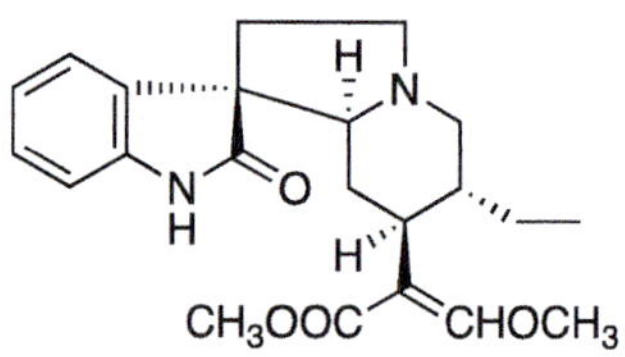

rhynchophylline

Rhynchophylline is composed of tryptamine and a C_{10} unit with the corynanthe skeleton. Thus, the partial structure of this alkaloid somewhat resembles those of reserpine (Section 2.10) and yohimbine (Section 2.11), although its biological activities are different. Unlike reserpine, rhynchophylline does not cause sedative activity, and, unlike yohimbine, no convulsive activity occurred when it was given to frogs and mice. In addition, unlike reserpine, which caused long-lasting hypotensive activity, rhyncophylline caused only a short period of hypotensive activity followed by a lasting hypertensive activity.

LITERATURE CITED

[1] T. Namba, Genshoku Wakanyaku Zukan ("Illustrated Reference Book of Crude Drugs"), p. 187, Hokuryukan Publishing Co. Ltd, Tokyo, Japan, 1980R.
[2] H. Kondo, T. Fukuda, M. Tomita, Yakugaku Zasshi 48 (1928) 54.
[3] J.C. Seaton, M.D. Nair, O.E. Edwards, L. Marion, Can. J. Chem. 38 (1960) 1035.
[4] Y. Ban, T. Oishi, Chem. Pharm. Bull. 11 (1963) 451.
[5] Y. Ban, M. Seto, T. Oishi, Tetrahedron Lett. 13 (1972) 2113.

2.14 NUX VOMICA AND STRYCHNINE

Strychnos nux-vomica (Loganiaceae) is a tree that grows wild in India, Sri Lanka, and the northern area of Australia. The seeds of this plant are known as "Machin-shi" (in Japanese) or vomica (Strychni Semen), and they are used for maintaining health of the stomach at a medicinal dose. The seeds are also the source of strychnine nitrate, and have found use for heart and lung diseases and for Raynaud's disease.

Strychnine and brucine were isolated as the main alkaloids of vomica (also known as nux vomica in other parts of the world). Strychnine was isolated at the beginning of the nineteenth century, but it was not until the middle of the twentieth century when the chemical structure of this alkaloid was clarified [1]. The now-classic total synthesis of strychnine was achieved by Woodward et al. [2].

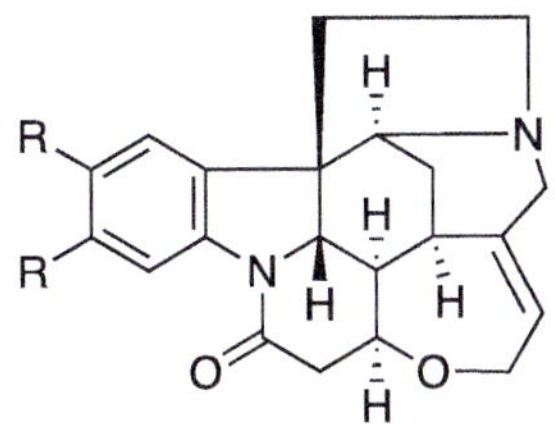

strychnine **R = H**
brucine **R = OCH_3**

Strychnine is a very poisonous substance, and its lethal dose for humans is in the range of 0.03–0.1 g. Therefore, a single seed of vomica may be lethal. On the other hand, the toxicity of brucine is about 1/20 to 1/30 that of strychnine. As a symptom of strychnine toxicity, typical convulsions occur. This convulsion caused by strychinine is repeated by slightly stimulating the bodies of experimental animals.

The seeds of *S. ignatii* are known as ignatius beans, and are used as a stimulant to restore body functions in both European and Chinese medicine. Ignatius beans also contain strychnine and brucine.

LITERATURE CITED

[1] V. Prelog, S. Szpiltogel, Helv. Chim. Acta 28 (1945) 1669.
[2] R.B. Woodward, M.P. Cava, W.D. Ollis, A. Hunger, H.U. Daeniker, K. Schenker, Tetrahedron 19 (1963) 247.

2.15 CALABASH CURARE AND C-CURARINE AND C-TOXIFERINE I

As described in Section 1.5, there are at least three kinds of arrow poisons used by the indigenous peoples of the Amazon basin: these are (1) tubocurare (tube of bamboo curare), (2) pot curare, and (3) calabash curare. As described in Section 1.5, the active principle of tubocurare is D-tubocurarine, which is derived from phenylalanine.

On the other hand, calabash curare is an arrow poison prepared from *Strychnos toxifera* (Loganiaceae) and related plants of the same genus. *S. toxifera* is also contained in the pot curare mentioned above.

The poisonous principle of calabash curare is C-curarine, a dimeric alkaloid derived from tryptophan. This alkaloid was first reported in 1937 [1], and total synthesis was achieved in 1964 [2]. C-Toxiferine I, an alkaloid with a structure similar to that of C-curarine, and with hydroxy groups at the 18 and 18′-positions, was isolated and characterized from the same plant [3,4]. These alkaloids possess two of the strychnine-like moieties described in the previous section. The "C" of these alkaloids indicates the origin from calabash.

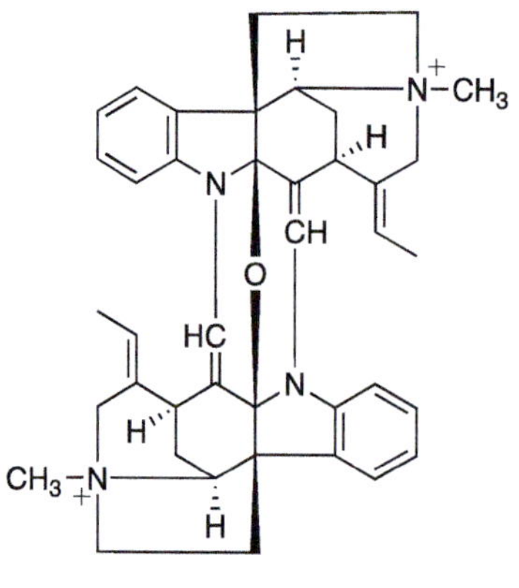

C-curarine

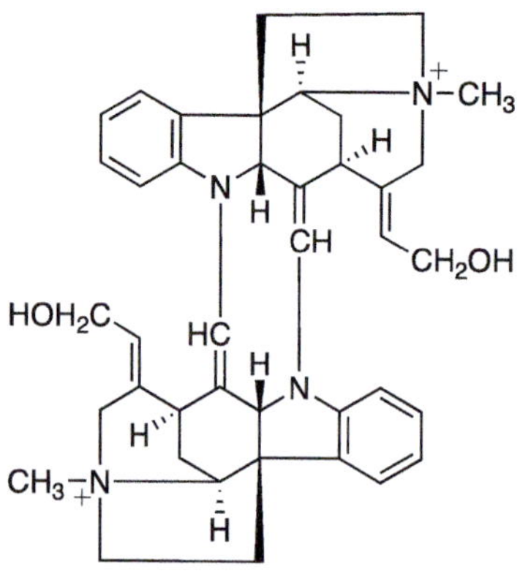

C-toxiferine I

C-Toxiferine possesses 250 times stronger muscle relaxant activity than D-tubocurarine, described in Section 1.5. However, it has drawbacks, in that it is too long acting to control muscle relaxant activity accurately, and it is also unstable in solution.

LITERATURE CITED

[1] H. Wieland, W. Konz, R. Sonderhoff, Liebigs Ann. Chem. 527 (1937) 160.
[2] M. Grdinic, D.A. Nelson, V. Boekelheide, J. Am. Chem. Soc. 86 (1964) 3357.
[3] H. Wieland, K. Bähr, B. Witkop, Liebigs Ann. Chem. 547 (1941) 156.
[4] W. Arnold, M. Hesse, H. Hiltebrand, A. Melera, W. von Philipsborn, H. Schmid, P. Karrer, Helv. Chim. Acta 44 (1961) 620.

2.16 ERGOT, ERGOT ALKALOIDS, AND LSD

When the parasitic fungus *Claviceps purpurea* lives on rye and other cereal crops, a sclerotium in the shape of horn and called ergot is produced. Ergot was the subject of great fear previously, because people who ate rye products infected with this, fungus experienced a strange and debilitating, and frequently lethal, disease. Ergot contains alkaloids that contract the blood vessels of arms and legs, preventing circulation of the blood, and gangrene results. Thus, people suffering from ergot poisoning lost their hands and legs without bleeding (ergotism), after they became darkened. Because, at first, the patients' felt heat in their arms and legs, the disease was called St. Anthony's fire in the Middle Ages. Many deaths occurred, particularly in France and Germany, in the period 900–1500 AD As an old record of ergot, a caution against ergot was found on a clay tablet from Assyria of 600 BC, and the records of St. Anthony's Fire extend to 1928.

Although it was well known that ergot was very dangerous, midwives in Europe were also using it for promotion of the contraction of the womb post-parturition. Subsequently, studies of the active principle(s) of ergot responsible for the contractions were initiated.

The first alkaloid isolated in crystalline form was ergotinine [1], but it did not possess the uterocontracting activity. Ergotoxine was the first biologically active alkaloid isolated [2], and was shown to be a mixture of ergocristine, ergocornine, and related alkaloids [3]. Ergotamine was isolated later in a pure form [4]. The common skeleton of these ergot alkaloids is known as lysergic acid, and ergotamine comprises a structure based on lysergic acid coupled with a peptide moiety comprising three amino acids.

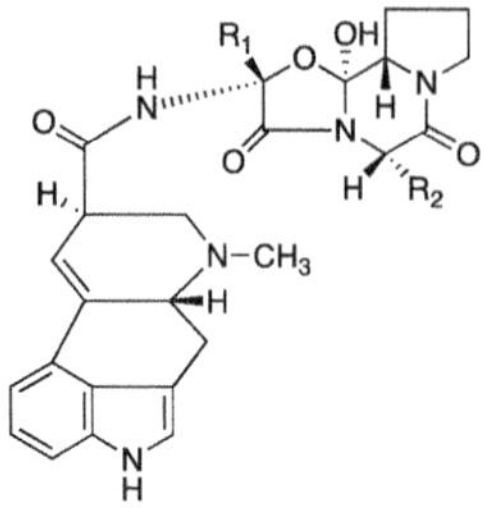

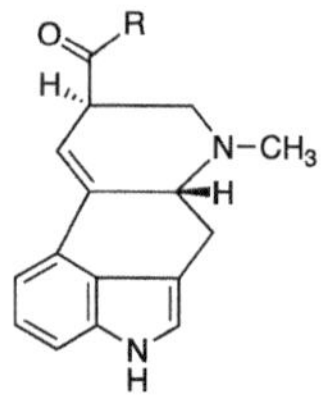

ergotamine $R_1 = CH_3$, $R_2 = CH_2C_6H_5$
ergocristine $R_1 = CH(CH_3)_2$, $R_2 = CH_2C_6H_5$
ergocornine $R_1 = R_2 = CH(CH_3)_2$

lysergic acid R = OH
LSD $R = N(CH_2CH_3)_2$
ergine $R = NH_2$

The universal skeleton of the ergot alkaloids is known as the ergoline nucleus. It was shown by using cultivated *Claviceps* that the ergoline skeleton was derived biosynthetically from tryptophan and a C_5 unit of mevalonic acid origin. Thus, when labeled D-tryptophan was fed, the labeled α-hydrogen (98% loss) and a nitrogen of the amino moiety (90% loss) were not incorporated into the ergoline skeleton. On the other hand, when labeled DL-tryptophan was fed, the losses of radioactivity were 57% and 50%, respectively. Consequently, it was deduced that for the biosynthesis of this nucleus, an intact L-tryptophan was incorporated. Interestingly, the stereochemistry of the C-5 position is opposite that in L-tryptophan. So, the ergoline skeleton was formed by incorporating L-tryptophan, and the stereochemistry of the α-carbon of the amino acid was inverted at some point in the pathway [5].

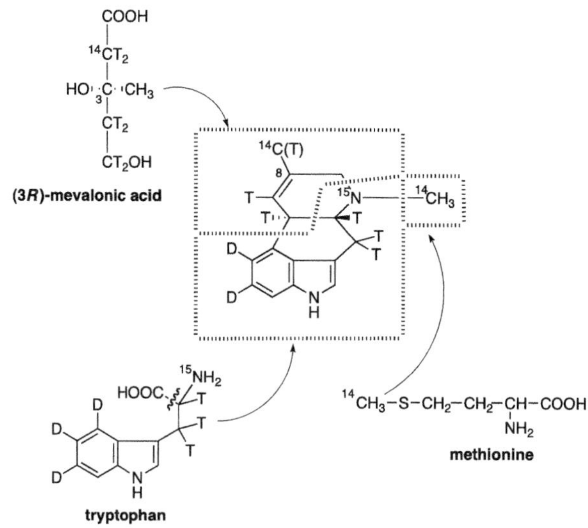

Incorporation of Labelled Mevalonic Acid, Tryptophan and Methionine Into the Ergoline Skeleton

Furthermore, it was found that (3*R*,*S*)-[2^{14}C]mevalonic acid was incorporated 50 times better than (3*S*)-[2-^{14}C]mevalonic acid. This indicates that as a precursor of the ergoloine nucleus, (3*R*)-mevalonic acid is 100 times superior to (3*S*)-mevalonic acid [5]. It was also clarified that the *N*-methyl moiety was derived from methionine [6].

Simple lysergic acid amides, such as ergine and lysergic acid hydroxyethylamide, are obtained from the ergot that lives on wild grass. These alkaloids are detected only as trace constituents in the ergot that lives on rye. These alkaloids are also obtained from the seeds of *Ipomoea violacea* and *Turbina corymbosa* (Convolvulaceae) [7]. The seeds of *I. violacea* are used as a hallucinogen under the name of ololiuqui. These alkaloids, which probably arise from a parasitic fungus [8], possess about 1/100 the hallucinatory activity of LSD, which will be described below.

LSD is a semi-synthetic compound prepared from lysergic acid by A. Hofmann of Sandoz Co. LSD is the diethylamide derivative of lysergic acid, and is named based on the initials of its German name, Lyserg Säure Diethylamid. LSD has provoked a number of social issues, and is in a group of such alkaloids, together with morphine and its diacetate derivative heroin (Section 1.10), cocaine (Section 3.3), and the stimulants related to ephedrine (Section 16.1).

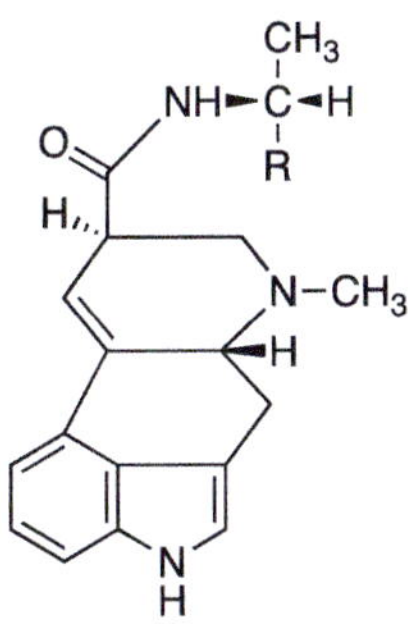

ergometrine **R = CH$_2$OH**
lysergic acid 2-hydroxyethylamide **R = OH**

A sudden and temporary hypertension (α-effect) and subsequent hypotension (β-effect) are observed from the intravenous injection of epinephrine. However, if ergotamine or ergotoxine was injected previously, the α-effect was not observed, and only the β-effect (hypotensive activity) occurred. This is known as the α-blocking effect of the ergot alkaloids. Such an effect is also shown by yohimbine, described previously (Section 2.11). On the other hand, through treatment with ergometrine, an immediate and strong uterine contraction occurred. Consequently, this alkaloid is used for the prevention of bleeding after childbirth, and for an incomplete miscarriage.

LITERATURE CITED

[1] M.C. Tanret, Compt. Rend 81 (1875) 896.
[2] G. Barger, F.H. Carr, J. Chem. Soc. 337 (1907).
[3] Z. Řeháček, P. Sajdl, Ergot Alkaloids, p. 41, Academia, Praha, 1990R.
[4] A. Stoll, Helv. Chim. Acta 28 (1945) 1283.
[5] H.G. Floss, Tetrahedron 32 (1976) 873.
[6] F. Weygand, H.G. Floss, Angew. Chem. 75 (1963) 783.
[7] R.E. Schultes, A. Hofmann, Plants of the Gods, p. 162, McGraw-Hill, New York, 1979R.
[8] R.E. Schultes, A. Hofmann, Plants of the Gods, p. 175, McGraw-Hill, New York, 1979R.

2.17 *Cinchona* AND QUININE

Cinchona ledgeriana and *C. succirubra* (Rubiaceae) are trees that grow wild in the Andes mountains of Peru and Bolivia. Crude drugs prepared from the trunk, stem, and root bark of these trees are the source of the anti-malarial agent quinine. These plants are cultivated in the Java islands, and most of them are *C. ledgeriana*. To prepare the crude drug, the 20- to 25-year-old, trees are dug out, and the bark of the trunks, as well as the stems and roots, are gathered. The alkaloid content is typically in the range of 5–8%, and the main alkaloid, quinine, accounts for two-thirds of the total alkaloids.

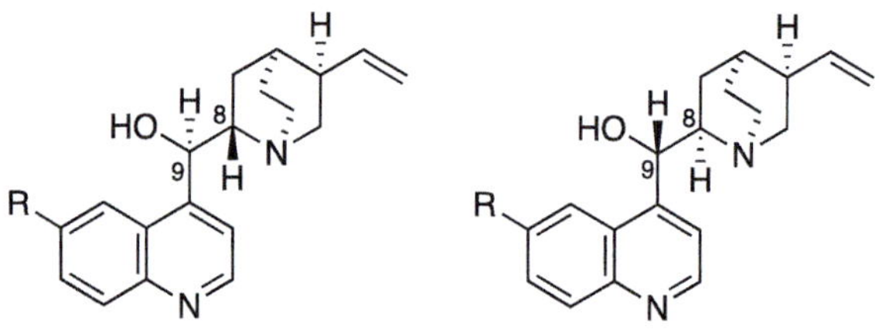

quinine **R = OCH$_3$** **quinidine** **R = OCH$_3$**
cinchonidine **R = H** **cinchonine** **R = H**

Other alkaloids co-occurring with quinine are cinchonine, quinidine, and cinchonidine. Among them, quinine and quinidine, as well as cinchonine and cinchonidine, are pairs of stereoisomers at the C-8 and C-9 positions. Quinine and cinchonidine possess the 8*S* and 9*R* geometry, whereas quinidine and cinchonine possess the 8*R* and 9*S* geometry.

These alkaloids possess quinoline and quinuclidine skeleta, and it was established that the precursor of the quinoline moiety of these alkaloids is tryptophan, as in the case of the indole alkaloids. It was further established that the remaining carbon framework was derived from the C$_9$ unit of corynanthe type C$_{10}$ unit (geraniol, section 2.9).

Thus, when DL-[2′-^{14}C]tryptophan was fed *to C. succirubra*, ^{14}C was incorporated into C-2′ of quinine [1]. On the other hand, when [1-^{15}N, 2-^{14}C]tryptophan was fed, the N and C-9 positions of quinine were labeled. Furthermore, the C-10 position of quinine was labeled by feeding [3-^{14}C] geraniol. From these observations, it was estimated that quinine was biosynthesized via a Corynanthe indole alkaloid, as figure shown below [2].

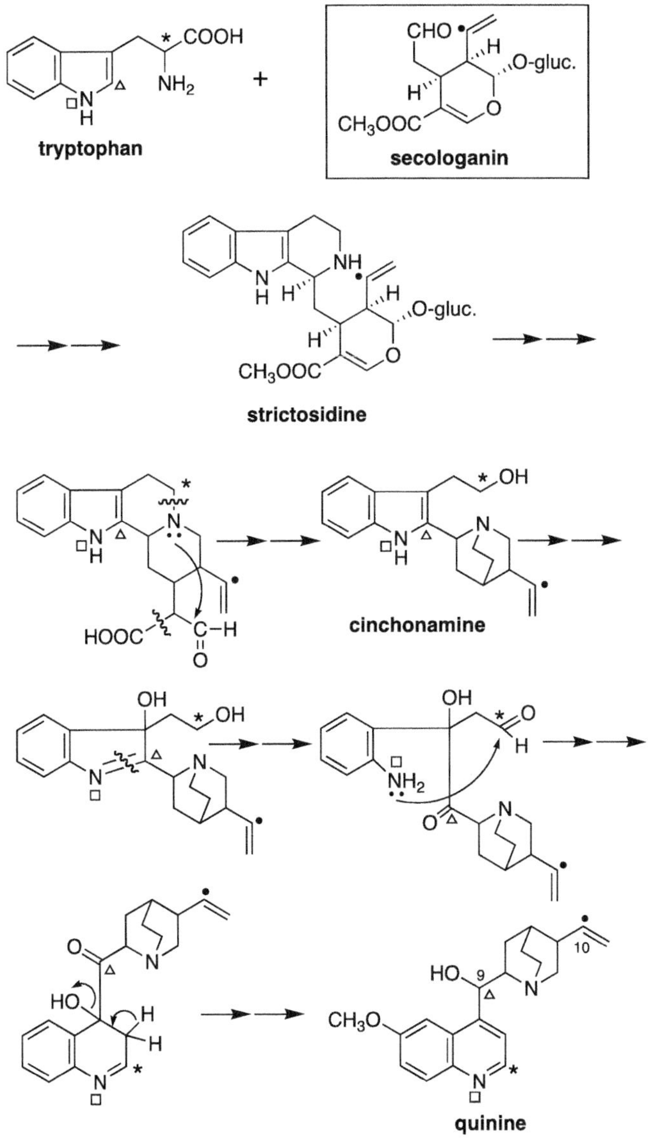

Biogenetic Scheme for Quinine

Quinine is used as an anti-malarial drug. Malaria is caused by a malarial parasite (*Plasmodium* sp.) with the mediation of malaria-bearing mosquito (*Anopheles*). There are five kinds of malarial parasites: *Plasmodium* sp., which are known to infect humans: *P. vivax*, *P. malariae*, *P. falciparum*, *P. ovale*, and *P. knowlesi*.

Malaria, particularly drug-resistant malaria, is increasing rapidly again, shortly after the discontinuance of the aerial spraying of insecticides for harmful insects, including mosquitos, in consideration of the environment. Other than quinine, a number of synthetic drugs, such as chloroquine and mefloquine, as well as artemisinin, a peroxy-sesquiterpenoid isolated from *Artemesia annua* (Asteraceae), are used for malaria treatment. However, quinine still occupies an important position as an anti-malarial drug; and is also present in drinks known as "tonic water."

Quinidine is used as an antiarrhythmic agent (class IA), and it is also used rarely as an anti-malarial drug.

LITERATURE CITED

[1] N. Kowanko, E. Leete, J. Am. Chem. Soc. 84 (1962) 4919.
[2] E. Leete, J.N. Wemple, J. Am. Chem. Soc. 91 (1969) 2698.

2.18 *Catharanthus roseus* AND VLB AND VCR

Svoboda and co-workers of Eli Lilly & Co. conducted screening tests of antineoplastic substances from various plant extracts and compounds of plant origin. As a result, it was found that an alkaloid isolated from the extract of the leaves of *Catharanthus roseus* (Apocynaceae) showed a remarkable increase in the lifespan of mice (DBA/2) implanted with P-1534 leukemia [1].

C. roseus grows wild in Madagascar, and is widely cultivated in both the tropical and temperate areas of the world as a decorative plant. This plant is also cultivated in Japan, and is known as "Nichi-nichi So" or "Vinca," and its English name is Madagascan periwinkle.

Some confusion exists about the botanical name of this plant, and names such as *Lochnera rosea*, *Vinca rosea*, and *Ammocallis rosea* are sometimes used; the correct name is *Catharanthus roseus* [2]. *Vinca* is another genus in the Apocynaceae, and *Vinca major* and *V. minor* are classified in this genus, but produce a very different range of monoterpenoid indole alkaloids.

Occasionally, the bisindole alkaloids vinblastine (vincaleukoblastine, VLB) and vincristine (leurocristine, VCR) are referred to as "Vinca" alkaloids, which is a misnomer; these bisindole alkaloids should be correctly referred to as *Catharanthus* alkaloids.

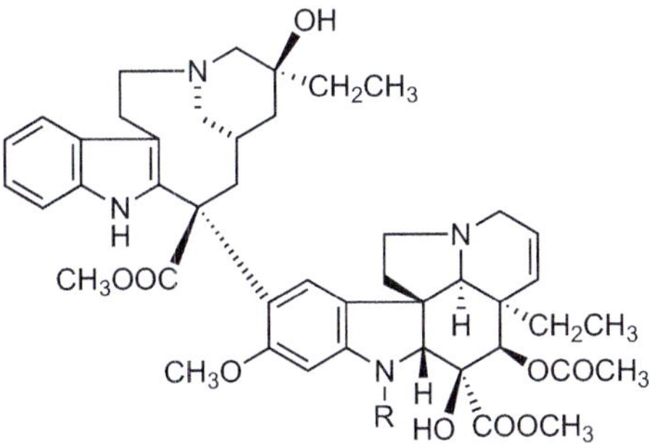

vinblastine (VLB) **R = CH_3**
vincristine (VCR) **R = CHO**

Before the discovery of VLB and VCR, the leaves of the Madagascan periwinkle were used as a remedy for diabetes in traditional medicine, and the activity was studied using experimental animals. Although hypoglycemic activity was found, during this study, Noble et al. of W. Ontario University [3] found that some of the fractions of the plant extracts showed a reducing activity of granules in the blood and a hematopoietic lowering activity in the bone marrow. They isolated a bisindole alkaloid vinblastine (VLB) as an active metabolite [3]. Later, through studies by Svoboda described below, it was found that vinblastine possessed strong antitumor activity, and vinblastine became a treatment for bone marrow leukemia. The chemical structure of vinblastine, including the absolute stereochemistry, was described in 1965 [4], and its partial synthesis [5] and total synthesis [6] were reported.

Vincristine (VCR) was isolated as a minor alkaloid from the leaves of the Madagascan periwinkle, in which the vindoline *N*-methyl group is oxidized to an *N*-formyl moiety. The isolation of vincristine was reported in 1961 [7] and the chemical structure, including stereochemistry, was reported at almost the same time as that of VLB [4].

The American Medical Association recommended that, as names and abbreviations of these alkaloids, vinblastine and VLB and vincristine and VCR, respectively, should be used [4]. In this connection, VLB and VCR are marketed, respectively, under the names of Velban® and Oncovin® (Eli Lilly & Co.).

These alkaloids are considered to be formed by the coupling of the iboga alkaloid catharanthine and the *Aspidosperma* alkaloid vindoline, both of them derived from tryptophan. The other parts of both alkaloids are derived from a secoiridoid (C_{10}) unit, and strictosidine is the primary precursor of both alkaloid units (Section 2.9) [8]. Mangeney et al. [9] succeeded in the syntheses of VLB and VCR by the coupling of catharanthine and vindoline. VLB can be oxidized under controlled conditions to VCR.

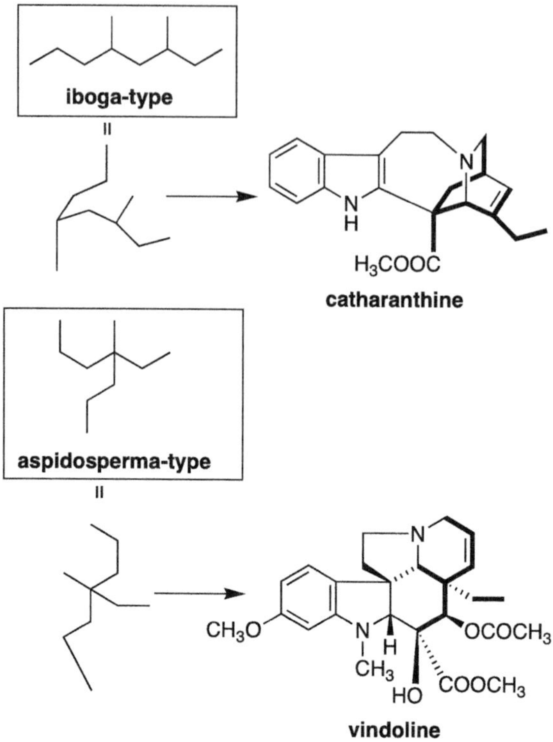

Incorporation Pattern of C_{10} Units Into Catharanthine and Vindoline

Sulfates of VLB and VCR are used as antineoplastic agents, and especially VCR sulfate is used for the treatment of leukemia, malignant lymphatic tumor, and childhood cancers, either as a single medicine or in several combination regimens with other antitumor agents. Side effects of these alkaloids are a decrease in white blood cells, a decrease in platelets, disorders of digestive organs, loss of hair, numbness, and muscle pains. On the other hand, VLB sulfate is not used frequently compared with VCR, although it shows better effects than VCR sulfate in some cases, and is used for the treatment of malignant lymphatic tumors, villous adenoma, and a cystic mole. The difference between the structures of VLB and VCR is slight, but the antitumor activities and side effects are different. Although VLB sulfate has side effects similar to those of VCR, the effects on the nervous system are weaker.

LITERATURE CITED

[1] G.H. Svoboda, D.A. Blake, in: W.I. Taylor, N.R. Farnsworth (Eds.), The Catharanthus Alkaloids, Marcel Dekker, New York, 1975R, p. 45.
[2] W.T. Stern, in: W.I. Taylor, N.R. Farnsworth (Eds.), The Catharanthus Alkaloids, Marcel Dekker, New York, 1975R, p. 9.
[3] R.L. Noble, C.T. Beer, J.H. Cutts, Ann. N.Y. Acad. Sci. 76 (1958) 882.
[4] J.W. Moncrief, W.N. Lipscomb, J. Am. Chem. Soc. 87 (1965) 4963.
[5] P. Potier, N. Langlois, Y. Langlois, F. Guéritte, Chem. Commun. 670 (1975).
[6] J.P. Kutney, J. Nat. Prod. 40 (1977) 107.
[7] G.H. Svoboda, J. Nat. Prod. 24 (1963) 173.
[8] R.J. Parry, in: W.I. Taylor, N.R. Farnsworth (Eds.), The Catharanthus Alkaloids, Marcel Dekker, New York, 1975R, p. 141.
[9] P. Mangeney, R.Z. Andriamialisoa, N. Langlois, Y. Langlois, P. Potier, J. Am. Chem. Soc. 101 (1979) 2243.

2.19 *Euodia rutaecarpa* AND EVODIAMINE AND RUTAECARPINE

Euodia rutaecarpa (Rutaceae) is a deciduous tree that grows wild in China and is dioecious. The tree was introduced into Japan around 1720 (Edo era), and is cultivated in various locations in Japan. The dried fruit of this tree is known as "Go-shuyu" in the Japanese crude drug market, and are used to maintain the health of the stomach, to stimulate the appetite, and to treat stomachache, headache, and vomiting caused by an excess of water in the body. It appears that currently the crude drug that is imported from China is derived from *E. officinalis* instead of *E. rutaecarpa*.

Evodiamine and rutaecarpine were isolated as the first alkaloids from this plant and their chemical structures determined [1,2]. Total syntheses of these compounds were achieved by several groups [3–8].

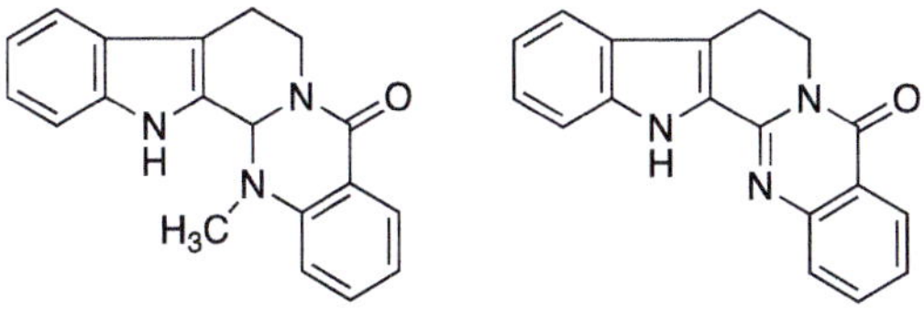

evodiamine **rutaecarpine**

There are always exceptions when one tries to classify alkaloids, and evodiamine and rutaecarpine are examples. As expected from the chemical structures, both of these compounds possess partial structures related to tryptamine, and these compounds can be classified as such. On the other hand, anthranilic acid, which is discussed in Chapter 9, is also involved in the biosynthesis of these alkaloids, and this biosynthetic unit supplies the third nitrogen atom [9–11]. Hence, these alkaloids can be also classified as being derived from anthranilic acid. Based on the nucleus, these alkaloids can also be classified as quinazoline alkaloids, as discussed in Section 9.5. Typically, evodiamine and rutaecarpine are classified as alkaloids derived from tryptophan, i.e., as indole alkaloids.

Other than evodiamine and rutaecarpine, various alkaloids isolated from Rutaceous plants, such as protoberberine (Section 1.8), quinoline (Section 9.1), and acridone (Section 9.3) alkaloids are sometimes referred to "Rutaceous alkaloids." This name, however, refers to a heterogeneous group of alkaloids with diverse biosynthetic origins.

The biosynthesis of evodiamine and rutaecarpine was studied by feeding of tryptophan labeled with ^{14}C, formic acid labeled with ^{14}C, and anthranilic acid labeled with ^{3}H [9]. Each of these biosynthetic precursors was incorporated into the two alkaloids. It was verified that the single carbon of formic acid was incorporated into the 13b position of evodiamine and rutaecarpine. This carbon was also labeled by [$^{14}CH_3$]methionine [10]. Furthermore, despite possessing an *N*-methyl moiety, the incorporation of [$^{14}CH_3$]methionine into evodiamine was lower than expected. This suggests that in the biosynthesis of evodiamine, previously existing unlabeled *N*-methylanthranilic acid was used mainly as the precursor material. In the biosynthesis of these alkaloids, dihydronorharman derived by adding a C_1 unit to a tryptamine unit originating from tryptophan, and anthranilic acid or *N*-methylanthranilic acid, are added to dihydronorharman to form rutaecarpine and evodiamine, respectively.

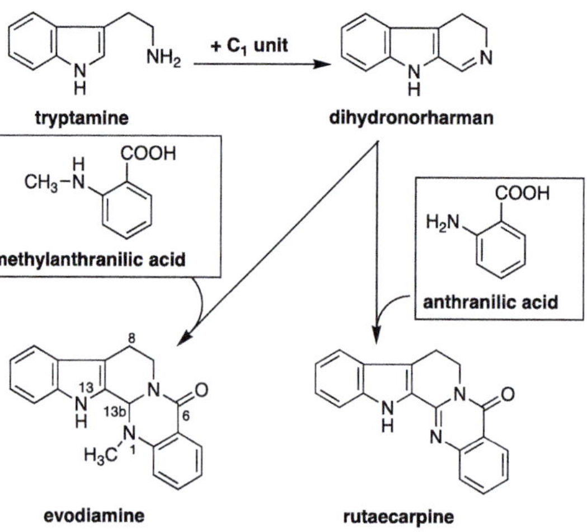

Biosynthetic Route to Evodiamine and Rutaecarpine

Rhetsinine is a yellow alkaloid isolated from the bark of *Zanthoxylum rhetsa* (Rutaceae) [12] and the fruits of *E. rutaecarpa* [11]. It appeared to be a biosynthetic precursor of evodiamine, and its existence in the plant was doubtful. Yamazaki and Kawana [11] reported that rhetsinine was obtained by illuminating a dioxane solution of evodiamine for a short time.

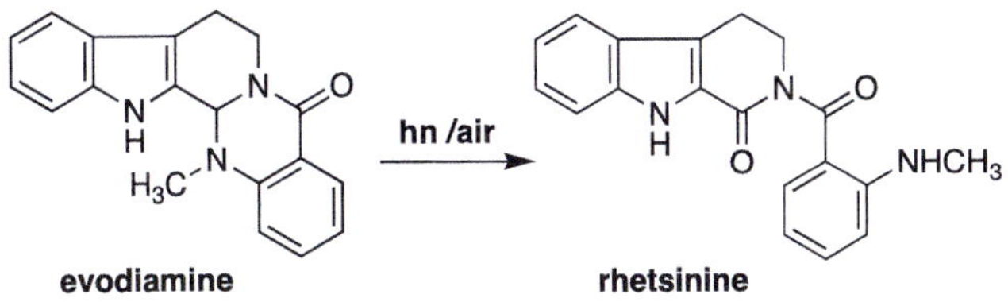

Transformation of Evodiamine into Rhetsinine by Photochemical Reaction

LITERATURE CITED

[1] Y. Asahina, K. Kashiwagi, Yakugaku Zasshi 35 (1915) 1273.
[2] Y. Asahina, Yakugaku Zasshi 44 (1924) 1.
[3] Y. Asahina, R.H.F. Manske, R. Robinson, J. Chem. Soc. 1708 (1927).
[4] Y. Asahina, T. Ohta, Chem. Ber. 61B (1928) 319.
[5] T. Kametani, T. Higa, C.V. Loc, M. Ihara, M. Koizumi, K. Fukumoto, J. Amer. Chem. Soc. 98 (1976) 6186.
[6] T. Kametani, C.V. Loc, T. Higa, M. Koizumi, M. Ihara, K. Fukumoto, J. Am. Chem. Soc. 99 (1977) 2306.
[7] H. Möhrle, C. Kamper, R. Schmid, Arch. Pharm. 313 (1980) 990.
[8] J. Kökösi, I. Hermecz, G. Szàz, Z. Mészàos, Tetrahedron Lett. 22 (1981) 4861.
[9] M. Yamazaki, A. Ikuta, Tetrahedron Lett. 7 (1966) 3221.
[10] M. Yamazaki, A. Ikuta, T. Mori, T. Kawana, Tetrahedron Lett. 8 (1967) 3317.
[11] M. Yamazaki, T. Kawana, Yakugaku Zasshi 87 (1967) 608.
[12] A. Chatterjee, S. Bose, C. Ghosh, Tetrahedron 7 (1959) 257.

2.20 ATHLETE'S FOOT AND PYRROLNITRIN

Pyrrolnitrin is an anti-fungal antibiotic produced by *Pseudomonas pyrrocinia* [1]. This antibiotic shows only weak antimicrobial activity against Gram-positive and Gram-negative bacteria (MIC = 12.5–100 mg/ml), but strong toxicity against fungi, especially trichophytes, and the MIC against *Trichophyton asteroides* is 0.05 mg/ml [1]. Pyrrolnitrin is used as a treatment for athlete's foot [2].

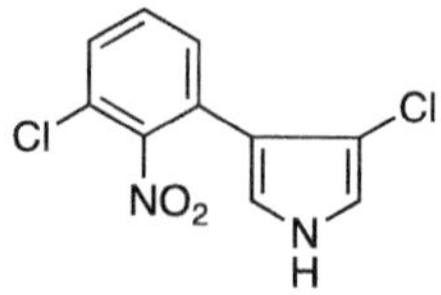

pyrrolnitrin

Pyrrolnitrin possesses a structure in which the benzene and pyrrole rings are attached and substituted by nitro and chlorine units [3]. Because of the unusual structure, the biosynthetic pathway became of interest, and Gorman and Lively proposed a biogenesis of pyrrolnitrin from tryptophan [2]. Studies using various stable and unstable isotopes were conducted using *P. aureofaciens*, which also produces pyrrolnitrin. As a result, it was shown that the skeleton of pyrrolnitrin originated from tryptophan, and that D-tryptophan was incorporated better than L-tryptophan. Namely, when [2-^{3}H, 3-^{14}C]tryptophan was fed, it was clarified that the incorporation rate of D-form was 15.8% and that of L-form was 9.6% [4,5]. The α-hydrogen and the nitrogen of the amino moiety of L-tryptophan were retained, indicating intact incorporation [5]. However, D-tryptophan probably was incorporated into pyrrolnitrin after it was transformed into the L-form via indole pyruvic acid.

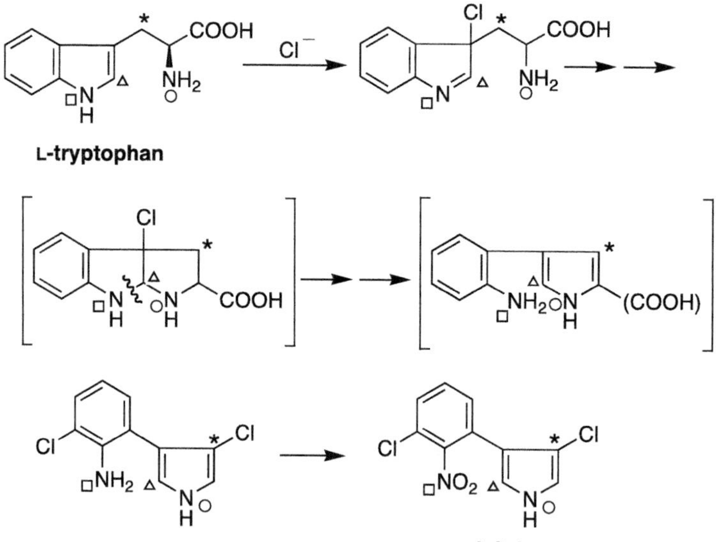

Biogenetic Pathway for Pyrrolnitrin

As a result of various incorporation studies, it was demonstrated that the nitro moiety of pyrrolnitrin originated from the nitrogen of the indole moiety of tryptophan and that the nitrogen of the pyrrole moiety originated from the α-amino moiety of tryptophan, as proposed [2]. Chang et al. suggested a 3-alkylidene indolenine derivative as a biosynthetic intermediate, rather than an intermediate with a chlorine atom in the structure as suggested previously [5]. It seems clear that a biosynthetic intermediate of pyrrolnitrin possesses a skeleton similar to that of physostigmine, described in Section 2.5.

LITERATURE CITED

[1] K. Arima, H. Imanaka, M. Kousaka, A. Fukuta, G. Tamura, Agric. Biol. Chem. 28 (1964) 575.
[2] M. Gorman, D.H. Lively, in: D. Gottlieb, P.D. Shaw (Eds.), Antibiotics, Vol. II, Springer-Verlag, New York (1967), p. 433.
[3] H. Imanaka, M. Kousaka, G. Tamura, K. Arima, J. Antibiot. (Ser. A) 18 (1965) 205.
[4] O. Salcher, F. Lingens, P. Fischer, Tetrahedron Lett. 19 (1978) 3097.
[5] C.-J. Chang, H.G. Floss, D.J. Hook, J.A. Mabe, P.E. Manni, L.L. Martin, K. Schröder, T.L. Shieh, J. Antibiot. 34 (1981) 555.

2.21 ANTHRAMYCIN AND TOMAYMYCIN

There is a group of antibiotics that possess the pyrrolo[1,4]benzodiazepine skeleton and show antimicrobial activity against Gram-positive bacteria, along with antiviral, antiphage, and antitumor activities. Anthramycin and tomaymycin are typical alkaloids in this group.

Anthramycin was isolated from the fermentation beer of *Streptomyces refuineus* [1], its chemical structure determined [2], and total synthesis achieved [3]. In contrast, tomaymycin was isolated from the fermentation broth of *S. achromogenes* [4].

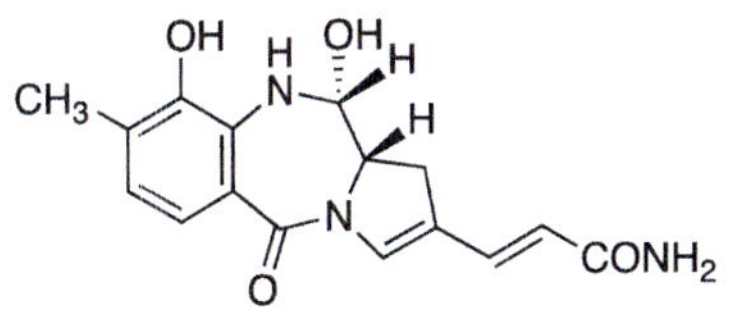

anthramycin

tomaymycin

Although it might be considered, based on their structures, that both antibiotics are derived from anthranilic acid, anthranilic acid does not serve as a precursor. Consequently, it was proposed, following incorporation experiments using radioisotopes, that the anthranilic acid–like moieties were derived from L-tryptophan via kynurenine, as shown in the Figure [5,6]. The other unit of these alkaloids is thought to be derived from L-tyrosine. The methyl and methoxyl methyl moieties are derived from methionine. The absolute stereochemistry of kynurenine was determined from the fact that this alkaloid was obtained by the chemical transformation of L-tryptophan [7].

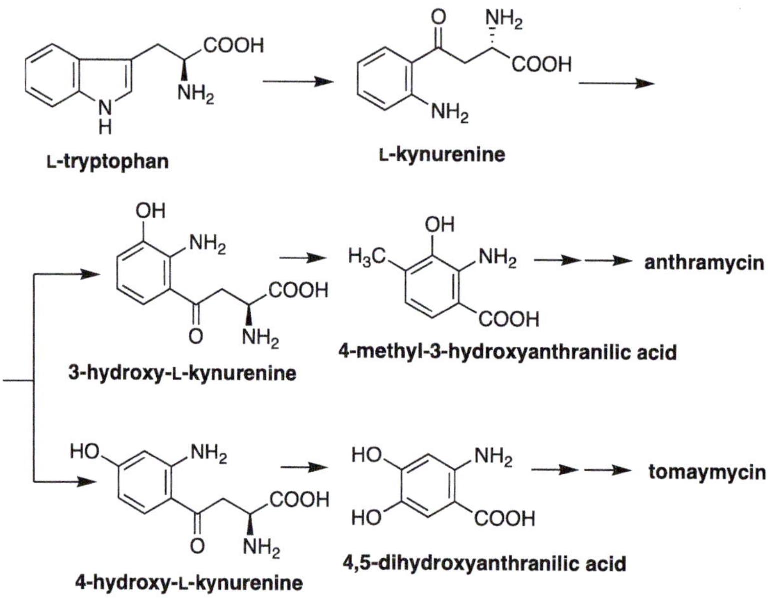

Biosynthetic Route to Anthramycin and Tomaymycin

LITERATURE CITED

[1] M.D. Tendler, S. Korman, Nature 199 (1963) 501.
[2] W. Leimgruber, A.D. Batcho, F. Schenker, J. Am. Chem. Soc. 87 (1965) 5793.
[3] W. Leimgruber, A.D. Batcho, R.C. Czajkowski, J. Am. Chem. Soc. 90 (1968) 5641.
[4] K. Arima, M. Kohsaka, G. Tamura, H. Imanaka, H. Sakai, J. Antibiot. 25 (1972) 437.
[5] L.H. Hurley, J. Antibiot. 30 (1977) 349.
[6] L.H. Hurley, C. Gairola, Antimicrob. Agents Chemoth. 15 (1979) 42.
[7] J.L. Warnell, C.P. Berg, J. Am. Chem. Soc. 76 (1954) 1708.

2.22 STREPTONIGRIN

Streptonigrin is an antibiotic produced by the actinomycetes, *Streptomyces flocculus* ATCC13257. Streptonigrin forms dark brown crystals and has shown strong cytocidal activity against various experimental animal tumors. The chemical structure of this antibiotic was determined by a combination of chemical degradation studies and physicochemical analyses [1]. The total synthesis of this alkaloid was achieved [2].

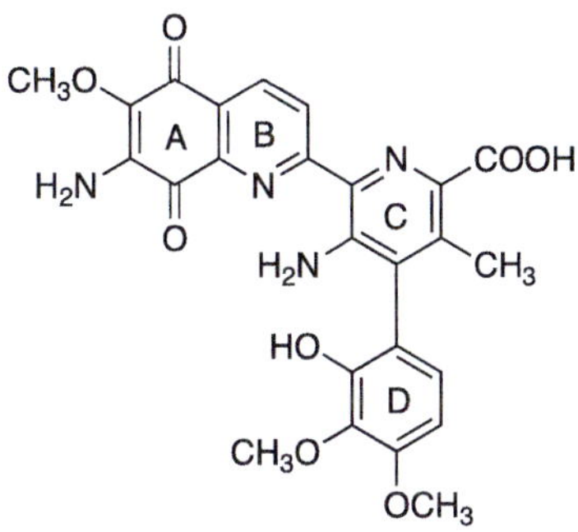

streptonigrin

If the aromatic rings of this alkaloid are tentatively named as ring A through ring D, it was shown that rings A and B are derived from anthranilic acid, and that rings C and D are derived from tryptophan [3,4]. The origin of rings C and D was proved by the observation of ^{15}N-^{13}C couplings in this alkaloid in the ^{13}C NMR spectrum [3]. A plausible mechanism of the transformation from tryptophan skeleton to the rings C and D through cleavage of a β-carboline has been shown [5]. Gould and Cane [4] also investigated the incorporation of [U-$^{13}C_6$]glucose into streptonigrin via tryptophan using ^{13}C NMR.

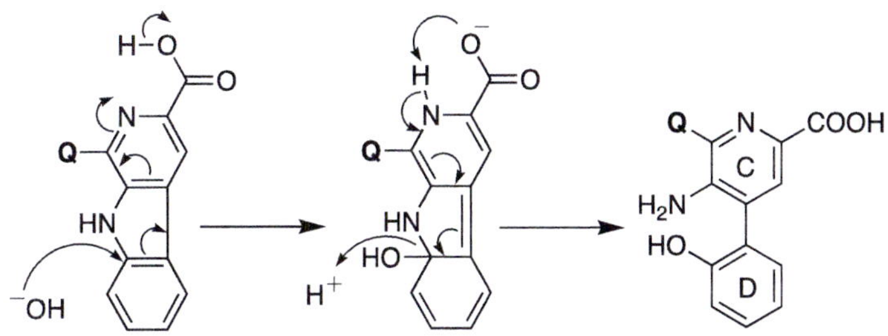

Q: quinolinequinone moiety

Biogenetic Origin of the Phenylpyridine System

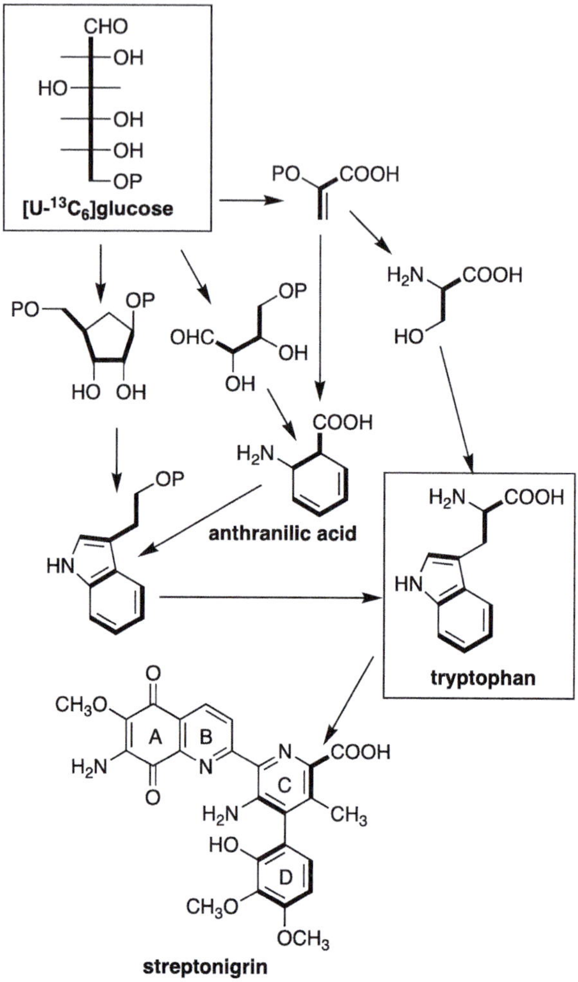

Labelling Pattern of the C and D Rings of Streptonigrin by the Incorporation of [U-$^{13}C_6$]glucose

LITERATURE CITED

[1] K.V. Rao, K. Biemann, R.B. Woodward, J. Am. Chem. Soc. 85 (1963) 2532.
[2] S.M. Weinreb, F.Z. Basha, S. Hibino, N.A. Khatri, D. Kim, W.E. Pye, T.-T. Wu, J. Am. Chem. Soc. 104 (1982) 536.
[3] S.J. Gould, C.-C. Chang, J. Am. Chem. Soc. 100 (1978) 1624.
[4] S.J. Gould, D.E. Cane, J. Am. Chem. Soc. 104 (1982) 343.
[5] A.J. Herlt, R.W. Rickards, J.-P. Wu, J. Antibiot. 38 (1985) 516.

2.23 STAUROSPORINE

Staurosporine, an indolo[2,3-a]carbazolole alkaloid that showed the potent protein kinase inhibiting activiry, was first isolated from *Streptomyces staurosporeus* (AM-2282) in 1977 [1] and subsequently from other actinomycetes, e.g., *S. actuosus* [2], and *Streptomyces* species strain M-193 [3]. The structure and stereochemistry of staurosporine were deduced by X-ray crystallography [4,5]. The acute toxicity (LD_{50}) of staurosporine hydrochloride on intraperitoneal administration in mice is 6.6 mg/kg. Although staurosporine has no significant effects on bacteria, it possesses inhibitory activity against fungi and yeasts [1], Preliminary evidence has also shown that staurosporine possesses strong antihypertensive activity [6,7], and pronounced in vitro activity against a number of experimental tumors, e.g., a human neuroblastoma cell line (NB-1) [2], HeLa S3 cells, B16 melanoma cells, and P-388 leukemia cells [2,8,9]. Most interestingly, it is a potent inhibitor of protein kinase C [8] and platelet aggregation [3].

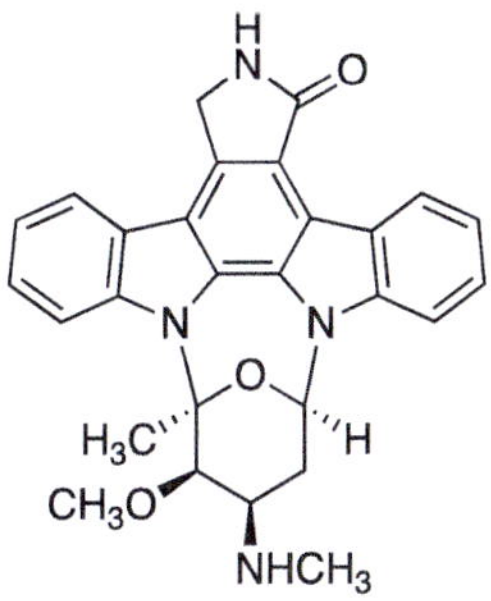

staurosporine

Regarding the biosyntheses of this alkaloid, it was established that the aglycone moiety was derived from two units of tryptophan, with the carbon skeleton incorporated intact, by examining the ^{13}C-NMR spectrum subsequent to stable isotope incorporation experiments [10,11]. It was found that during staurosporine biosynthesis, the nitrogen atom on the side chain of tryptophan is cleaved [12]. In contrast, it was established that the aminosugar moiety of staurosporine is derived from glucose, based on the observation of the direct incorporation of uniformly labeled ^{13}C glucose ([U-$^{13}C_6$]-D-glucose) [13]. It was also found that the origin of 3′-*O*- and 4′-*N*-methyl groups of this alkaloid was methionine, by the feeding of ^{13}C- and ^{2}H-enriched methionine to *S. staurosporeus* [14].

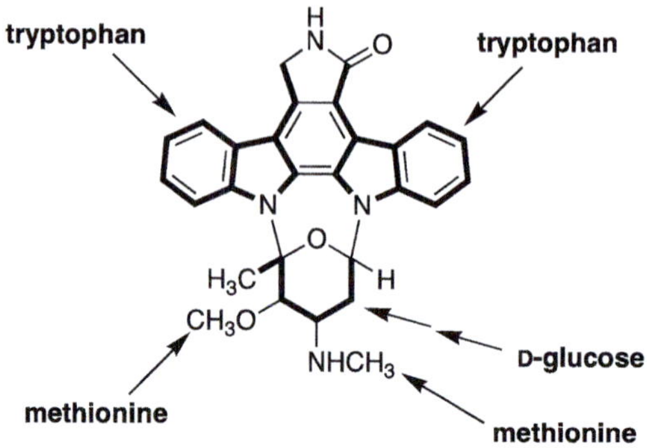

Biosynthetic Origins of Staurosporine

LITERATURE CITED

[1] S. Ōmura, Y. Iwai, A. Hirano, A. Nakagawa, J. Awaya, H. Tsuchiya, Y. Takahashi, R. Masuma, J. Antibiot. 30 (1977) 275.
[2] H. Morioka, M. Ishihara, H. Shibai, T. Suzuki, Agric. Biol. Chem. 49 (1985) 1959.
[3] S. Oka, M. Kodama, H. Takeda, N. Tomizuka, H. Suzuki, Agric. Biol. Chem. 50 (1986) 2723.
[4] A. Furusaki, N. Hashiba, T. Matsumoto, A. Hirano, Y. Iwai, S. Ōmura, J. Chem. Soc. Chem. Commun. 800 (1978).
[5] A. Furusaki, N. Hashiba, T. Matsumoto, A. Hirano, Y. Iwai, S. Ōmura, Bull. Chem. Soc. Jpn. 55 (1982) 3681.
[6] S. Ōmura, Y. Iwai, A. Hirano, Ger. Offen. 2,754,326 [Chem. Abstr., 89, 58348y (1978)].
[7] S. Ōmura, Y. Iwai, A. Hirano, Jpn. Kokai 78 73,501 [Chem. Abstr., 89, 178086h (1978)].
[8] T. Tamaoki, H. Nomoto, I. Takahashi, Y. Kato, M. Morimoto, F. Tomita, Biochem. Biophys. Res. Commun. 135 (1986) 397.
[9] H. Morioka, H. Shibai, Y. Yokogawa, M. Ishihara, T. Kida, T. Suzuki, Jpn. Kokai Tokkyo Koho JP 60,185,719 [Chem. Abstr., 104, 18649u (1986)].
[10] D. Meksuriyen, G.A. Cordell, J. Nat. Prod. 51 (1988) 884.
[11] D. Meksuriyen, G.A. Cordell, J. Nat. Prod. 51 (1988) 893.
[12] S.-W. Yang, G.A. Cordell, J. Nat. Prod. 60 (1997) 788.
[13] S.-W. Yang, G.A. Cordell, J. Nat. Prod. 59 (1996) 828.
[14] S.-W. Yang, L.-J. Lin, G.A. Cordell, P. Wang, D.G. Corley, J. Nat. Prod. 62 (1999) 1551.

CHAPTER 3

Alkaloids Derived from Ornithine and Arginine

Datura metel (Solanaceae)

As will be described later, ornithine is an amino acid that is biosynthetically derived from arginine. Consequently, alkaloids derived from these amino acids are combined and described in this chapter.

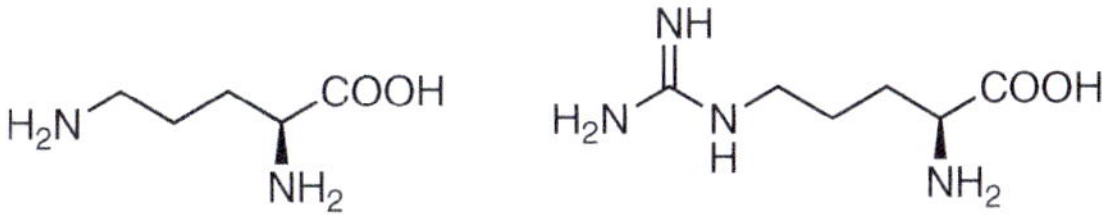

Also, in this chapter, the polyamine alkaloids, such as ephedradine and kukoamine, which contain a polyamine unit derived from putrescine and are biosynthesized by decarboxylation of ornithine, will be discussed. Furthermore, atropine and cocaine, which are biosynthesized via the *N*-methyl-1-pyrrolinium ion and formed by methylation of putrescine followed by cyclization, and tetrodotoxin, which was once considered to be biosynthesized from arginine, will be discussed.

3.1 TOBACCO AND NICOTINE

At the end of fifteenth century, Columbus (about 1446–1506) observed that the natives of the Caribbean were smoking tobacco. Sailors learned about the plant and its usage, and this knowledge was brought back to Europe. The cultivation of tobacco spread to Europe, Africa, and Asia in the sixteenth century.

Alkaloids
http://dx.doi.org/10.1016/B978-0-12-417302-6.00003-9

Tobacco is derived from the leaves of *Nicotiana tabacum* (Solanaceae), and it was used to treat headache and toothache. However, it is not used as a medicine now, and smoking tobacco is now a global addictive habit. Tobacco leaf contains a large amount of nicotine (2–8%), and the nicotine extracted as nicotine sulfate is used as an insecticide in agriculture. Tobacco leaf contains more than ten related alkaloids other than nicotine, and all of these alkaloids possess a pyridine skelcton with 3-substitution. The main alkaloids other than nicotine, anabasine and nor-nicotine, are isolated from the leaf material, and these alkaloids also possess insecticidal activity.

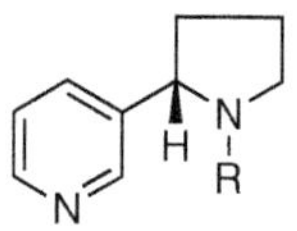

nicotine **R = CH_3**
nornicotine **R = H**

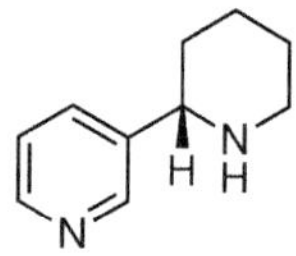

anabasine

Nicotine was isolated by Posselt and Reimann in 1828 [1], and subsequently the chemical structure was clarified [2–4]. As described later, the pyridine ring of nicotine is derived from nicotinic acid, which is biosynthesized from aspartic acid. Therefore, nicotine can also be described as an alkaloid derived from aspartic acid. On the other hand, the pyrrolidine ring is biosynthesized from ornithine. When DL-[5-^{14}C]-pyrroline-5-carboxylic acid, a postulated biosynthetic precursor derived from ornithine, was incorporated into *Nicotiana rustica*, the incorporated rate (0.04%) was very low, and the 2′- and 5′-positions were equally labeled with ^{14}C (Figure). Consequently, it was estimated that the biosynthetic intermediate was not this compound but another symmetrical structure [5].

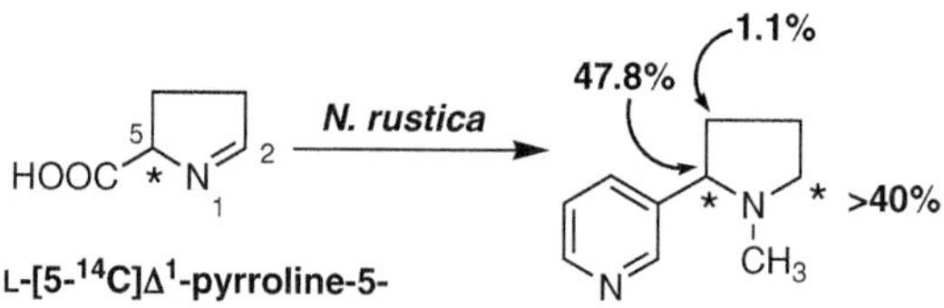

D,L-[5-^{14}C]Δ^1-pyrroline-5-carboxylic acid

nicotine

Incorporation of D,L-[5-^{14}C]Δ^1-pyrroline-5-carboxylic acid into Nicotine in *Nicotiana rustica*

Thus, it was elucidated that ornithine was decarboxylated to form putrescine (Figure), and this intermediate was transformed into the *N*-methyl-1-pyrrolinium species via 4-aminobutyraldehyde or 4-(*N*-methyl)-aminobutyraldehyde. *N*-Methyl-1-pyrrolinium then combines with a nicotinic acid moiety to give nicotine. The fact that a symmetrical intermediate (putrescine) is involved when ornithine is incorporated into the pyrrolidine moiety of nicotine was demonstrated by feeding DL-[2,3-$^{13}C_2$]ornithine. In the ^{13}C NMR spectrum of the labeled nicotine obtained, couplings between C-2′ and C-3′, and between C-4′ and C-5′, were observed [6]. It was considered that the two nitrogens at the α- and δ-positions of ornithine function differently, because the nitrogen at the δ-position is selectively incorporated, whereas the nitrogen at the α-position is not well incorporated. It is probable that some of the nitrogen of the α-position is captured by a transaminase [7].

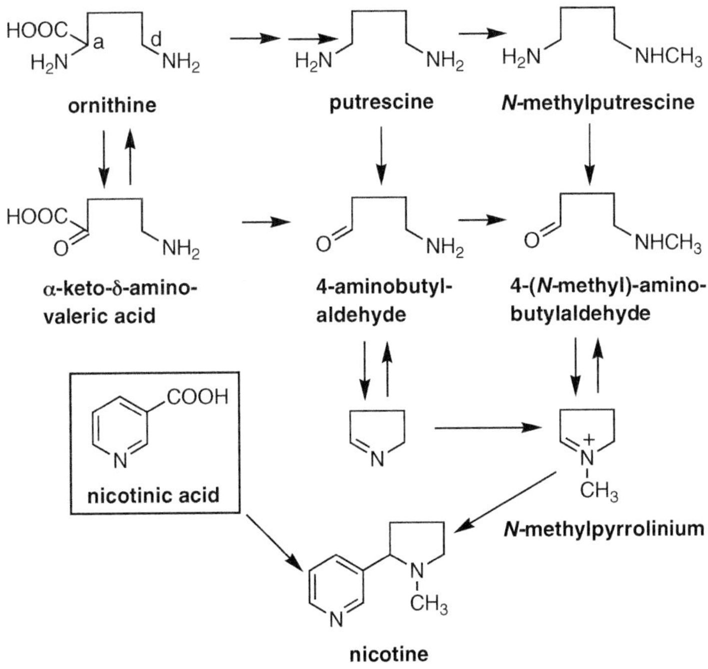

Biogenetic Route of Nicotine

The piperidine moiety of anabasine is biosynthesized via lysine. The uptake of lysine into anabasine is unsymmetrical, as discussed in the next chapter.

Nicotine possesses excitation and blocking effects on the autonomic nervous system. In the autonomic nervous system, there is a ganglion between two nervous fibers, before the information reaches the target organ. Nicotine shows a dichotomic action for the ganglion. First, the ganglion is excited initially, and it will be suppressed subsequently. Large doses of nicotine show a suppressing effect from the beginning. Such a dichotomic action as shown by nicotine is referred to as a nicotinic effect.

LITERATURE CITED

[1] H. Itokawa, A. Ohta, Y. Nishikawa (Translated): Steinegger and Hänsel, Lehrbuch der Pharmacognossie, Hirokawa Pub. Co., Tokyo (1976), p. 285.
[2] A. Pinner, Chem. Ber. 26 (1893) 292.
[3] A. Pictet, A. Rotschy, Chem. Ber. 37 (1904) 1225.
[4] L.C. Craig, J. Am. Chem. Soc. 55 (1933) 2854.
[5] V. Krampl, H.R. Zielke, R.U. Byerrum, Phytochemistry 8 (1969) 843.
[6] E. Leete, M.L. Yu, Phytochemistry 19 (1980) 1093.
[7] E. Leete, E.G. Gros, T.J. Gilbertson, Tetrahedron Lett. 5 (1964) 587.

3.2 BELLADONNA AND ATROPINE

The leaves and roots of *Atropa belladonna* (Solanaceae) are known as Belladonna leaves and Belladonna roots, and are used as medicine in the form of extracts. These materials are also used as materials for the manufacture of atropine sulfate. Other alkaloids isolated from belladonna include (−)-hyoscyamine and (−)-scopolamine. Atropine is derived from (−)-hyoscyamine by racemization of its tropic acid ester moiety.

Belladonna means "beautiful (bella) grande dame (donna)." It is said that, in the old times, society ladies of Italy used the diluted extract of this plant material as an eye lotion to expand the pupil to enhance the appearance of their eyes.

Datura tatula and *Datura stramonium* are Solanaceae plants from the North American continent, and now grow naturally in various places in Japan. Leaves of these plants are known as Datura leaves or "Mandala" leaves and contain (−)-hyoscyamine. They are used as materials for producing atropine sulfate.

Datura metel (=*Datura alba*) and *Datura innoxia* are naturalized in the tropical zone of Asia. (−)-Hyoscyamine is present in the leaves, and

(−)-scopolamine is present in the seeds of these plants; consequently, they are used for the manufacture of (−)-scopolamine sulfate.

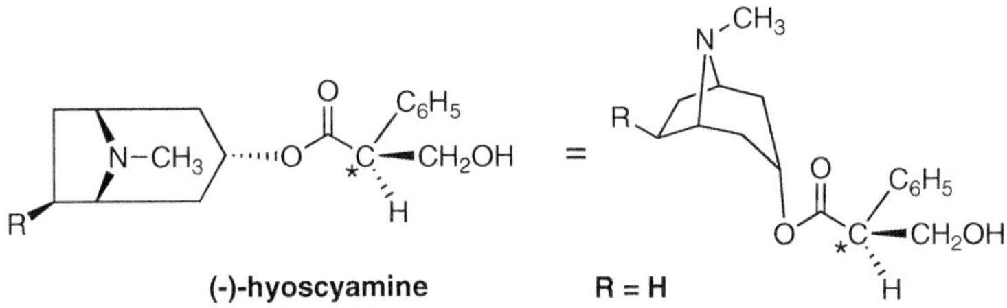

(-)-hyoscyamine R = H
atropine (*dl*-hyoscyamine) R = H
6β-hydroxyhyoscyamine R = OH

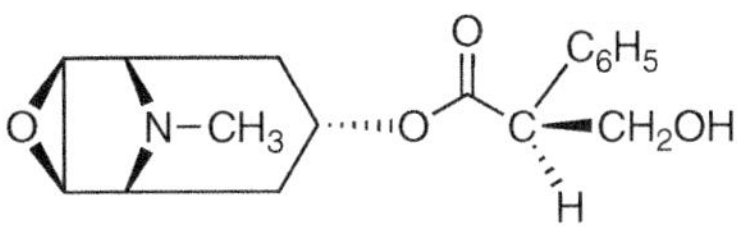

(-)-scopolamine

Hyoscyamus niger (Solanaceae) grows wild in Europe and is cultivated in many countries. Because (−)-hyoscyamine is present in the leaves, they are used as material for the manufacture of atropine sulfate.

Scopolia japonica grows wild in Japan, and this plant contains (−)-hyoscyamine and (−)-scopolamine. The rhizomes and roots of *S. japonica* are known as "Roto" root, and the extract of this material is used as an analgesic, for spasmolysis, and as a digestive juice secretion suppressor. Originally, "Roto" was prepared from the rhizome and root of *Hyoscyamus niger* var. *chinensis*, which grows wild in China.

Atropine and (−)-scopolamine are representative suppressors of the parasympathetic nervous system (cholinolytic drug). The mechanism of action of these alkaloids is not to suppress the generation of acetylcholine (ACh) in the terminal of the cholinergic nerve, but to affect the receptor in the control organ side. Thus, atropine shows a competitive antagonism action vs. ACh at the parasympathetic ganglions in the juncture. The representative action of atropine is to open the pupil by relaxing the musculus sphincter pupillae under parasympathetic innervation. Therefore, it is widely used in ophthalmology to facilitate refined examination. The parasympathetic nervous system inhibitory effect of (−)-scopolamine resembles that of atropine; however, the mydriatic action of (−)-scopolamine is much stronger than that of atropine.

Although atropine shows almost no central nervous action at a therapeutic dose (1 mg), large doses cause excitement of the motor area and enhanced vitality, producing hallucinations and delirium.

In 1826, an oculist Genseki Habu (1768–1854) visited P. F. von Siebold, (1796–1866), a medical doctor who had come from Europe to Edo (the present Tokyo), because he had heard that von Siebold possessed a medicine (belladonna) that opened the pupil of the eye. Habu asked for the medicine. von Siebold gave the medicine to Habu with pleasure, and Habu immediately used it in ophthalmology to demonstrate that the medicine really did dilate the pupil.

However, the medicine ran out, and Habu again asked von Siebold for more belladonna. At that time, Habu presented a robe bearing a crest presented by the Shogun (the general). von Siebold could not give the medicine to Habu because his stock of belldonna was insufficient at that time. Instead, von Siebold told him that a similar plant grew in Japan, and the plant was *S. japonica*. In fact, von Siebold showed a sketch of *S. japonica* drawn by the medical botanist Toyobumi Mizutani (1779–1833) of Owari (present Nagoya), and he estimated it to be belladonna. It was the first time that *S. japonica* was used in place of belladonna in Japan.

After 2 years, in 1828, the Dutch boat in which von Siebold planned to return to his country while anchored in the harbor in Nagasaki city was wrecked because of a typhoon. As a result, the cargo that von Siebold was planning to bring to Europe was unloaded, and an inspection by a governmental official was done. Among the materials, the crested robe presented by Habu was found, as well as the counterpart of a map of Japan (made by Tadataka Inoo (1745–1818) and presented by Kageyasu Takahashi (1785–1829), astronomy scientist). Because there was a national ban on such exports at that time, the matter became a big affair. Eventually, G. Habu and K. Takahashi were arrested, and later the corpse of K. Takahashi, who had died in prison, was beheaded. G. Habu was dismissed and his property was forfeited. In the end, more than 50 people including the students of von Siebold were subjected to punishment. von Siebold was banished from Japan and his return prohibited. In 1996, Japan issued an 80-yen postage stamp commemorating the 200th anniversary of the birth of von Siebold.

(−)-Hyoscyamine is considered to be biosynthesized from a C_4 unit derived from two molecules of malonyl CoA attached to the *N*-methyl-Δ^1-pyrrolinium ion, followed by decarboxylation, cyclization, reduction, and esterification [1,2]. The biosynthetic route through which ornithine becomes putrescine by decarboxylation and is transformed to *N*-methyl-Δ^1-pyrrolinium during the biosynthesis of (−)-hyoscyamine was described in the section of the biosynthesis of nicotine.

Through the incorporation studies using *D. innoxia*, it was found that a β-ketothioester was the intermediate in the biosynthesis of (−)-hyoscyamine and (−)-scopolamine [3–7]. Tropinone is formed from the β-ketothioester, and is stereoselectively reduced by a NADPH-dependent oxidoreductase TR-1 to give tropine. Then littorine is formed by adding a phenyllactic acid moiety derived from phenylalanine via phenylpyruvic acid. Regarding the incorporation of phenyllactic acid into tropine and its transformation, it was found that [1,3-$^{13}C_2$](−)-hyoscyamine was obtained when [1,3-$^{13}C_2$] phenyllactic acid was incorporated into *D. innoxia* [8]. A transformation therefore occurred on littorine to form (−)-hyoscyamine, and scopolamine was biosynthesized from 6β-hydroxyhyoscyamine by oxidation as described below.

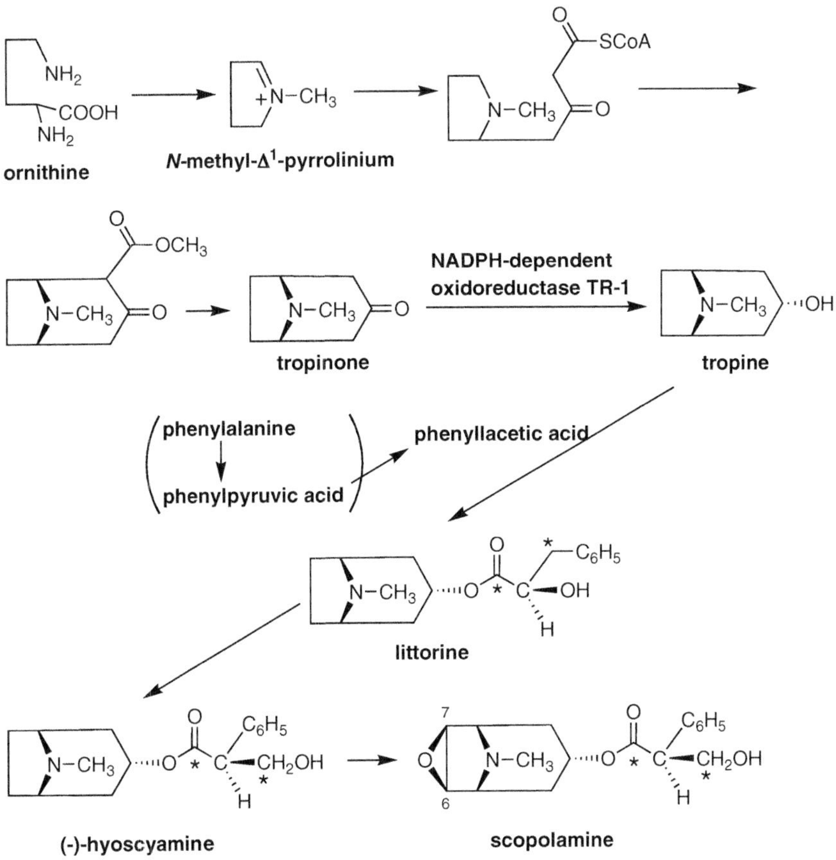

Biosynthetic Pathway to (-)-Hyoscyamine and (-)-Scopolamine

It was estimated that (−)-scopolamine was biosynthesized from 6β-hydroxyhyoscyamine formed via (−)-hyoscyamine. In this biosynthetic pathway, the two routes (a) and (b) could be considered. When [6-^{18}O]6β-hydroxyhyoscyamine was fed to the cultured stems and leaves of *Duboisia myoporoides*, it was found that (−)-scopolamine retained 100% of the ^{18}O [9]. The experimental result indicated that biosynthetic route (b) was taken in the pathway from the hydroxy precursor to the epoxy product [10].

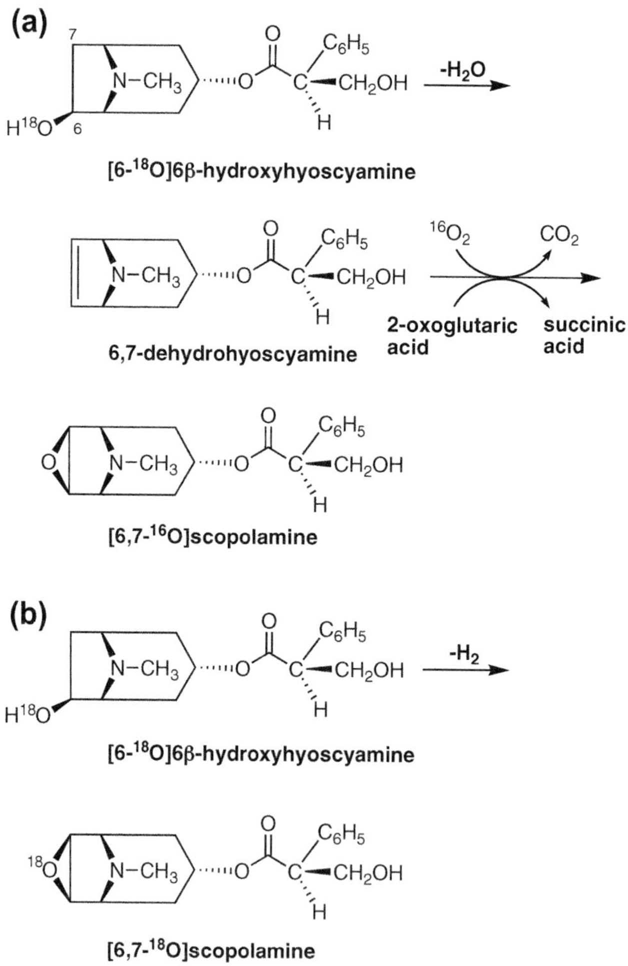

Two Biogenetic Routes for [6-^{18}O]6β-Hydroxyhyoscyamine in *Duboisia myoporoides* Stem

The biosynthesis of cuscohygrine, one of the main alkaloids of belladonna, was examined using ^{14}C-labeled compounds [11,12]. According to the results, the labeling pattern of ^{14}C of cuscohygrine obtained 15 days after feeding C_1 or C_2 labeled sodium acetate to the cultivated belladonna is as shown in the Figure. Namely, the skeleton of this alkaloid is biosynthesized from *N*-methyl-Δ^1-pyrrolidine and acetoacetic acid, as in the case of hyoscyamine biosynthesis.

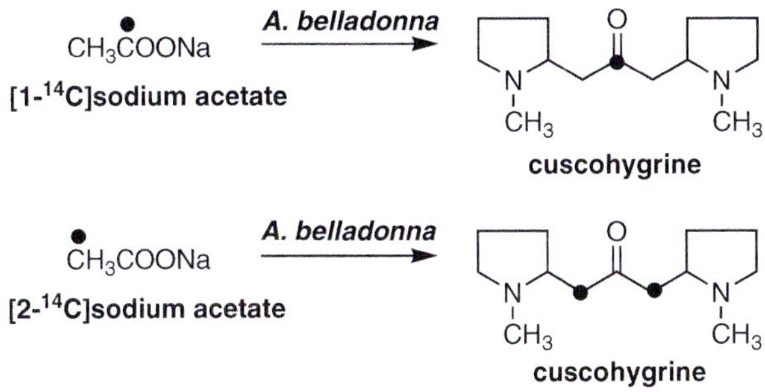

Incorporation of ^{14}C Labeled Sodium Acetate into Cuscohygrine

LITERATURE CITED

[1] E. Leete, J. Am. Chem. Soc. 104 (1982) 1403.
[2] E. Leete, Phytochemistry 22 (1983) 699.
[3] T. Hashimoto, Y. Yamada, E. Leete, J. Am. Chem. Soc. 111 (1989) 1141.
[4] U. Sankawa, H. Noguchi, T. Hashimoto, Y. Yamada, Chem. Pharm. Bull. 38 (1990) 2066.
[5] B.A. McGaw, J.G. Woolley, Phytochemistry 17 (1978) 257.
[6] E. Leete, S.H. Kim, J. Am. Chem. Soc. 110 (1988) 2976.
[7] M.N. Huang, T.W. Abraham, S.H. Kim, E. Leete, Phytochemistry 41 (1996) 767.
[8] T. Hashimoto, K. Nakajima, G. Ongena, Y. Yamada, Plant Physiol. 100 (1992) 836.
[9] A. Lanoue, M. Boitel-Conti, J.-C. Portais, J.-C. Laberche, J.-N. Barbotin, P. Christen, B. Sangwan-Norreel, J. Nat. Prod. 65 (2002) 1131.
[10] T. Hashimoto, J. Kohno, Y. Yamada, Plant Physiol. 84 (1987) 144.
[11] F.E. Baralle, E.G. Gros, Phytochemistry 8 (1969) 849.
[12] F.E. Baralle, E.G. Gros, Phytochemistry 8 (1969) 853.

3.3 COCA LEAVES AND COCAINE

Cocaine is an alkaloid isolated from the leaves of *Erythroxylum coca* or *Erythroxylum novogranatense* (Erythroxylaceae). The former shrub grows wild in Bolivia, and the latter grows in Peru. The mean cocaine content of cocaine leaves is said to be 0.7–2.5%, and the cocaine content of the latter species is considered to be higher. The latter plant is also cultivated in Java, and the total amount of the alkaloids is higher than that which grows wild. However, the content of cocaine in these plants is low, and the levels of cinnamoylcocaine and truxilline are high. These alkaloids are also isolated from *E. coca*, and ecgonine is obtained by hydrolysis. Ecgonine obtained in this way is transformed into cocaine by methylation, followed by benzoylation.

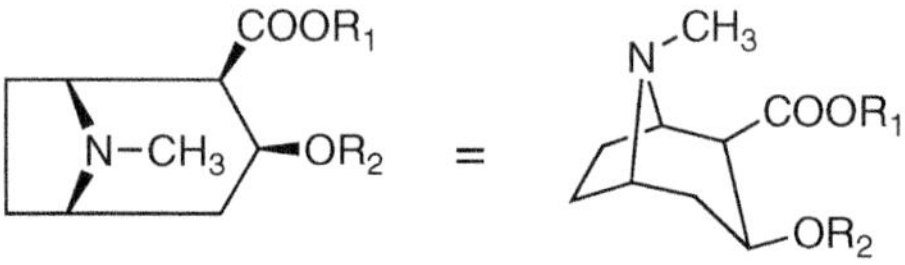

cocaine	$R_1 = CH_3$, $R_2 = C_6H_5CO$-
cinnamoylcocaine	$R_1 = CH_3$, $R_2 = C_6H_5$-CH=CHCO-
ecgonine	$R_1 = R_2 = H$

Cocaine possesses the tropane skeleton similar to (−)-hyoscyamine and (−)-scopolamine. Therefore, these alkaloids are known as tropane alkaloids collectively. However, the stereochemistry at the C-3 position in cocaine is epimeric to those of (−)-hyoscyamine and (−)-scopolamine. Cocaine is still used as a local anesthetic. Unfortunately, this alkaloid causes various social problems because of its neurotropic effect, and is now a notorious illicit drug.

Extract of Coca leaves was formerly mixed in the soft drink Coca-Cola. Pharmacist J. S. Pemberton (1831–1888) of Atlanta (Georgia, USA) mixed the extracts of Coca leaves and cola (kola) nuts with syrup and named it "Coca-Cola". He added water and began to sell the carbonated drink in 1886 [1]. In 1888, A. G. Chandler (1851–1929) handed over the Coca-Cola's production right, and he began to advertise it as a soft drink without completely advertising the biological effects of this drink. Until 1903, this drink still contained cocaine.

It is considered that the biosynthesis of cocaine resembles that of (−)-hyoscyamine and (−)-scopolamine. It was shown, through the feeding experiments using *E. coca* [2–5], that the β-ketothioester was the biosynthetic intermediate of cocaine, as in the case of (−)-hyoscyamine and (−)-scopolamine. It was also demonstrated that a thioester of benzoic acid derived from phenylalanie was the precursor of the benzoyl moiety of cocaine [6].

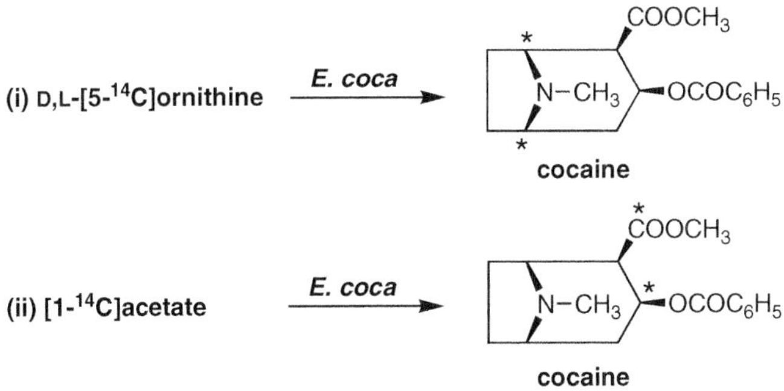

Distribution of Radio Isotopes in Cocaine by feeding D,L-[5-^{14}C]ornithine and [1-^{14}C]acetate into *E. coca*

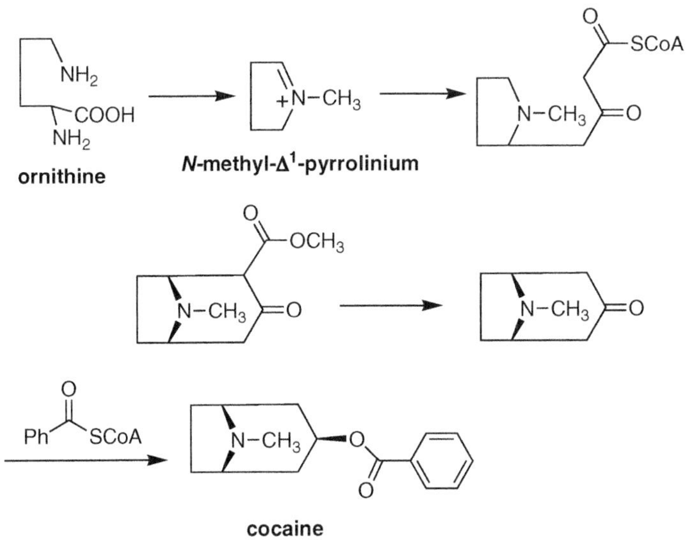

Biosynthetic Pathway to Cocaine

LITERATURE CITED

[1] J.L. Phillips, R.D. Wynne, Cocaine: The Mystique and the Reality, pp. 52–53, Avon Books, New York, 1980.
[2] E. Leete, J. Am. Chem. Soc. 104 (1982) 1403.
[3] E. Leete, Phytochemistry 22 (1983) 699.
[4] E. Leete, S.H. Kim, J. Am. Chem. Soc. 110 (1988) 2976.
[5] M.N. Huang, T.W. Abraham, S.H. Kim, E. Leete, Phytochemistry 41 (1996) 767.
[6] E. Leete, J.A. Bjorklund, S.H. Kim, Phytochemistry 27 (1988) 2553.

3.4 *Senecio* AND PYRROLIZIDINE ALKALOIDS

Senecio vulgaris (Asteraceae) is a perennial plant that came to Japan in the Meiji Era (1868–1912) and spread as a naturalized plant all over the country. In contrast, *Senecio pierotii* (Asteraceae) is a perennial plant that grows in the swamps of the mountains in Japan. Pyrrolizidine alkaloids, which possess a pyrrolizidine nucleus, were isolated from a number of *Senecio* plants, includeing *S. vulgaris* and *S. pierotti*. These alkaloids are distributed in *Senecio* plants, and several other genera of plants in the Asteraceae, Boraginaceae, and Fabaceae. Because their principal source is *Senecio* plants, these pyrrolizidine alkaloids are also known as Senecio alkaloids.

Most pyrrolizidine alkaloids are esters constructed from an aminoalcohol called a necine unit (pyrrolizidine derivative), and straight or branched chain fatty acid, called a necic acid. Necine is the derivative of pyrrolizidine, i.e., azabicyclo[3,3,0]octane, and a hydroxymethyl ($-CH_2OH$) moiety is always attached at the C-1 position. Necines can be classified into two groups, either with or without a double bond between the C-1 and C-2 positions. The stereochemistry of the H-8 position is usually H-8α. When the C-8 position is oxidized, as in the case of otonecine, it becomes an eight-membered ring by decyclization.

Necines are classified into the following three categories by the number of hydroxyl groups attached: (1) no other hydroxyl moieties, except a $-CH_2OH$ at the C-1 position, e.g., isoretronecanol and supinidine; (2) a secondary alcohol is attached at the C-7 position, e.g., platynecine and retronecine; and (3) an additional alcohol is attached at the C-2 or C-6 position, e.g., rosmarinecine and crotanecine. The alkaloids of the latter group are rare.

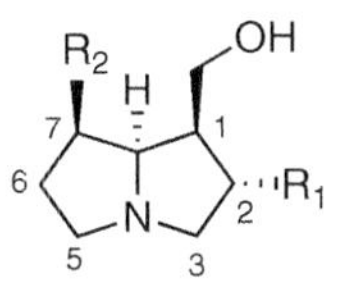

isoretronecanol $R_1 = R_2 = H$
platynecine $R_1 = H, R_2 = OH$
rosmarinecine $R_1 = R_2 = OH$

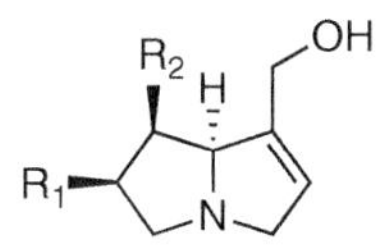

supinidine $R_1 = R_2 = H$
retronecine $R_1 = H, R_2 = OH$
crotanecine $R_1 = R_2 = OH$

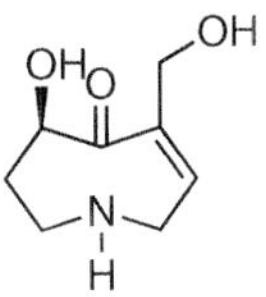

otonecine

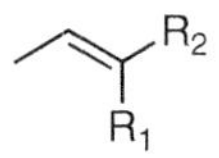

angelic acid $R_1 = COOH, R_2 = CH_3$
tiglic acid $R_1 = CH_3, R_2 = COOH$

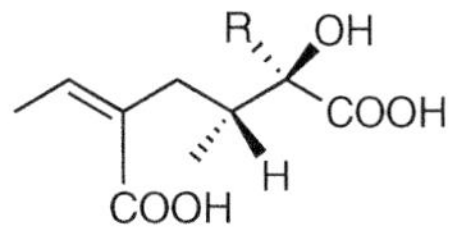

senecic acid $R = CH_3$
isatinecic acid $R = CH_2OH$

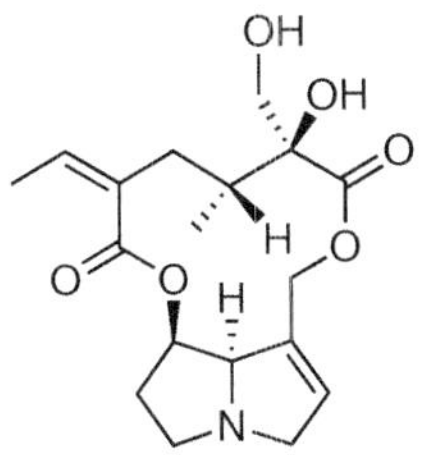

retrorsine

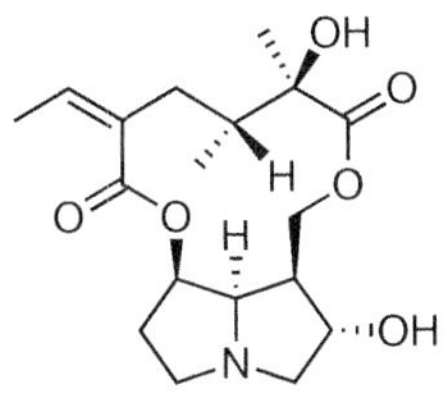

rosmarinine

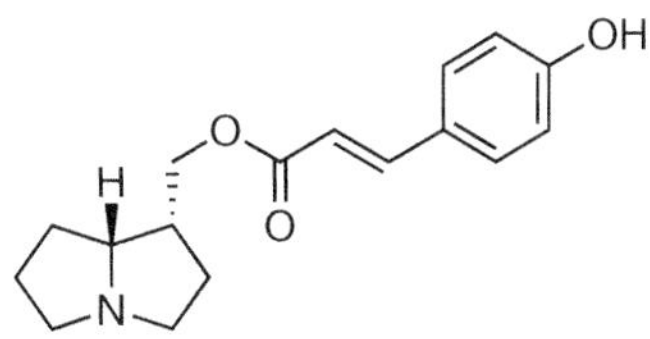

thesinine

The necic acids are linear or branched compounds constructed usually with 5, 7, 8, or 10 carbons. One or two necic acid(s) are combined with the necine as an ester. For example, angelic acid and tiglic acid with five carbons, and senecic acid and isatinecic acid with ten carbons, are frequent acylating units.

Various pyrrolizidine alkaloids are produced by the combination of the necines and necic acids. For example, the combination of retronecine, a necine, and isatinecic acid, a necic acid, forms retrorsine, which was isolated from *Senecio retrorsus* [1]. The combination of rosmarinecine and senecic acid forms rosmarinine, which is the main alkaloid of *Senecio pleistocephalus* [2].

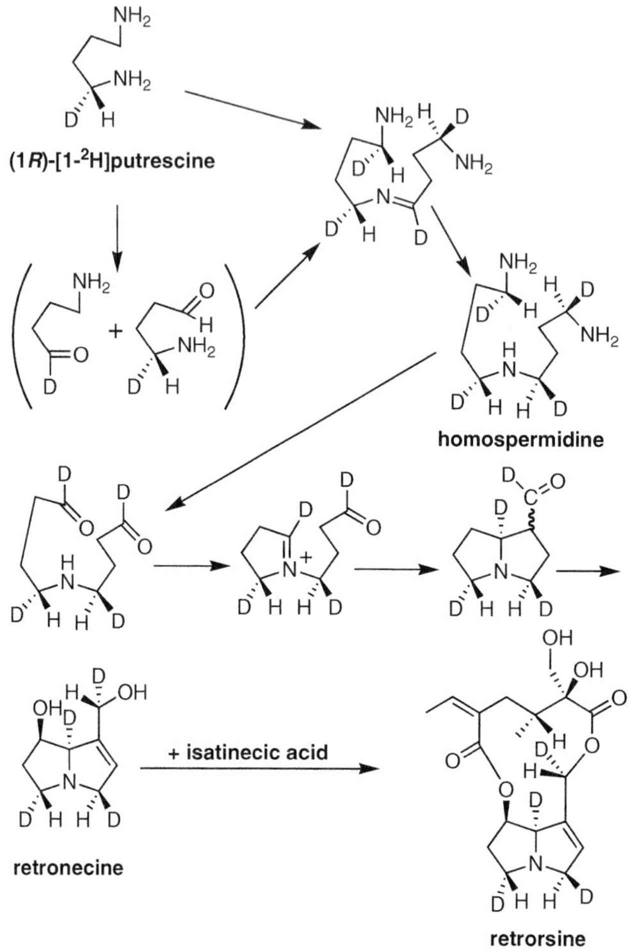

Incorporation of (1*R*)-[1-^{2}H]putrescine into Retronecine and Retrorsine

Some more unusual pyrrolizidine alkaloids possess an organic acid with an aromatic ring instead of a necic acid. For example, thesinine is formed by the combination of a pyrrolizidine derivative and coumaric acid. Thesinine was isolated from the flowers of *Borago officinalis* (Boraginaceae).

The necine part of a pyrrolizidine alkaloid is biosynthesized from the intermediate homospermidine via L-ornithine, followed by putrescine. It has been established, using labeled putrescine in the biosynthesis of the retrorsine in *Senecio isatideus*, that a biosynthetic intermediate with a symmetrical structure is formed.

That is to say, 3β,5α,8α,9-*pro-S* deuterated retrorsine was obtained by feeding (1*R*)-[1-^{2}H] putrescine. On the other hand, 3α,5β-deuterated retrorsine was obtained when (1*S*)-[1-^{2}H] putrescine was administered. In this case, no label was observed at the C-8 and C-9 positions [3,4]. The methylamino groups of labeled putrescine are oxidized to either one of the two aldehydes, shown in parentheses, and these combine with another molecule of putrescine. Stereospecific reduction of the imine double bond forms homospermidine. The labeled homospermidine is then stereospecifically oxidized to a dialdehyde, during which the amino group and the ^{1}H are lost and the D (^{2}H) remains. Furthermore, the aldehyde undergoes stereospecific reduction to form retronecine after the pyrrolizidine ring is formed through a Mannich reaction. Retrorsine is then formed if isatinecic acid combines to form the ester.

Plants containing 1,2-dehydro-pyrrolizidine alkaloids may cause liver necrosis to domestic animals and humans. Therefore, care must be taken not to consume plants that are known to contain pyrrolizidine alkaloids, such as the flower stalks of *Petasites japonicus* (Asteraceae), comfrey *Symphytum officinale* (Boraginaceae), and *Borago officinalis* (Boraginaceae). Lycopsamine was isolated as a main alkaloid of Russian comfrey (a hybrid of *S. officinale* and *Symphytum asperum*) [5].

In mammals, including humans, 1,2-dehydropyrrolizidine alkaloids incorporated internally are hydrolyzed to form retronecine. A pyrrole derivative, dehydroretronecine, is derived from retronecine by introducing a second double bond through oxidative didehydrogenation. It is this dehydroretoronecine that manifests the liver toxicity [6,7].

In the meantime, the male butterfly of *Danaid hamatus hamatus*, unlike Mammalia, transforms retronecine into hydroxydanaidal, danaidal, and danaidone, and these alkaloids are stored in the organ known as a hair pencil. The male butterfly spreads the hair pencil when he invites the female for mating, and the alkaloids act as an aphrodisiac-like pheromone at that time. Although it is observed that the male butterflies visit the Boraginaceae and *Senecio* genus plants in which the pyrrolizidine alkaloids are contained, there is a mystery as to how they take in these alkaloids, because they do not possess chewing mouthparts. It is possible that the alkaloids are contained in the honey of the flower [8–10].

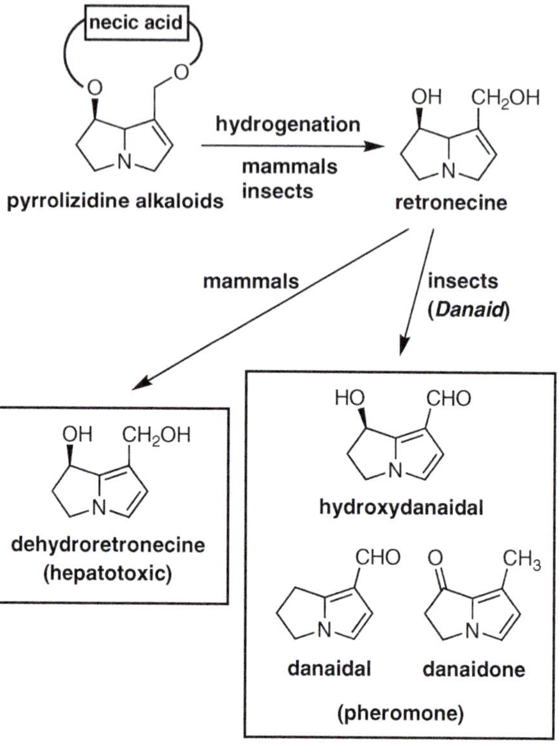

Metabolism of Pyrrolizidine Alkaloids in Mammals and Insects (*Danaid*)

The isolation and synthesis of pyrrolizidine alkaloids originating from plants and animals have been reviewed [11,12].

LITERATURE CITED

[1] R.H.F. Manske, Can. J. Res. 14B (6) (1936), [*Chem. Abstr.*, 30, 25713(1936)].
[2] H.A. Kelly, D.J. Robins, J. Chem. Soc. Perkin Trans. 1 (1987) 177.
[3] H.A. Khan, D.J. Robins, Chem. Commun. 1981 (1981) 554.
[4] J. Rana, D.J. Robins, Chem. Commun. 1984 (1984) 517.
[5] C.C.J. Culvenor, M. Clarke, J.A. Edgar, J.L. Frahn, M.V. Jago, J.E. Peterson, L.W. Smith, Experientia 36 (1980) 377.
[6] J. Meinwald, Y.C. Meinwald, P.H. Mazzocchi, Science 164 (1969) 1174.
[7] J.A. Edgar, C.C.J. Culvenor, L.W. Smith, Experientia 27 (1971) 761.
[8] J. Meinwald, W.R. Thompson, T. Eisner, Tetrahedron Lett. 12 (1971) 3485.
[9] J.A. Edgar, C.C.J. Culvenor, Nature 248 (1974) 614.
[10] J.A. Edgar, C.C.J. Culvenor, T.E. Pliske, Nature 250 (1974) 646.
[11] T. Furumae (Ed.), Kagakuseitaigakueno Shoutai, Sankyo Pub, 1996R, p. 106.
[12] J.R. Liddell, Nat. Prod. Rep. 19 (2002) 773.

3.5 ALKALOIDS DERIVED FROM POLYAMINES

Anton van Leeuwenhoek, known for inventing the microscope, observed various things through his self-made microscope, and one of them was human sperm. In 1678, he reported the existence of crystals in semen, which were later named Ladenburg spermine. However, it was not until 1924 that Rosenheim reported an efficient method for removing the crystals of spermine phosphate from semen for the first time. The yield depends on the purification method, and can vary in the range of 13–28 mg/10 ml [1]. The chemical structure of spermine was clarified in 1926, almost 250 years after the first report by van Leeuwenhoek. It was later established that spermine exists in various animal tissues and bacteria and in all eukaryotic cells. A material isolated from tissues that was effective against *Mycobacterium tuberculosis* was demonstrated to be spermine [2].

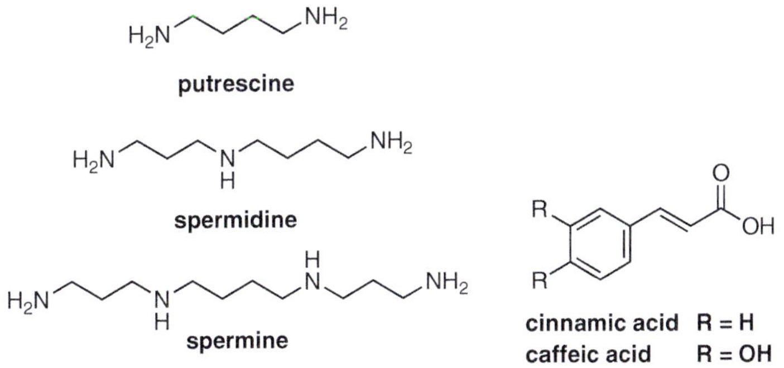

Chemical compounds such as spermine are called polyamines, and among them are spermidine and putrescine, in addition to spermine. Spermidine is also present in semen. Various biological activities are known for these alkaloids [3,4].

On the other hand, alkaloids in which cinnamic acid and various phenylpropanoids like caffeic acid and other acids combine with polyamines as amides are also isolated from higher plants. Among them, about eighty alkaloids contain spermidine as a base, and about thirty alkaloids contain spermine as a base [5]. These alkaloids are known as polyamine alkaloids. Some representatives of these alkaloids are described in the next section.

Most alkaloids exist in limited plant families. The distribution of the polyamine alkaloids is not limited to a few families but rather in broad range of families, including the Poaceae, Liliaceae, Fabaceae, Amaranthaceae, Asteraceae, and Solanaceae, etc. [6].

Among these polyamines, putrescine is biosynthesized from ornithine by decarboxylation with ornithine decarboxylase. Putrescine receives a propylamino unit (C_3N unit) from decarboxylated SAM (*S*-adenosylmethionine) to form spermidine. SAM is derived from methionine. Spermidine synthase catalyzes this biosynthetic process. Spermine is formed from spermidine through the addition of a C_3N unit from a decarboxylated SAM unit under the catalysis of spermine synthase [3].

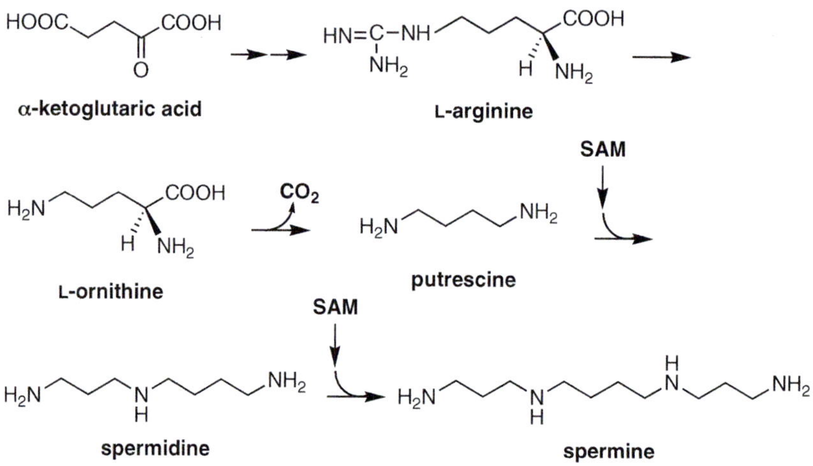

Biosynthetic Route of Putrescine, Spermidine and Spermine

The putrescine, spermidine, spermine, and related polyamine alkaloids have been reviewed [7].

LITERATURE CITED

[1] O. Rosenheim, Biochem. J. 18 (1924) 1253.
[2] J.G. Hirsh, R.J. Dubos, J. Exp. Med. 95 (1952) 191.
[3] Y. Takeda, K. Igarashi, Poriamin No Kiso (S. S. Cohen), Saine Pub. Co, 1976R.
[4] K. Igarachi, Seimei No Shimpi-Busshitu - Poriamin, Kyoritu Pub. Co, 1993R.
[5] A. Guggisberg, M. Hesse, Alkaloids 22 (1983R) 85.
[6] J. Martin-Tanguy, F. Cabanne, E. Perdrizet, C. Martin, Phytochemistry 17 (1978) 1927.
[7] S. Bienz, R. Detterbeck, C. Ensch, A. Guggisberg, U. Hausermann, C. Meisterhans, B. Wendt, C. Werner, M. Hesse, Alkaloids 58 (2002R) 83.

3.6 PUTRESCINE AND PIRIFERINE

From the virus-infected leaf of tobacco, alkaloids in which phenylpropanoids, such as *p*-coumaryl or ferulyl radical combined with putrescine were obtained. It was reported that these alkaloids possessed antiviral activity [1].

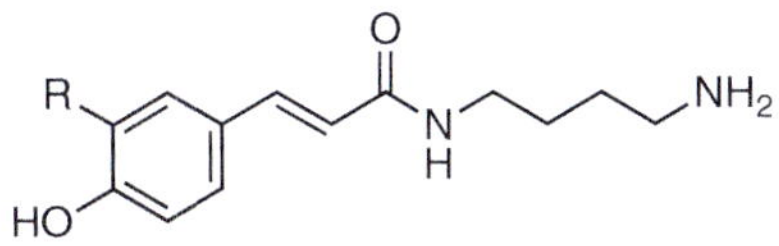

***p*-coumaroylputrescine** **R = H**
feruloylputrescine **R = OCH_3**

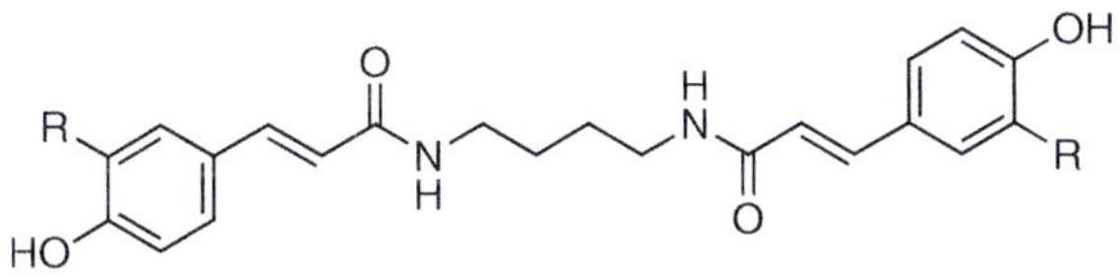

di-*p*-coumaroylputrescine **R = H**
diferuloylputrescine **R = OCH_3**

In contrast, it was reported that piriferine, isolated from *Aglaia pirifera* (Meliaceae), and odorine and 5′-*epi*-odorine isolated from *Aglaia odorata* and *Aglaia roxburghiana*, respectively, showed the ability to overcome resistance in vinblastine-resistant KB cells [2].

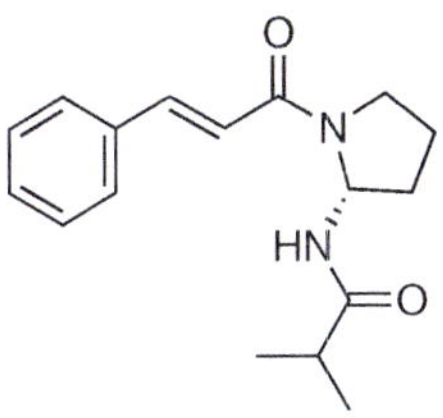

piriferine

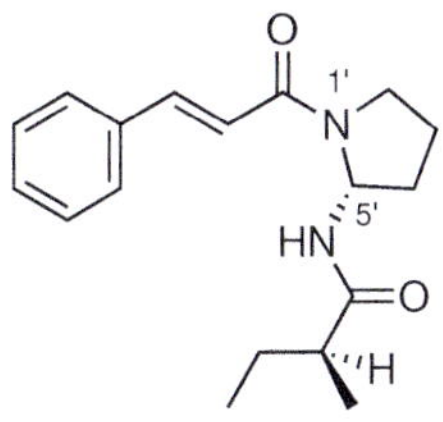

odorine
5'-*epi*-odorine (5'-eimer)

LITERATURE CITED

[1] J. Martin-Tanguy, C. Martin, M. Gallet, R.Vernoy, C. R. Acad. Sci. (Ser. D) 282 (1976) 2231.
[2] E. Saifah, J. Puripattanavong, K. Likhitwitayawuid, G.A. Cordell, H. Chai, J.M. Pezzuto, J. Nat. Prod. 56 (1993) 473.

3.7 SPERMIDINE AND AGROBACTIN

As an example of alkaloids with a spermidine moiety in the structure, maytenin was isolated from *Maytenus chuchuhuasha* (Celastraceae). This alkaloid possesses a structure in which two *trans*-cinnamoyl moieties are combined at each end of spermidine [1]. Among the polyamine alkaloids, those that possess a spermidine moiety in their structure are the most abundant [2].

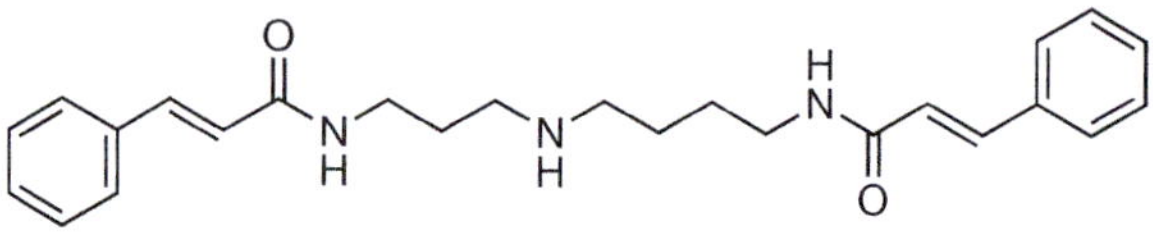

maitenine

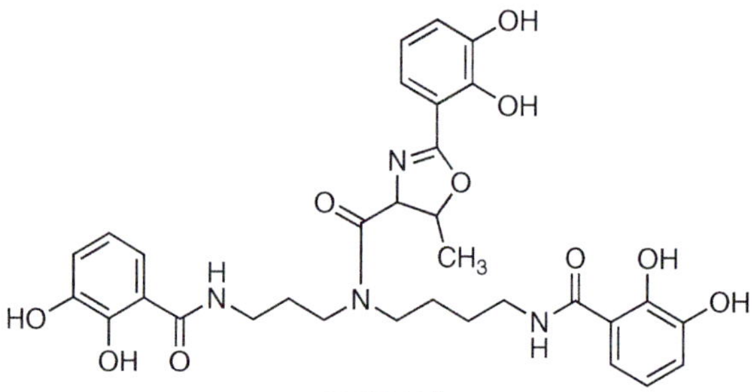

agrobactin

Agrobactin is an alkaloid formed when the microorganism *Agrobacterium tumefaciens* B6 is cultivated in a medium that lacks iron. Agrobactin is a siderophore; it forms a chelate with ferrous ion and is useful for its transportation [3].

LITERATURE CITED

[1] G. Englert, K. Klinga, R. Hamet, E. Schlittler, W. Vetter, Helv. Chim. Acta 56 (1973) 474.
[2] A. Guggisberg, M. Hesse, Alkaloids 22 (1983) 85.
[3] S.A. Ong, T. Peterson, J.B. Neilands, J. Biol. Chem. 254 (1979) 1860.

3.8 SPERMINE AND EPHEDRADINE

The number of alkaloids that contain spermine as a partial structure (spermine alkaloids) is less than the number of alkaloids that contain spermidine in their structure (spermidine alkalolids). Even so, these alkaloids are distributed in eleven plant families (i.e., Acanthaceae, Amaranthaceae, Bromeliaceae,

Cruciferae. Ephedraceae, Flacourtiaceae, Fabaceae, Liliaceae, Scrophulariaceae, Solanaceae, and Verbenaceae) [1].

The history of the isolation and structure elucidation of the spermine alkaloids dates from 1968, when homaline was isolated from the leaf of *Homalium pronyense* (Flacourtiaceae). The total synthesis of this alkaloid was reported in 1982, and the chemical structure, including absolute stereochemistry, was confirmed [2].

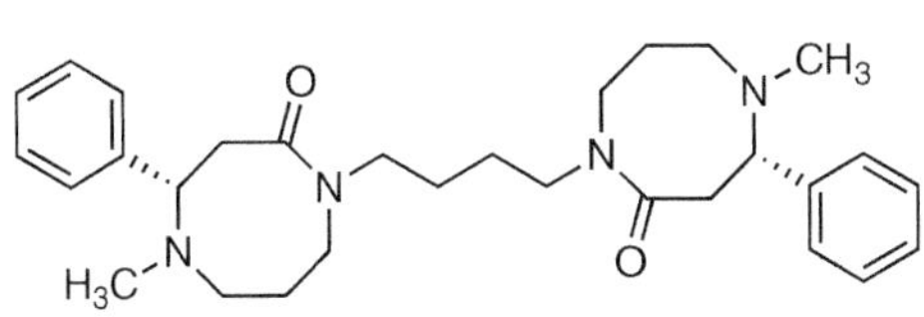

homaline

The crude drug "Mao" is prepared from the above-ground parts of *Ephedra* species (Ephedraceae). Mao is used for the manufacture of ephedrine hydrochloride, and is also used to prepare the Kampo medicine "Sho-seiryu-to," etc.

In Kampo medicine, gnarls and the subterranean part ("Mao-kon") of *Ephedra* species are regarded as contradictory to the action of Mao (the above-ground part). It is known that an extract of the subterranean part lowers blood pressure, whereas the extract of Mao raises blood pressure. Examination of the active constituents afforded ephedradines A–D [3]. Among these active constituents, the structure of ephedradine A was determined by X-ray crystallography of the brominated derivative [3]. Chemical structures of ephedradines B–D were determined by comparison of their spectroscopic data with those of ephedradine A [3,4].

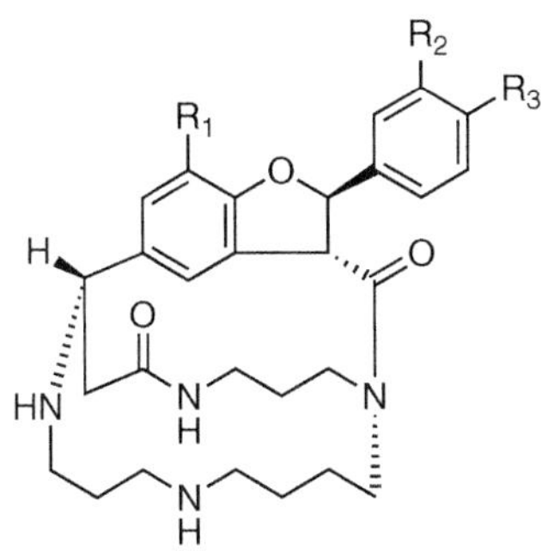

ephedradine A $R_1 = R_2 = H, R_3 = OH$
ephedradine B $R_1 = H, R_2 = OCH_3, R_3 = OH$
ephedradine C $R_1 = H, R_2 = R_3 = OCH_3$
ephedradine D $R_1 = OCH_3, R_2 = H, R_3 = OH$

The fruit, leaves, and root bark of *Lycium chinenese* (Solanaceae) are used mainly as a tonic. The methanol extract of the root bark of *L. chinense* is used as an antipyretic and has shown remarkable hypotensive activity in animals. The active components were isolated and were determined to be alkaloids, named kukoamines A and B [5,6]. Kuko is the Japanese name for *L. chinense*. These alkaloids are composed of a spermine unit and two molecules of dihydrocaffeic acid, as in the case of the ephedradines described above. The total synthesis of kukoamine A was independently achieved by two groups [7,8].

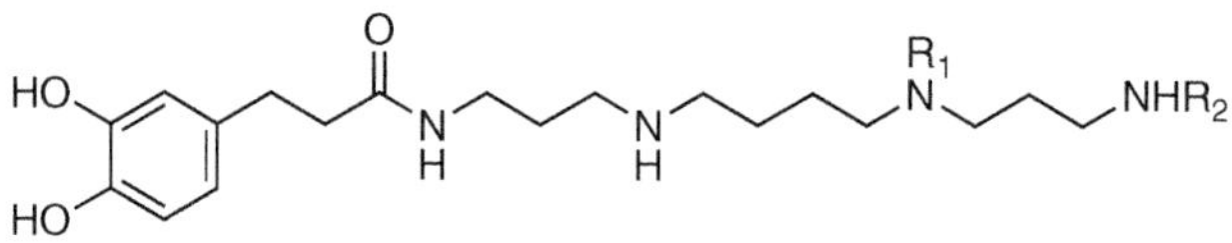

kukoamine A R_1 = H, R_2 = dihydrocaffeoyl
kukoamine B R_1 = dihydrocaffeoyl, R_2 = H

Several plants of the genus *Aphelandra* (Acanthaceae) also produce spermine alkaloids, and among them, aphelandrine is the main alkaloid. When radioisotope-labeled potential precursors of aphelandrine were fed to *Aphelandra tetragona*, it was shown that putrescine, spermidine, and cinnamic acid were incorporated into aphelandrine. It was not clear whether spermine was incorporated directly into aphelandrine. Spermine might have been metabolized to putrescine and spermidine before incorporation. It was clarified in this experiment that methionine was the precursor of the 3-aminopropyl moiety of spermidine and spermine [9].

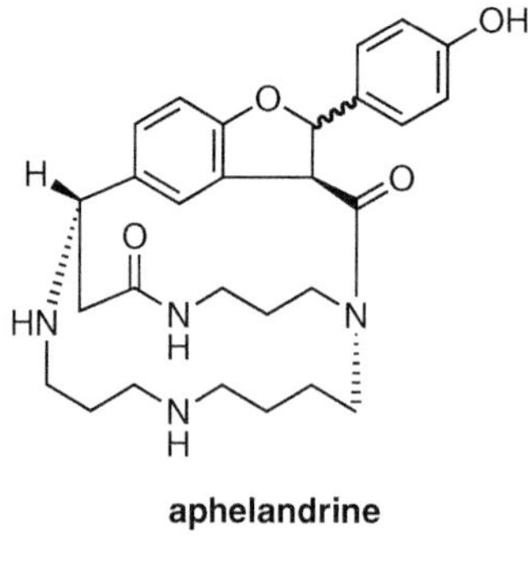

aphelandrine

LITERATURE CITED

[1] A. Guggisberg, M. Hesse, Alkaloids 22 (1983) 85.
[2] H.H. Wasserman, G.D. Berger, K.R. Cho, Tetrahedron Lett. 23 (1982) 465.
[3] M. Tamada, K. Endo, H. Hikino, C. Kabuto, Tetrahedron Lett. 20 (1979) 873.
[4] H. Hikino, M. Ogata, C. Konno, Heterocycles 17 (1982) 155.
[5] S. Funayama, K. Yoshida, C. Konno, H. Hikino, Tetrahedron Lett. 21 (1980) 1355.
[6] S. Funayama, G.-R. Zhang, S. Nozoe, Phytochemistry 38 (1995) 1529.
[7] K. Chantrapromma, B. Ganem, Tetrahedron Lett. 22 (1981) 23.
[8] T. Moriwake, S. Saito, H. Tamai, H. Mitsuda, M. Inaba, Heterocycles 23 (1985) 277.
[9] G. Papazoglou, J. Sierra, K. Homberger, A. Guggisberg, W. Woggon, M. Hesse, Helv. Chim. Acta 74 (1991) 565.

3.9 FUGU AND TETRODOTOXIN

"Fugu" is the generic name of the fish classified to the Plectognath of the Tetraodontidae. In the sea near Japan, about forty kinds of Fugu are present, and many kinds of Fugu have been served as a food in Japan. It seems that Japanese people have been eating Fugu since the earliest historic times, and evidence of this has been found.

The fact that Fugu possesses a poison that might intoxicate and kill humans has been known for some time. Research on the poison of Fugu began in the Meiji era (1868–1912), when modern science was introduced into Japan from Europe. Although the research at that time was mostly limited to pharmacological and toxicological studies, some chemical studies on Fugu toxin were done by Yoshizumi Tahara (1855–1935) [1–3]. In a research paper published in 1909, Tahara named this toxin as tetrodotoxin, but the purity of the toxin was found subsequently to be only about 0.2% [4]. It was in 1964, at the International Symposium of Natural Products Chemistry held in Kyoto, Japan, that the purification and structure determination of tetrodotoxin (TTX) were reported. At that symposium, the same chemical structure of this alkaloid was reported by three groups, i.e., two groups from Japan (Tsuda's and Hirata's groups) and the group of Woodward [4]. Among the eleven carbons of TTX, nine of the carbons are asymmetric.

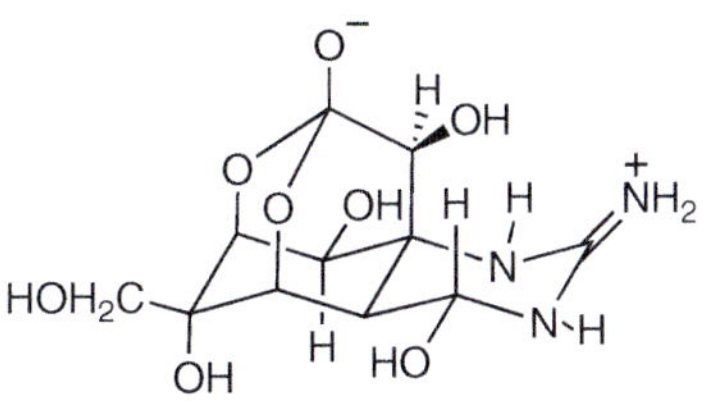

tetrodotoxin

Subsequently, it became clear that some of the animal toxins that had been known under different names were identified as TTX. Among them are tarichatoxin, isolated from the egg of California newt (*Taricha torosa*) of the United States [5,6], and maculotoxin, a neurotoxin secreted by the blue-ringed octopus (*Haplochlaena maculosa*) of Australia [7]. TTX was obtained mainly from the skin of several frogs from Costa Rica, namely, *Atelopus varius ambulatorius*, *A. varius varius*, and *Atelopus chiriquiensis* [8]. In addition, TTX was obtained from the cultured broth of a *Pseudomonas* sp. [9] and a *Vibrio* [10] strain.

The concern that TTX might be derived from a microorganism arose among researchers of the pharmacology of marine products around 1983, because TTX was distributed among various kinds of animal species. Regarding the TTX of Fugu, the content of this alkaloid varied among individual fish. Nontoxic Fugu could be produced by culturing it from the egg, and nontoxic Fugu became toxic by co-culturing with toxic Fugu. Thus, at least it is considered that the origin of the TTX of Fugu is extracorporeal.

TTX is a specific blocking agent of the voltage-dependent sodium (Na) channel, and occupies a unique position as a toxin for the Na channel. By applying TTX labeled with radioisotope, it became possible to isolate the channel molecule itself [11]. As a result, the primary structure of the Na channel was clarified by applying gene recombination technology [12].

With respect to the biosynthesis of TTX, although there are other possibilities, it was considered that TTX is formed by the combination of an arginine molecule and a C_5 isoprenoid unit [13,14]. The biogenetic units are shown in the Figure. If this is the biosynthesis of TTX, an arginine molecule is incorporated without decarboxylation.

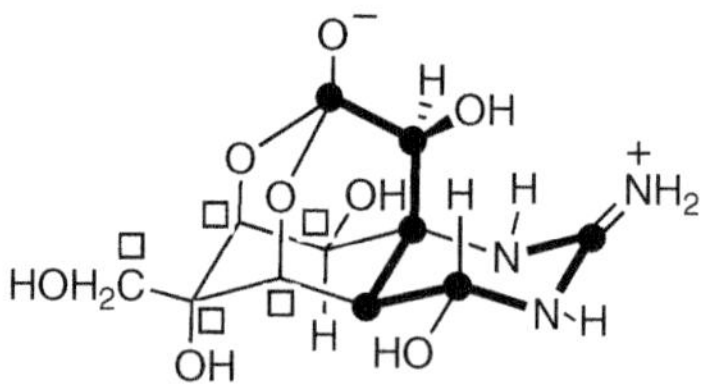

● arginine
□ isoprene (C_5 unit)

Biogenesis of Tetrodotoxin

However, this biogenetic pathway was subsequently re-examined because 5,6,11-trideoxy-TTX and its 4-epimer were isolated in equivalent yield from the Fugu [14,15]. If TTX is synthesized by the combination of an arginine unit and a C_5 unit, it is considered that oxidation of the C_4 position occurs enzymatically after cyclization. In this case, the C-4 position of TTX must be more stereoselective, and it seems relevant to consider that the C-4 position is epimerized before cyclization. Therefore, the C-4 position is derived from the anomer of a sugar, and it was proposed that the moiety that was formerly thought to be derived from arginine was actually derived from guanidine and 2-deoxy-3-oxopentose. According to this new hypothesis, the existence of 4-*epi*-TTX, in addition to 6-*epi*-TTX, 11-deoxy-TTX, 11-oxo-TTX, and 11-nor-TTX-6-ol, is explicable [15,16].

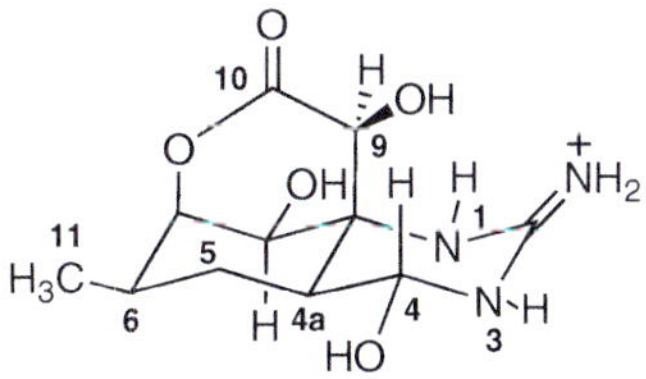

5,6,11-trideoxytetrodotoxin

If it is demonstrated that there is no involvement of arginine in the biosynthesis of TTX, this discussion should be moved to another chapter. Yet, if TTX is considered to be an alkaloid derived from an amino acid and a C_5 unit, there is an another example similarly derived. Namely, kainic acid (6.1) is shown to be derived from a glutamic acid and a C_5 unit. It is noteworthy that both of these very interesting, biologically active alkaloids, TTX and kainic acid, obtained from marine organisms in Japan, are derived from an amino acid and a C_5 unit.

Yokoo [17] and Tachikawa and Sakai [18] have provided reviews on the extraction and separation of Fugu poison. The latter review also summarizes the studies of the structure of TTX and related alkaloids, and their distribution.

LITERATURE CITED

[1] Y. Tahara, Yakugaku Zasshi 14 (1894) 732.
[2] Y. Tahara, Yakugaku Zasshi 15 (1895) 409.
[3] Y. Tahara, Yakugaku Zasshi 29 (1909) 587.
[4] K. Tuda, Yakushigaku Zasshi 29 (1994R) 424.

[5] H.D. Buchwald, L. Durham, H.G. Fischer, R. Harada, H.S. Mosher, C.Y. Kao, F.A. Fuhrman, Science 143 (1964R) 474.
[6] H.S. Mosher, F.A. Fuhrman, H.D. Buchwald, H.G. Fischer, Science 144 (1964R) 1100.
[7] D.D. Sheumack, M.E.H. Howden, I. Spence, R.J. Quinn, Science 199 (1978) 188.
[8] Y.H. Kim, G.B. Brown, H.S. Mosher, F.A. Fuhrman, Science 189 (1975) 151.
[9] T. Yasumoto, D. Yasumura, M. Yotsu, T. Michishita, A. Endo, Y. Kotaki, Agr. Biol. Chem. 50 (1986) 793.
[10] M.-J. Lee, D.-Y. Jeong, W.-S. Kim, H.-D. Kim, C.-H. Kim, W.-W. Park, Y.-H. Park, K.-S. Kim, H.-M. Kim, D.-S. Kim, Appl. Environ. Microbiol. 66 (2000) 1698.
[11] J.A. Miller, W.S. Agnew, S.R. Levinson, Biochemistry 22 (1983) 462.
[12] M. Noda, S. Shimizu, T. Tanabe, T. Takai, T. Kayano, T. Ikeda, T. Takahashi, H. Nakayama, Y. Kanaoka, N. Minamino, K. Kangawa, H. Matsuo, M.A. Raftery, T. Hirose, S. Inayama, H. Hayashida, T. Miyata, S. Numa, Nature 312 (1984) 121.
[13] T. Yasumoto, M. Yotsu, M. Murata, J. Am. Chem. Soc. 110 (1988) 2344.
[14] J. Kobayashi, M. Ishibashi, Alkaloids 41 (1992R) 42.
[15] M. Yotsu-Yamashita, Y. Yamagishi, T. Yasumoto, Tetrahedron Lett. 36 (1995) 9329.
[16] M. Miyakoshi, Farumasia 32 (1996R) 960.
[17] A. Yokoo, Tennen Yukikagobutu Toriatukai-ho (Jikken Kagaku Koza 22), p. 381, Maruzen Pub. Co., 1966R.
[18] R. Tachikawa, J. Sakai, Tennen Yukikagobutu Jikken-ho, p. 575, Kodansha Pub. Co., 1977R.

CHAPTER 4

Alkaloids Derived from Lysine

Sophora flavescens (Fabaceae)

Among the alkaloids derived from lysine, some are alkaloids in which lysine was incorporated intact, and others are alkaloids in which cadaverine, the symmetrical precursor derived from lysine by decarboxylation, was incorporated.

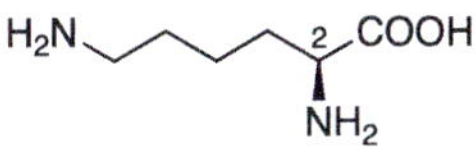

L-lysine

In the biosynthesis of pelletierine, lysine was incorporated without the involvement of a symmetrical intermediate such as cadaverine. On the other hand, it was established that in the biosynthesis of matrine, a symmetrical intermediate is incorporated, implying that cadaverine is involved.

Among the alkaloids obtained from *Nicotiana tabacum* (Solanaceae), the main component, nicotine, is biosynthesized through the combination of the five-membered ring of pyrrolidine, derived from ornithine, and nicotinic acid (3.1). On the other hand, the related alkaloid anabasine possesses a structure in which a six-membered piperidine ring derived from lysine is attached to nicotinic acid. Incorporation of ornithine into the pyrrolidine ring of nicotine occurs through a symmetrical intermediate, putrescine. However, lysine is incorporated into the piperidine ring of anabasine, through an asymmetrical precursor without involving a symmetrical intermediate such as cadaverine.

http://dx.doi.org/10.1016/B978-0-12-417302-6.00004-0

4.1 PEPPER (*Piper nigrum*) AND PIPERINE

Many members of the Piperaceae are of Indian origin, and are vine-like evergreens which entwine other trees. *Piper nigrum* is cultivated in India, the West Indies, and South America. "Ko-sho" is the Japanese name of pepper, and means foreign "Sho (spice)". Capsicum, which will be described in Chapter 16, is known as "Ban-sho", the fruit of which is also essential for the production of spices.

Black pepper is prepared from the unripe fruits of *Piper nigrum* with their pericarp, and white pepper is prepared from the ripe fruits after removal of their pericarp. Although the hot taste of white pepper is weaker than that of black pepper, the aroma of the former is stronger than that of the latter.

Piperine is recognized as the predominant pungent principle of pepper. Piperine possesses a structure in which a piperidine moiety is combined with piperinic acid through an amide function. Piperine was isolated at the the end of the nineteenth century, and the molecular formula determined [1,2]. It was some time afterward that the chemical structure was elucidated and its total synthesis accomplished [3]. Although the existence of a pungent substance, chavicine, a *cis,cis*-isomer of piperine was also reported, it was found by Grewe *et al.* [3] that chavicine was a mixture of piperine and piperyline, and other related species. Piperyline is identical with trichostachine [4], isolated from the leaves of *P. trichostachyon*, and it also possesses the same hot taste as piperine.

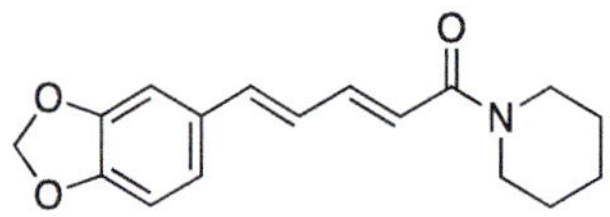

piperine

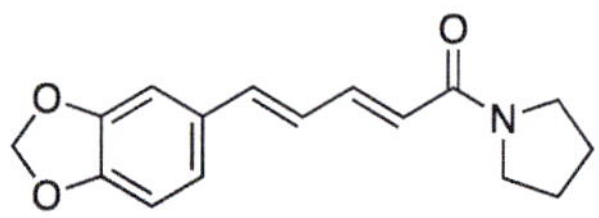

piperyline (trichostachine)

In the meantime, if the hot taste of *trans,trans*-piperine and piperyline is displayed with +++++, and that of chavicine is displayed with ++++, it was found that *cis,cis*-piperine showed a weak hot-taste (++) among the chemically synthesized isomers of piperine, *cis,cis*-piperine, *cis,trans*-piperine, and *trans,cis*-piperine. The remaining two isomers possess only a slight hot taste (+) [3], and, no hot taste was found for piperoleines A and B isolated from *P. nigrum* as minor constituents. It was also reported [3] that *cis,cis*-piperine, which corresponds to chavicine, was not found in black pepper. From these results, it was established that chavicine was an impure form of piperine derivatives.

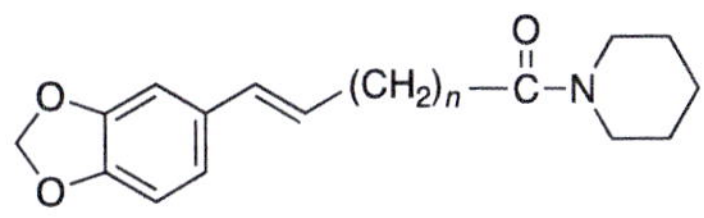

piperoleine A ***n*** **= 4**
piperoleine B ***n*** **= 6**

Among the alkaloids described above, the piperidine moieties of piperine and the piperoleines A and B are considered to be biosynthetically derived from lysine.

LITERATURE CITED

[1] P. Cazeneuve, O. Caillol, Bull. Soc. Chim. 27 (1877) 290.
[2] L. Rügheimer, Chem. Ber. 15 (1882) 1390.
[3] R. Grewe, W. Freist, H. Neumann, S. Kersten, Chem. Ber. 103 (1970) 3752.
[4] J. Singh, K.L. Dhar, C.K. Atal, Tetrahedron Lett. 10 (1969) 4975.

4.2 *Punica granatum* AND PELLETIERINE

Punica granatum (Punicaceae) is a deciduous tree native to Asia Minor, and is cultivated as a garden tree or a fruit tree in many places of Japan. The bark, branches, and roots of *P. granatum* are called "Zakuro-hi", "Sekiryu-hi", and "Sekiryu-kanpi", among other names in Japanese, and they are used as a taeniacide. A single daily dose is 30 - 40 g as a decoction, and because the active component of this crude drug is volatile, use of the fresh material is preferable.

Pelletierine was isolated [1,2] as an active ingredient and its chemical structure was determined [3]. It was found that the structure of pelletierine corresponded to that of isopelletierine synthesized previously [4].

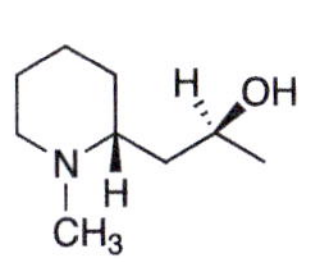

***N*-methylallosedridine**

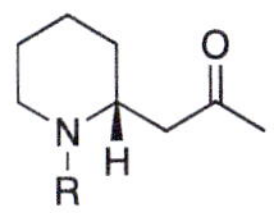

pelletierine **R = H**
***N*-methylpelletierine** **R = CH_3**

Pelletierine is biosynthesized through the incorporation of lysine. Among the several types of alkaloid possessing the piperidine nucleus, lobeline involves a similar biosynthetic pathway (Section 4.4) as that of pelletierine, whereas arecoline and coniine are biosynthesized through completely different pathways. The former alkaloid is derived from nicotinic acid (Section 10.3), and the latter alkaloid is biosynthesized via the polyketide pathway (Section 15.1).

Alkaloids possessing a piperidine moiety in their chemical structure are sometimes collectively referred to as piperidine alkaloids, regardless of their biosynthetic origin.

The biosynthetic pathway of the pelletierine type alkaloids was studied using *Sedum sarmentosum* (Crassulaceae). Namely, the NMR spectra of *N*-methylpelletierine and *N*-methylallosedridine, obtained following the administration of [1,2-$^{13}C_2$] acetic acid and [1,2,3,4-$^{13}C_4$] acetoacetic acid, were examined.

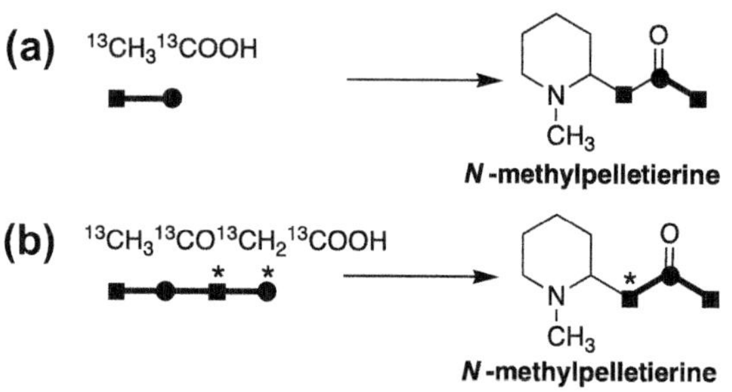

Incorporation Patterns of (a) [1,2-$^{13}C_2$]Acetic Acid and (b) [1,2,3,4-$^{13}C_4$]Acetoacetic Acid into *N*-Methylpelletierine

As a result, it was found that [1,2-$^{13}C_2$] acetic acid was incorporated into the C-2′ and C-3′ positions of the alkaloid. On the other hand, when [1,2,3,4-$^{13}C_4$] acetoacetic acid was fed, carbons 2 - 4 of acetoacetic acid were incorporated intact into carbons 1′ - 3′ of the alkaloid (figs. a and b) [5].

It was clarified by 1970 that pelletierine was biosynthesized from lysine and acetic acid, and that the C-2 position in pelletierine was labeled when C-2 labeled lysine was fed.

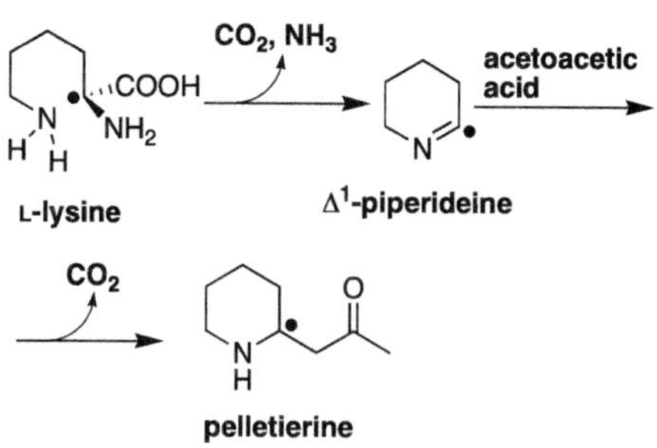

Biogenetic Route to Pelletierine

That is to say, in the biosynthesis of pelletierine, lysine was transformed into an asymmetric alkaloid Δ^1-piperideine without forming a symmetrical intermediate, such as cadaverine. The C_4 unit derived from acetic acid was then combined with this moiety, and pelletierine was formed by decarboxylation of this intermediate [6–8] However, there are four different pathways for the introduction of the two acetic acid moieties, as shown in the fig. (A - D).

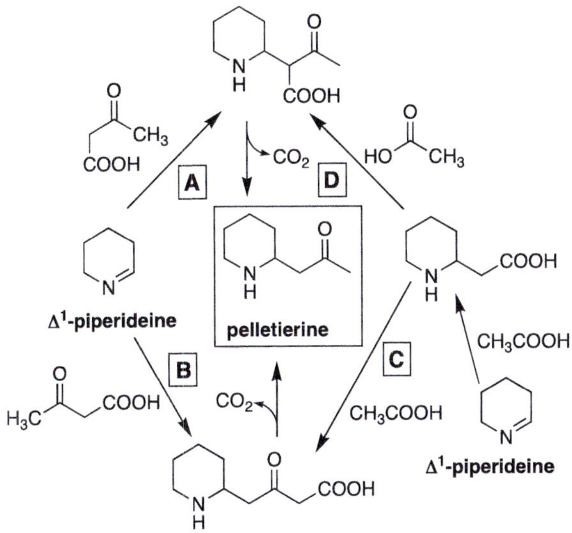

Four Biogenetic Routes (A - D) from Δ^1-Piperideine to Pelletierine

It was only recently that the exact pathway was determined through the experiment described above. [1,2-$^{13}C_2$] Acetic acid was introduced intact into the C-2′ and C-3′ positions of the resulting alkaloid, clarifying that pathway A or D was taken. Since [1,2,3,4-$^{13}C_4$] acetoacetic acid was introduced at C-1″, C-2″, and C-3″ in the alkaloid, it was concluded that pelletierine was biosynthesized by route A [5].

LITERATURE CITED

[1] M.Tanret, Compt. Rend. Acad. Sci. 86 (1878) 1270.
[2] R.E. Gilman, L. Marion, Bull. Soc. Chim. Fr. (1993) 1961.
[3] G. Drillien, C.Viel, Bull. Soc. Chim. Fr. (1963) 2393.
[4] E. Anet, G.K. Hughes, E. Ritchie, Nature 164 (1949) 501.
[5] T. Hemscheidt, I.D. Spenser, J. Am. Chem. Soc. 112 (1990) 6360.
[6] R.N. Gupta, I.D. Spenser, Chem. Commun. (1968) 85.
[7] R.N. Gupta, I.D. Spenser, Phytochemistry 8 (1969) 1937.
[8] M.F. Keogh, D.G. O'Donovan, J. Chem. Soc., Sect. C (1970) 1792.

4.3 TOBACCO AND ANABASINE

The main alkaloid of tobacco, *Nicotiana tabacum* (Solanaceae), is (-)-nicotine, as described in Section 3.1. (-)-Nicotine is an alkaloid composed of nicotinic acid (which will be described later) and a pyrrolidine ring derived from ornithine.

Also obtained from tobacco are the alkaloids (-)-anabasine and (-)-anatabine and related derivatives. (-)-Anabasine is biosynthesized by combining nicotinic acid and a piperidine ring derived from lysine.

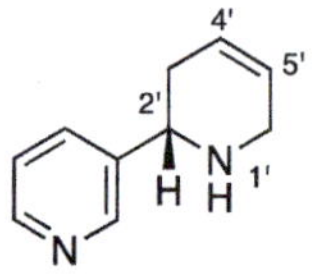

(-)-anabasine **4',5'-H$_2$**
(-)-anatabine

As well as nicotine, (-)-anabasine possesses some contact poisonous activity for insects, so that it can be used as an insecticide. Although anabasine is a minor constituent of tobacco, in *Anabasis aphylla* (Chenopodiaceae), it is the main alkaloid [1].

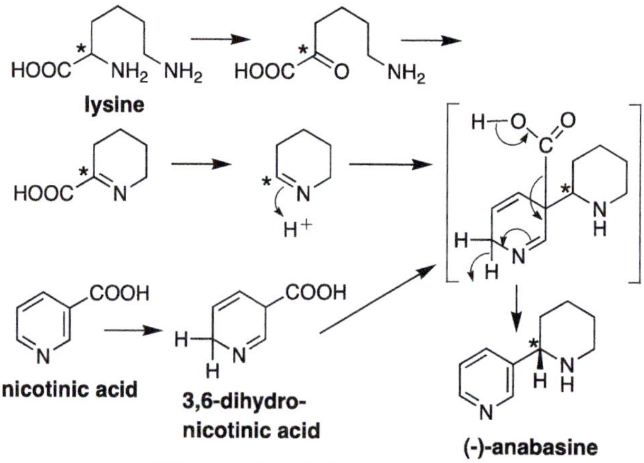

Biogenesis of (-)-Anabasine

It was already described in Section 3.1 that ornithine is incorporated into nicotine in the form of a symmetrical intermediate, putrescine. Although the last stages in the biosynthesis of (-)-anabasine are potentially quite similar to those of (-)-nicotine, during this biosynthetic pathway, it is considered that a symmetrical intermediate, such as cadaverine, is not involved. Thus, when [2-^{14}C]lysine was incorporated into (-)-anabasine, ^{14}C was introduced only into the C-2 position of the piperidine ring of

(-)-anabasine. Consequently, it seems that the biosynthesis of the piperidine ring of (-)-anabasine is formed in a similar process to that of pelletierine, as shown in the figure. The rationale for the involvement of 3,6-dihydronicotinic acid as an intermediate during the biosynthesis from nicotinic acid [2,3] will be described in Section 10.1.

Although the chemical structure of (-)-anatabine is quite similar to that of (-)-anabasine, remarkably, the biosynthetic pathways of these alkaloids are considerably different. Thus, (-)-anabasine is biosynthesized from nicotinic acid and lysine as described above, whereas, (-)-anatabine is biosynthesized from two molecules of nicotinic acid [2,3], as described in Chapter 10.

LITERATURE CITED

[1] A. Orechoff, G. Menschikoff, Chem. Ber. 64 (1931) 266.
[2] R.F. Dawson, D.R. Christman, A. D'Adamo, M.L. Solt, A.P. Wolf, J. Am. Chem. Soc. 82 (1960) 2628.
[3] E. Leete, S.A. Slattery, J. Am. Chem. Soc. 98 (1976) 6326.

4.4 *Lobelia inflata* (INDIAN TOBACCO) AND LOBELINE

Lobelia inflata (Campanulaceae) grows wild in the U.S.A., the eastern and central part of Canada, and Kamchatka. The leaf has been used by Native Americans as an emetic since ancient times. From this plant, a large number of piperidine and *N*-methylpiperidine derivatives were obtained, and the main alkaloids are lobeline, lobelanine, and lobelanidine [1,2].

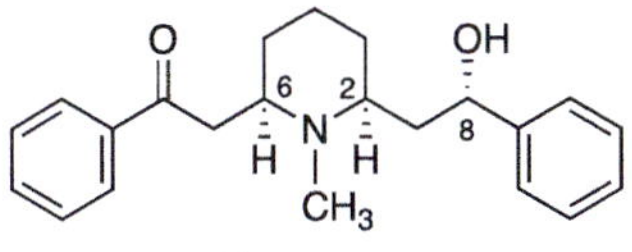

(-)-lobeline

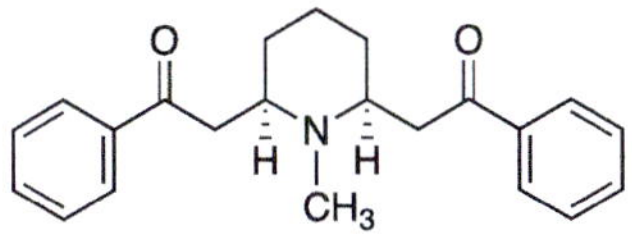

lobelanine

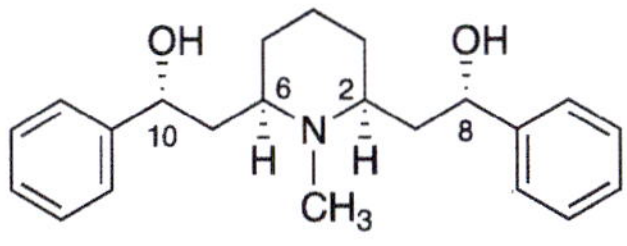

lobelanidine

In 1965, the absolute structures of (-)-lobeline and lobelanidine were reported [3]. The absolute configurations of the C-2 and C-8 positions of the former alkaloid are *S*, and that of C-6 is *R*, respectively. The absolute configurations of C-2, C-6, and C-8 of lobelanidine are the same as those of lobeline, and that of C-10 is *R*.

These alkaloids are used clinically as respiratory stimulants in the case of non-responding newborn infants, for collapse in cases of gas poisoning and anaesthetic poisoning, for smoking cessation, and for suffocation caused by drowning.

LITERATURE CITED

[1] H. Wieland, Chem. Ber. 54 (1921) 1784.
[2] H. Wieland, C. Schöpf, W. Hermsen, Liebigs Ann. Chem. 444 (1925) 40.
[3] C. Schöpf, E. Müller, Liebigs Ann. Chem. 687 (1965) 241.

4.5 *Sophora flavescens* AND MATRINE

The crude drug prepared from the roots of *Sophora flavescens* (Fabaceae) is known as "Kujin" in Traditional Chinese Medicine. It possesses a strong bitter taste, and has been used as a medicine from ancient times in China, where it was recorded in the very old text book of herbal medicine known as "Jinno-Honzo-Kyo" in Japanese. "Jinno-Honzo-Kyo" was said to be first written in China at the end of 5th century, but the original has not survived until now.

The crude drug is used for the treatment of peptic ulcers, diuresis, as an antipyretic, an analgesic, and an antiparasitic. In Japan, it was formerly used for killing maggots, and for prickly heat. An extract of the whole plant was applied as a topical insecticide for cattle, and as an insecticide for plants.

The Japanese name of this plant is "Kurara", and it is thought that this name may be derived from its very bitter taste. Namely, because a person who tastes an extract of the roots may get dizzy ("Kuramu" in Japanese).

"Kurara" grows wild in Japan, and was initially named as *Sophora angustifolia*. Subsequently, this plant was considered to be a variety of *S. flavescens* growing wild in China, and was named as *S. flavescens* var. *angustifolia*. However, this plant is now regarded as the same species as that of Chinese origin, and the botanical name *S. flavescens* is used.

Matrine was obtained as the main alkaloidal component of this plant. In addition, oxymatrine, anagyrine, methylcytisine, and sophocarpine were also isolated, together with other alkaloids. The name matrine is derived from one of the Japanese plant names of *S. flavescens*, where this plant is known as "Matori-gusa" in some areas [1,2].

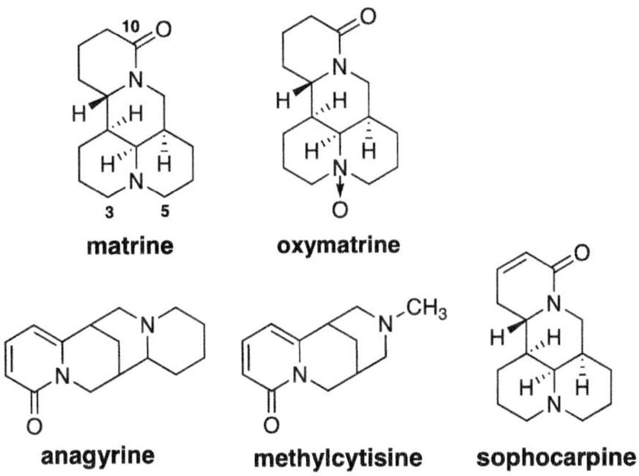

Matrine was isolated at the beginning of the twentieth century [1], and its plane structure was reported about 30 years later. However, it was a further 30 years before the absolute configuration of matrine was deduced by the total synthesis of (+)-matrine [3–5].

Regarding the biosynthesis of matrine, from an incorporation experiment using [1,5-$^{14}C_2$]-cadaverine, it was concluded that three cadaverine moieties were introduced into a molecule of matrine [6]. Because it was known that cadaverine was derived from lysine, it was implied that the nitrogen atoms of the quinolizidine units of matrine were derived from lysine.

Shibata *et al.* [7] obtained ^{14}C-labelled matrine by administering [2-^{14}C]-lysine, [1,5-$^{14}C_2$]-cadaverine, and [6-^{14}C]-Δ^1-piperideine to *S. flavescens*, and the location of the ^{14}C established an overall biosynthetic pathway as shown in the figure.

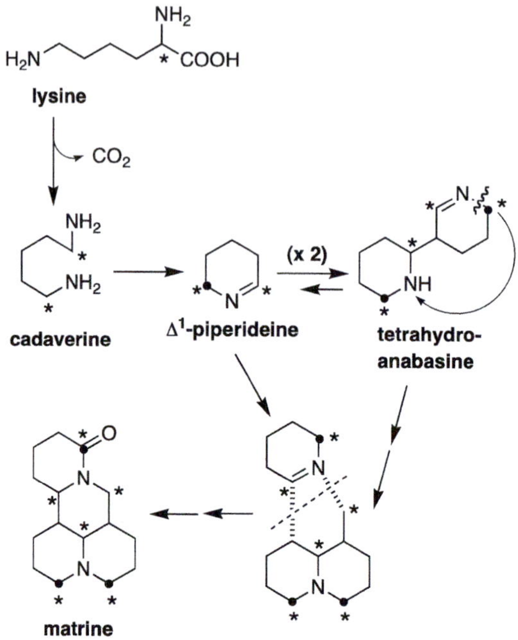

Biogenetic Pathway of Matrine

Thus, the tetrahydroanabasine derived from two molecules of Δ^1-piperideine was cleaved and was closed to form a ring resulting in a lupinane skeleton. To this lupinane moiety was attached another Δ^1-piperideine moiety to form matrine. Actually, when [6-^{14}C]-Δ^1-piperideine was administered, an unequal result was obtained, namely that ^{14}C (black circle) was introduced to the extent of 90% at the C-3 and C-5 positions of matrine, whereas only 10% of ^{14}C was introduced at the C-10 position. This observation suggests that the pool size of tetrahydroanabasine is larger than that of Δ^1-piperideine.

The total synthesis of (±)-matrine was accomplished [8], and the antitrichomonal activity of matrine was reported [9].

LITERATURE CITED

[1] N. Nagai, H. Kondo, Yakugaku Zasshi 23 (1903) 993.
[2] K. Tsuda: Tennen Yukikagobutsu Jikkenho, Jikken Kagaku Koza 22 (Experimental Methods for Natural Products Chemistry, Experimental Chemistry 22), p. 397, Maruzen Pub. Co. (1958R).
[3] K. Tsuda, Chem. Ber. 69 (1936) 429.
[4] S. Okuda, M. Yoshimoto, K. Tsuda, Chem. Pharm. Bull. 14 (1966) 275.
[5] S. Okuda, M. Yoshimoto, K. Tsuda, N. Utzugi, Chem. Pharm. Bull. 14 (1966) 314.

[6] H.R. Schütte, H. Aslanow, C.H. Schäfer, Arch. Pharm. 295 (1962) 34.
[7] S. Shibata, U. Sankawa, Chem. & Ind. (London) (1963) 1161.
[8] J. Chen, L.J. Browne, N.C. Gonnella, Chem. Commun. (1986) 905.
[9] H.-H. Wang, J.-T. Cheng, Phytother. Res. 8 (1994) 70.

4.6 COMMON BROOM (*Cytisus scoparius*) AND SPARTEINE

Sparteine, isolated from common broom (*Cytisus scoparius* = *Spartium scoparium*) (Fabaceae), also possesses a quinolizidine nucleus [1].

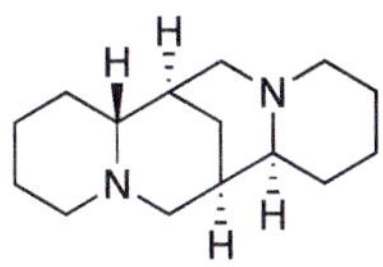

(-)-sparteine

Sparteine sulfate possesses several important biological activities, such as improving the tuning failure (a rhythmic disorder) of the heart, and produces a unique and periodic oxytocic action [2]. Consequently, as an injection, it is applied in cases of tachycardia, arrhythmia, uterine contraction disorder, inertia uteri, etc.

Sparteine is considered to be constructed from three molecules of cadaverine derived from lysine decarboxylation, as in the case of matrine [1]. Thus, when a Δ^1-piperideine moiety is attached to the lupinane skeleton, if route (a) is taken a matrine type alkaloid is formed, and a sparteine type alkaloid is formed by taking route (b).

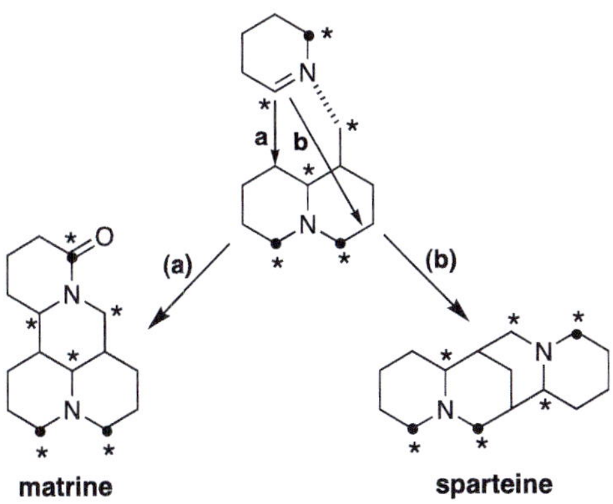

Biogenesis of Matrine and Sparteine

Alkaloids possessing a quinolizidine skeleton in their structure, such as sparteine and matrine and their related alkaloids, are frequently referred to as quinolizidine alkaloids.

LITERATURE CITED

[1] F. Binnig, Arzneim.-Forsch. 24 (1974) 752.
[2] M. Raschak, Arzneim.-Forsch. 24 (1974) 753.

CHAPTER 5

Alkaloids Derived from Proline

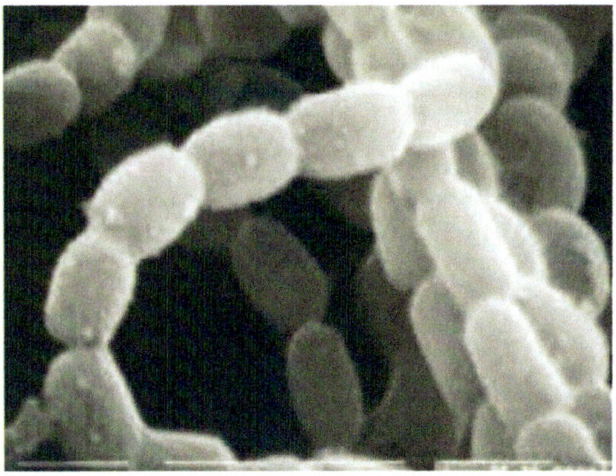

Streptomyces sp. 82-85

Proline is an amino acid that possesses a pyrrolidine moiety, and relatively few alkaloids are known, or considered, to be derived directly from proline. These alkaloids possess either a pyrrolidine or a pyrrole moiety in their structures.

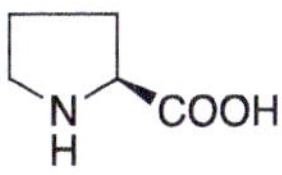

L-proline

Pyrrole derivatives are quite rare in plants, except for the chlorophylls present in plants, pigments in the blood (heme), and the bile pigments derived from heme. Therefore, one may initially regard chlorophyll and heme as alkaloids containing a pyrrolidine or pyrrole ring that might possibly be derived from proline in their chemical structures.

However, chlorophyll and heme are not derived from proline. Instead, part of the porphyrin ring of these alkaloids is derived from glutamic acid or glycine; these alkaloids are described in Chapter 12 (Section 12.1).

Other examples of alkaloids containing proline-like skeleta (for example, kainic acid and its related alkaloids) are also known. However, it was shown that the pyrrolidine rings of these alkaloids are derived from a glutamic acid and a C_5 unit; these alkaloids are described in Chapter 6 (Sections 6.1–6.2), where the alkaloids derived from glutamic acid are discussed. In addition, a pheromone of the butterfly of *Danaus genusas* is an example of a naturally occurring compound containing a pyrrole ring. However, this alkaloid is

Alkaloids
http://dx.doi.org/10.1016/B978-0-12-417302-6.00005-2

derived from a pyrrolizidine alkaloid of plant origin, and the pyrrolizidine nucleus originates from ornithine, as described in Chapter 3 (Section 3.4). The pyrrolidine ring of nicotine from tobacco (*Nicotiana* sp.) is also derived from ornithine (Section 3.1). Consequently, there are a number of alkaloids that appear to be derived from proline but that in actuality are derived through other biosynthetic pathways.

In this chapter, prodigiosin, obtained from the cultured broth of *Serratia marcescens*; stachydrine, isolated from Fabaceae plants; and pyrrole-2-carboxylic acid, isolated from the culture broth of *Streptomyces* sp. No. 82–85 are described. Regarding these alkaloids, it was established that labeled proline is incorporated into these alkaloids. It was also determined that alanine and glycine are incorporated into prodigiosin, in addition to proline [1]. However, intact incorporation of proline into prodigiosin as a unit retaining the nitrogen atom was observed, and consequently this alkaloid is described in this chapter.

In the meantime, it is known that proline is a consistent component of part of the tripeptide unit of the peptide ergot alkaloids, such as ergotamine, ergocristine, and ergocornine. These ergot alkaloids were described in Chapter 2 (Section 2.16).

LITERATURE CITED

[1] D.V. Lim, S.M.H. Qadri, C. Nichols, R.P. Williams, J. Bacteriol. 129 (1977) 124.

5.1 *SERRATIA* AND PRODIGIOSIN

Serratia marcescens is a small, Gram-negative bacillus in the intestinal bacterium family (Enterobacteriaceae), and colonies of this microorganism show a deep red color. This microorganism is sometimes found in feces cultures. A red pigment, prodigiosin, isolated from the culture broth shows antifungal, antileukemic, and antimalarial activity [1]. Subsequently, prodigiosin was synthesized chemically, where it also showed strong toxicity, and thus it has not been developed clinically [2].

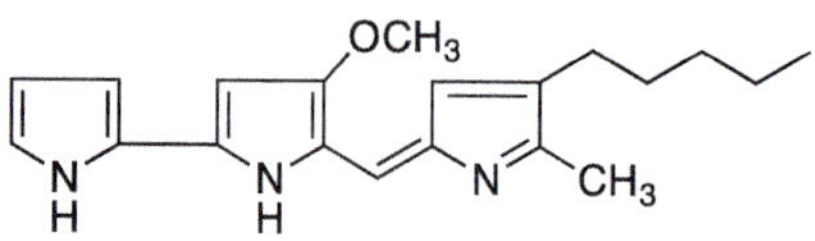

prodigiosin

Three pyrrole nuclei exist in prodigiosin, and it was shown that one of these units is derived from proline. In addition, a C_2 unit originated from alanine, five C_2 units were of acetic acid or alanine origin, and two C_1 units were derived from glycine or serine in the complex biosynthetic pathway. The methyl moiety of the methoxyl group of this alkaloid originated from methionine [3,4]. These derivations were determined by examining the ^{13}C nuclear magnetic resonance spectra of prodigiosin obtained by feeding each amino acid labeled with ^{13}C. When [1–^{14}C]-, [2–^{14}C]-, or [3–^{14}C]-labeled alanine was administered, the uptake of the radioisotope of the last two precursors was high. On the other hand, it was established that [1–^{14}C]alanine was not incorporated [4]. From these observations, it was shown that the C-2 and C-3 carbons of alanine were incorporated into prodigiosin, and that the carbonyl carbon of alanine was not incorporated.

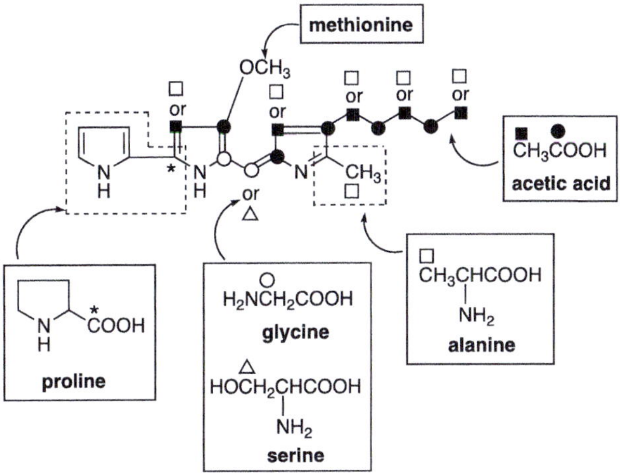

Biosynthetic Precursors of Prodigiosin

Prodigiosin showed significant cytotoxic activity against P388 mouse leukemia cells with IC_{50} 0.37 ng/ml, and it also showed cytotoxic activity against L-1210 mouse lymphocytic leukemia, B16 mouse melanoma, and human epidermoid nasopharynx carcinoma (KB) cells with IC_{50} values of 20–40 ng/ml [5]. The structures, biological activity, and biosynthesis of prodigiosin and related alkaloids are reviewed [6].

LITERATURE CITED

[1] F. Wrede, O. Hettche, Chem. Ber. 62 (1929) 2678.
[2] H. Rapoport, K.G. Holden, J. Am. Chem. Soc. 84 (1962) 635.
[3] H.H. Wasserman, R.J. Sykes, P. Peverada, C.K. Shaw, R.J. Cushley, S.R. Lipsky, J. Am. Chem. Soc. 95 (1973) 6874.

[4] D.V. Lim, S.M.H. Qadri, C. Nichols, R.P. Williams, J. Bacteriol. 129 (1977) 124.
[5] D.L. Boger, M. Petel, J. Org. Chem. 53 (1988) 1405.
[6] J.W. Bennett, R. Bentley, Adv. Appl. Microbiol. 47 (2000) 1.

5.2 STACHYDRINE

Stachydrine was isolated from *Medicago sativa* (Fabaceae), and it was established that this alkaloid was biosynthetically derived from proline.

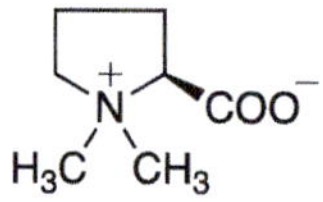

When either proline or hygric acid (*N*-methyl proline) labeled with ^{14}C at the carbonyl carbon was incorporated, stachydrine labeled with ^{14}C at the carboxyl carbon position was obtained [1]. It was also shown that proline was transformed into stachydrine in mature plants.

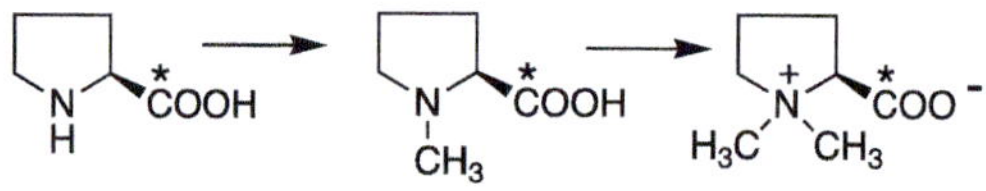

Biosynthetic Route of Stachydrine in *Medicago sativa.*

Namely, when labeled hygric acid was incorporated into young plants three weeks after germination, it was transformed into labeled stachydrine, whereas no labeled stachydrine was obtained by feeding labeled proline into the same plant. It seems that an enzyme that converts proline into hygric acid is absent or inactive in the young plant.

LITERATURE CITED

[1] J.M. Essery, D.J. McCaldin, L. Marion, Phytochemistry 1 (1962) 209.

5.3 PYRROLE-2-CARBOXYLIC ACID

It appears that pyrrole-2-carboxylic acid may be derived from proline, although there is no direct evidence to support this. This alkaloid was isolated from the culture broth of *Streptomyces* sp. 82–85 (Actinomycetaceae), and it shows platelet aggregation inhibition activity [1]. It is also a potential starter unit for prodigiosin biosynthesis (Section 5.1).

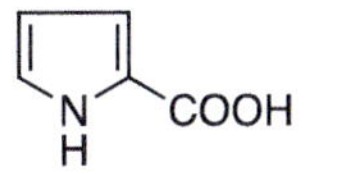

pyrrole-2-carboxylic acid

LITERATURE CITED

[1] K. Komiyama, C. Tronquet, Y. Hirokawa, S. Funayama, O. Sato, I. Umezawa, S. Oishi, Jpn. J. Antibiot. 39 (1986) 746.

CHAPTER 6

Alkaloids Derived from Glutamic Acid

Clitocybe acromelalga (Tricholomataceae)

Alkaloids derived from glutamic acid, other than γ-aminobutyric acid (GABA), include kainic acid isolated from *Digenea simplex*, domoic acid isolated from *Chondria armata*, acromelic acids isolated from *Clitocybe acromelalga*, ibotenic acid isolated from *Amanita strobiliformis* (*Amanita pantherina*), and tricholomic acid isolated from *Tricholoma musucaria*, and are described in this chapter.

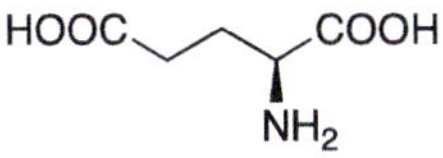

L-glutamic acid

Until recently, these alkaloids were considered to be unusual amino acids rather than alkaloids. These alkaloids are biosynthesized without passing through the process of decarboxylation, except for GABA and muscimol, which is secondarily produced from ibotenic acid.

6.1 γ-AMINOBUTYRIC ACID

GABA is an amino acid discovered in the encephalon of animals. This compound can also be regarded as a nitrogen-containing compound derived from L-glutamic acid by the action of glutamate decarboxylase and pyridoxal phosphate. GABA functions as a transmitter in the brain, and the blood pressure is lowered when GABA is injected intravenously, whereupon it acts on the vasomotor center of the medulla oblongata.

http://dx.doi.org/10.1016/B978-0-12-417302-6.00006-4

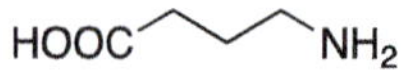

γ-aminobutyric acid (GABA)

GABA was formerly known as piperidic acid, because it was synthesized by the oxidation of piperidine at the end of the nineteenth century [1].

However, it was in the middle of the twentieth century that GABA was isolated from natural sources [2] for the first time. Dent et al. confirmed that GABA existed in the extract of potato by two-dimensional paper partition chromatography (PPC) [3]. In the meantime, in the animal kingdom, it was found that small amounts of GABA existed in human blood and urine by two-dimensional PPC, and GABA was isolated from the fresh encephalon of cattle [4]. Since then, it was found that GABA is widely distributed in the brains and nerves of animals and in various plants and microorganisms [5,6].

Until recently, GABA was regarded as one of the bioamines, such as dopamine (1.2), serotonin (2.1), and histamine (7.1). However, GABA should be regarded as one of the important alkaloids to be included in this chapter in view of its biological activity and process of formation.

GABA was isolated as a hypotensive principle of red-mold rice [7], and it was reported that a single dose of GABA (0.5 mg/kg, oral) and GABA-enriched fermented milk product (0.5 ml/kg, oral) significantly decreased the blood pressure of spontaneously hypertensive rats (SHR) [8]. It was also found that GABA had a hypotensive effect in SHR by low-dose (0.3–1.0 mg/kg, i.d.) [9]. *Lactobacillus* sp. Y-3, which produces GABA and is useful for manufacturing hypotensive health foods, is patented [10].

LITERATURE CITED

[1] C. Shotten, Chem. Ber. 16 (1883) 643.
[2] C.E. Dent, W. Stepka, F.C. Steward, Nature 160 (1947) 682.
[3] C.E. Dent, Biochem. J. 43 (1948) 169 [Chem. Abstr. 45 (1951) 5216e].
[4] J. Awapara, A.J. Landua, R. Fuerst, B. Seale, J. Biol. Chem. 187 (1950) 35.
[5] D. Jinnai, A. Mori, Nissin Igaku 45 (1958) 1.
[6] H. Hikino, S. Funayama, K. Endo, Planta Med. 30 (1976) 297.
[7] Y. Kohama, S. Matsumoto, T. Mimura, N. Tanabe, A. Inada, T. Nakanishi, Chem. Pharm. Bull. 35 (1987) 2484.
[8] K. Hayakawa, M. Kimura, K. Kasaha, K. Matsumoto, H. Sansawa, Y. Yamori, Brit. J. Nutr. 92 (2004) 411.
[9] M. Kimura, K. Hayakawa, H. Sansawa, Jpn J. Pharmacol. 89 (2002) 388.
[10] H. Matsushita, M. Furubayashi, R. Tsukiyama, Jpn Kokai Tokkyo Koho (2004), JP 2003-157713.

6.2 *Digenea simplex* AND KAINIC ACID

Digenea simplex (Rhodomelaceae) is a form of red algae. It is distributed in areas including the southern area of the Shiono Cape in Japan, the Indian Ocean, the Red Sea, and the Mediterranean. Dried whole algae are used as a crude drug under the name "Kaininso" in Japan, and are the material for the isolation of the roundworm ascaricide, kainic acid.

This crude drug has been used as an anthelmintic in Japan for a long time, and "Kaininso" was recorded as an anthelmintic in the old Japanese herbal text "Kinrampo," written in 866. However, the active component of this crude drug was uncertain until the twentieth century.

Finally, kainic acid was isolated as an active principle of "Kaininso" [1,2], and allokainic acid [3] was also isolated as an additional active component. The latter alkaloid is the stereoisomer of kainic acid at the C-4 position. Initially, kainic acid was named digenic acid, after the generic name *D. simplex*, and the name was subsequently changed to kainic acid [4].

The plane structure of kainic acid was determined by three independent research groups by three different methods, and resulted in identical chemical structure [5–7]. The stereochemistry of L-α-kainic acid and of L-α-allokainic acid was determined by the total synthesis of the respective alkaloids [8,9].

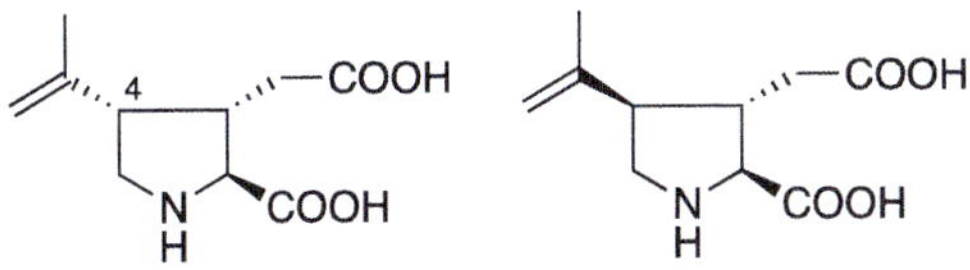

L-α-kainic acid **L-α-allokainic acid**

In Tokunoshima in the Amami Islands, *C. armata* (Rhodomelaceae) is used as an anthelmintic, and it was said that *C. armata* is more effective than *D. simplex*. From this seaweed, domoic acid was isolated as an active component [10–15]. Domoic acid is named after the local name, "Domoi," for *C. armata* in Tokunoshima.

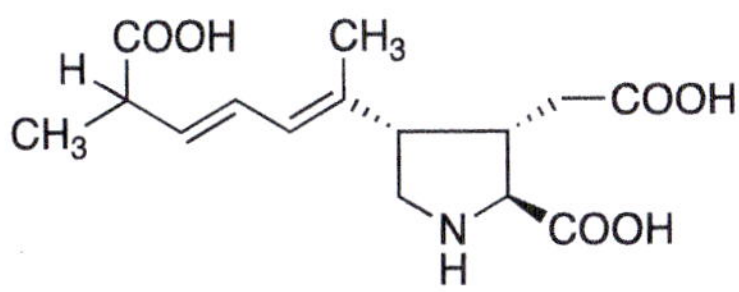

domoic acid

Though the name "Shakosai" is used sometimes for *D. simplex*, Shakosai is not *D. simplex*, but is actually *Caloglossa leprieurii* (Delesseriaceae). In

China, *C. leprieurii* is also used as a roundworm ascaricide, and it is known that this seaweed contains kainic acid.

Although kainic acid contains a proline-like skeleton, proline is not a biosynthetic precursor. It is thought that kainic acid is formed through a molecule of L-glutamic acid condensing with a C_5 unit. On the other hand, in the biosynthesis of domoic acid it is thought that a monoterpene unit combines with L-glutamic acid [15].

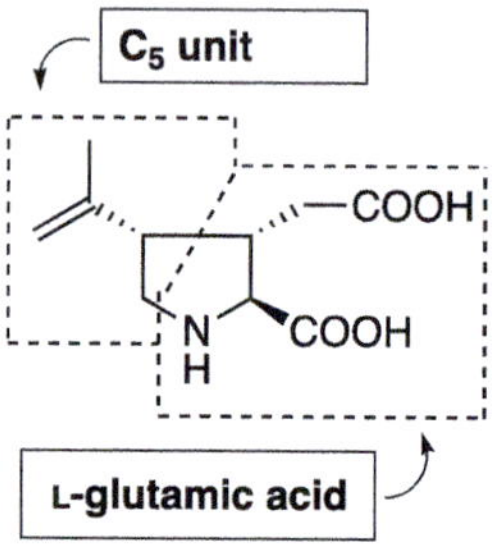

Biosynthetic Precursors of L-α-Kainic acid

Kainic acid has become an important drug in research associated with the nervous system [16].

LITERATURE CITED

[1] S. Murakami, T. Takemoto, Z. Shimizu, Yakugaku Zasshi 73 (1953) 1026.
[2] T. Takemoto, Handling Methods of Naturally Occurring Organic Compounds (Experimental Chemistry Series No. 22), p. 325, Maruzen Co., 1958R.
[3] S. Murakami, T. Takemoto, Z. Tei, K. Daigo, Yakugaku Zasshi 75 (1955) 869.
[4] S. Murakami, T. Takemoto, Z. Tei, K. Daigo, Yakugaku Zasshi 74 (1954) 560.
[5] M. Miyazaki, Yakugaku Zasshi 75 (1955) 695.
[6] Y. Ueno, H. Nawa, J. Ueyanagi, H. Morimoto, R. Nakamori, T. Matuoka, Yakugaku Zasshi 75 (1955) 840.
[7] S. Murakami, T. Takemoto, Z. Tei, K. Daigo, N. Takagi, Yakugaku Zasshi 75 (1955) 1252.
[8] M. Miyamoto, T. Sukawa, H. Morimoto, M. Uchibayashi, K. Tanaka, S. Tatsuoka, Yakugaku Zasshi 77 (1957) 580.
[9] J. Ueyanagi, H. Nawa, M. Honjo, R. Nakamori, K. Tanaka, Y. Ueno, S. Tatsuoka, Yakugaku Zasshi 77 (1957) 618.
[10] T. Takemoto, K. Daigo, Chem. Pharm. Bull. 6 (1958) 578.
[11] K. Daigo, Yakugaku Zasshi 79 (1959) 350.

[12] K. Daigo, Yakugaku Zasshi 79 (1959) 353.
[13] K. Daigo, Yakugaku Zasshi 79 (1959) 356.
[14] K. Daigo, Yakugaku Zasshi 79 (1959) 360.
[15] T. Takemoto, K. Daigo, Y. Kondo, K. Kondo, Yakugaku Zasshi 86 (1966) 874.
[16] J.W. Ferkany, R. Zaczek, J.T. Coyle, Nature 298 (1982) 757.

6.3 *Clitocybe acromelalga* AND ACROMELIC ACID

Clitocybe acromelalga (Tricholomataceae), "Dokusasako" in Japanese, is a toxic mushroom native to Japan. It is gregarious in bamboo and *Quercus serrata* forests in the flatland, and it grows wild mainly in western prefectures, such as Fukui, Toyama, and Niigata prefectures. It is distributed on Honshu island from Yamagata and Miyagi prefectures in the northern area to Shiga, Kyoto, and Wakayama prefectures in the southern area.

The toxic symptoms of this poisonous mushroom are extremely unpleasant. Discomfort starts several hours after a meal, followed by severe pain at the extremities, which occurs after several days following digestion. This pain continues for more than a month (sometimes for three months). Whether the new alkaloids, acromelic acids A and B, which were isolated and characterized, are the main toxic compounds is uncertain [1]. These alkaloids share the same pyrrolidine ring and the side chains at the C-2 and C-3 positions with kainic acid. In addition, kainic acid was transformed into acromelic acid A by changing the side chain at the C-4 position without changing the stereochemistry at C-4. Consequently, the absolute stereochemistry of acromelic acid was concluded to be the same as that of kainic acid [2].

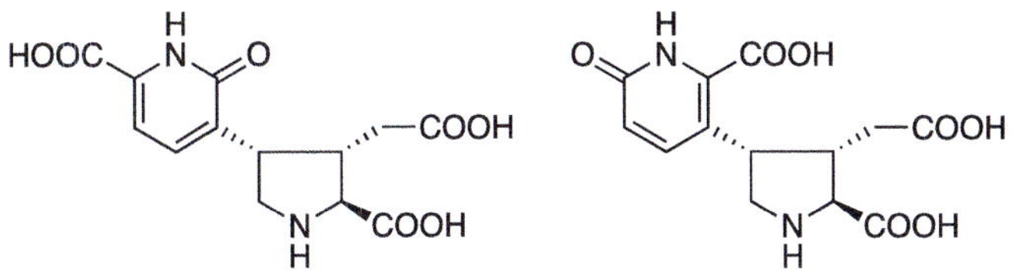

acromelic acid A **acromelic acid B**

It seems that glutamic acid is introduced into acromelic acids A and B during biosynthesis, as in the case of kainic acid.

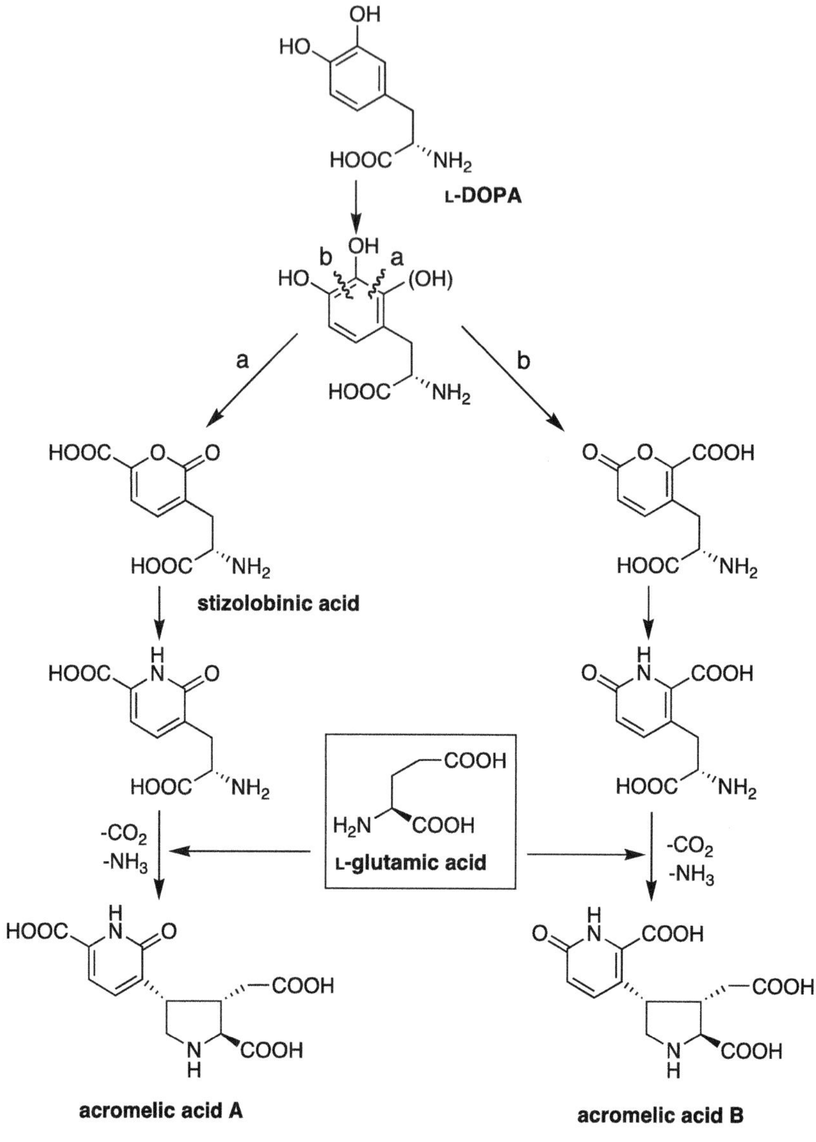

Biogenesis of Acromelic Acids A and B

In the biological activity test using the second ambulatory leg open chela muscle of the crayfish, acromelic acid A showed a strong depolarizing action. It is said that the activity shown by acromelic acid A is the strongest among the glutamic acid–related compounds that show such activity [3]. The biogenetic pathways of acromelic acids A and B are shown in the figure [4,5].

LITERATURE CITED

[1] K. Konno, H. Shirahama, T. Matsumoto, Tetrahedron Lett. 24 (1983) 939.
[2] K. Konno, K. Hashimoto, Y. Ohfune, H. Shirahama, T. Matsumoto, Tetrahedron Lett. 27 (1986) 607.
[3] M. Horikawa, K. Hashimoto, H. Shirahama, Tetrahedron Lett. 34 (1993) 331.
[4] K. Yamano, K. Konno, H. Shirahama, Heterocycles 35 (1993) 125.
[5] K. Yamano, H. Shirahama, Tetrahedron 49 (1993) 2427.

6.4 *Amanita pantherina* AND IBOTENIC ACID

Amanita strobiliformis (Japanese name: "Ibotengutake"; syn. *A. pantherina*) and *Amanita muscaria* (Amanitaceae) contain a fly-killing component [1–3], and it is known that *A. pantherina* and *A. muscaria* exhibit parasympathetic nerve excitatory action (muscarine receptor).

Muscarine was formerly considered to be the main active ingredient. However, the content of muscarine in these *Amanita* species was low, and although the parasympathetic nerve system excitement activity of muscarine is very high, the activity could not be explained solely by this component. As a result, ibotenic acid and muscimol, the decarboxylation product of ibotenic acid, were isolated as active components and characterized [2,3]. Muscimol was identified as the same constituent that had been named pantherine or agarine previously [4]. These alkaloids possess an isoxazole skeleton, and it seems that glutamic acid or glutamine is involved in their biosynthesis.

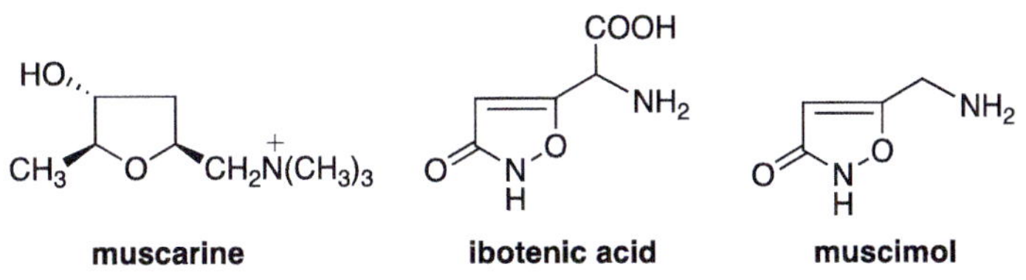

Potent central nervous system inhibitory effects and GABA antagonism were reported for muscimol [5], and this alkaloid was synthesized [4,6,7]. On the other hand, it was shown that ibotenic acid, like kainic acid (see Section 6.2), possessed more potent neuronal excitement effects than glutamic acid, and the activities of these alkaloids were 2–7 times and 8–80 times that of glutamic acid, respectively [8]. It can be considered that these alkaloids are derivatives of glutamic acid in which the conformation was fixed. These alkaloids are used as biological tools in neuroexcitatory experiments [9].

In the Tohoku area (northeast area of Honshu island) in Japan, there was a custom in which grilled *Tricholoma muscarium* (Tricholomataceae) was used for the purpose of killing flies. Takemoto et al. conducted research to isolate the active component(s) of this mushroom, and they succeeded in isolating an active component named tricholomic acid and elucidated its chemical structure [10]. The active component was identified as a dihydro derivative of ibotenic acid, and it possessed powerful fly-killing activity.

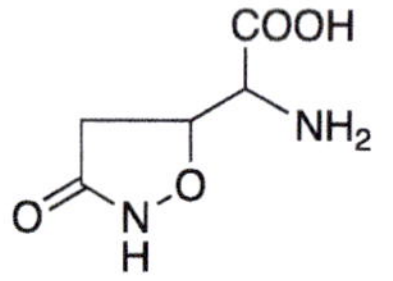

tricholomic acid

In addition, it was shown by Takemoto and Nakajima [11] that tricholomic acid and ibotenic acid possess peculiar palatability. In the meantime, it was established that kainic acid and domoic acid, which possess similar chemical structures, also showed fly-killing activity (see earlier Section 6.2).

LITERATURE CITED

[1] M. Onda, H. Fukushima, M. Akagawa, Chem. Pharm. Bull. 12 (1964) 751.
[2] T. Takemoto, T. Nakajima, T. Yokobe, Yakugaku Zasshi 84 (1964) 1232.
[3] T. Takemoto, T. Nakajima, R. Sakuma, Yakugaku Zasshi 84 (1964) 1233.
[4] A.R. Gagneux, F. Häfliger, J.R. Geigy, S.A. Basle, C.H. Eugster, R. Good, Tetrahedron Lett. 6 (1965) 2077.
[5] F.V. DeFeudis, Neurochem. Res. 5 (1980R) 1047.
[6] H. Göth, A.R. Gagneux, C.H. Eugster, H. Schmid, Helv. Chim. Acta 50 (1967) 137.
[7] K. Bowden, G. Crank, W.J. Ross, J. Chem. Soc. Sect. C 1968 (1968) 172.
[8] G.A.R. Johnston, D.R. Curtis, J. Davies, R.M. McCulloch, Nature 248 (1974) 804.
[9] A. Contestabile, P. Migani, A. Poli, L. Villani, Experientia 40 (1984R) 524.
[10] T. Takemoto, T. Nakajima, Yakugaku Zasshi 84 (1964) 1230.
[11] T. Takemoto, T. Nakajima, Hakko to Kogyo 34 (1976) 759.

CHAPTER 7

Alkaloids Derived from Histidine

Phytolacca americana (Phytolaccaceae)

Histamine is obtained by the decarboxylation of histidine. It is distributed in the tissues and blood, and is also produced by fermentation through the activity of microorganisms. Because histamine is produced by decarboxylation of histidine, it can be regarded as one of the simplest alkaloids derived from histidine.

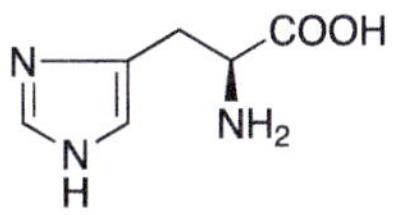

L-histidine

Only a very limited number of alkaloids derived from histidine are known. As an example of these alkaloids, (+)-pilocarpine is described in this chapter, as well as histamine. Alkaloids derived from histidine are also called imidazole alkaloids because of the imidazole nucleus in the histidine skeleton.

7.1 HISTAMINE

As described above, the alkaloid histamine originates from histidine. Histamine also exists as a common alkaloid in the body and is, like dopamine (Section 1.2) and serotonin (Section 2.1), called a bioamine. In some instances this alkaloid is discussed separately from the main alkaloids as one of the bioamines.

Histamine causes contraction of the smooth muscles, such as those of the bronchi, stomach, and intestine. The activity differs between the species of animal and between the various organs. For example, the guinea pig is particularly sensitive to this alkaloid, and among the human organs, the

http://dx.doi.org/10.1016/B978-0-12-417302-6.00007-6

bronchi are quite sensitive. Histamine does not affect the usual blood vessels; however, it causes a remarkable expansion of capillary vessels. In addition, this alkaloid promotes the secretion of gastric juice, pancreatic juice, and saliva.

It is clear that histamine is not the only mediator of an allergic reaction, but it is unquestionable that this alkaloid plays a very important role in the reaction. That is why an antihistaminic agent is active against the expansion of capillary vessels and the contraction of bronchi. Clinically it is used for nettle rash and allergic inflammation of the nose. On the other hand, an antihistaminic agent cannot suppress bronchial spasms or the secretion of gastric juice. These phenomena are explained in that a different compound exhibits the same biological activity as histamine, and because of the different receptors of histamine.

It is now clarified that there are two different receptors, H_1 and H_2, for histamine [1]. Cimetidine was synthesized as a typical antagonist of the H_2 receptor of histamine [2]. This compound possesses an imidazole skeleton like histamine, and it occupies the H_2 receptor of histamine to prevent the attachment of histamine, and finally it prevents the secretion of gastric juice. As a result, this compound became widely used for the treatment of stomach ulcers.

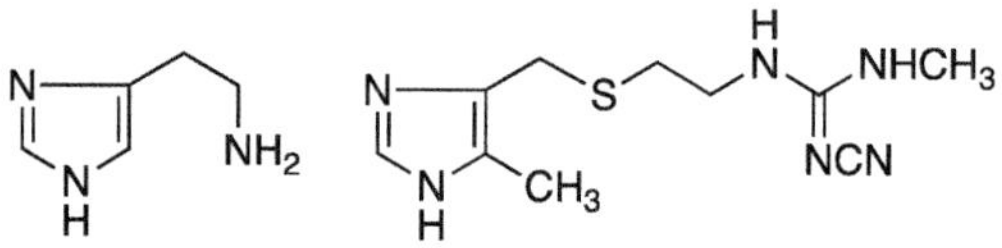

histamine **cimetidine**

There are few reports concerning the isolation of free histamine from higher plants. Among them, it was reported that from the dried roots of *Phytolacca americana*, as much as 1.3–1.6 mg/g of histamine was present [3]. Histamine was not detected, or was detected as only a trace amount from extracts of the dried roots of *Phytolacca esculenta* and *Phytolacca japonica*. The dried roots of *P. esculenta* and/or *P. japonica* are used as the material for the crude drug "Sho-riku" in Chinese herbal medicine. Two out of four samples of marketed "Sho-riku" in Japan contained almost the same amount and about half the amount of histamine as that of *P. americana*, respectively. So it is considered that some of the marketed "Sho-riku" in Japan are prepared from the dried roots of *P. americana*, or are increased in quantity with the root of *P. americana*, which is naturalized widely in Japan and which possesses a larger root than those of *P. esculenta* and *P. japonica*.

LITERATURE CITED

[1] J.W. Black, W.A.M. Duncan, C.J. Durant, C.R. Ganellin, E.M. Parsons, Nature 236 (1972) 385.
[2] Durant, C.J., Emmet, J.C., Ganellin, C.R. US. Patent, 3,950,333 (1976) [Chem. Abstr. 85, 33015s (1976)].
[3] S. Funayama, H. Hikino, J. Nat. Prod. 42 (1979) 672.

7.2 JABORANDI AND PILOCARPINE

"Jaborandi" is the name of the crude drug prepared from the evergreen plants *Pilocarpus jaborandi* and *Pilocarpus pinnatifolius*, which grow wild in South America, especially in Brazil. In South America, 13 plants in this genus are known, and all are used as the material for "Jaborandi," which is important as the material for the extraction of pilocarpine.

Pilocarpine is used as its hydrochloride and possesses excitatory activity on the parasympathetic nerve system, like physostigmine (Section 2.5) and arecoline (Section 10.3). Thus, this alkaloid acts as an antagonist of atropine (Section 3.2), and it promotes the secretion of sweat, saliva, and tears and causes myosis. It is reported that subcutaneous injection of 10 mg of pilocarpine HCl causes violent sweating (0.5–1.0 l) and salivation (1 l). As an eye lotion, a 1% solution of pilocarpine HCl is used for recovery from the mydriasis caused by atropine, or for the treatment of glaucoma.

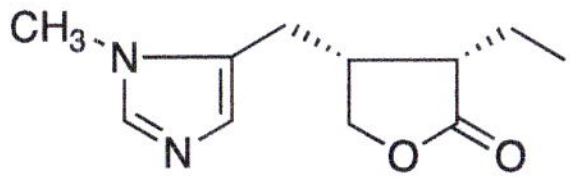

pilocarpine

The isolation of pilocarpine was reported in 1875 [1], separately by Byasson and Gerrard, and some chemical information was obtained subsequently [2]. The two-dimensional structure of the alkaloid was reported at the beginning of the twentieth century [3]. The absolute chemical structure and the total synthesis were reported subsequently [4,5].

Pilocarpine possesses an imidazole nucleus and the biosynthesis of this alkaloid is regarded as beginning with histidine, although the details are lacking.

LITERATURE CITED

[1] M.E. Hardy, Bull. Soc. Chim. Paris 1875 (1875) 497.
[2] A. Petit, M. Polonovski, Bull. Soc. Chim. Paris 1897 (1897) 554.
[3] H.A.D. Jowett, J. Chem. Soc. 83 (1903) 438.
[4] R.K. Hill, S. Barcza, Tetrahedron 22 (1966) 2889.
[5] H. Link, K. Bernauer, Helv. Chim. Acta 55 (1972) 1053.

CHAPTER 8

Alkaloids Derived from 2,3-Diamino-propionic Acid

Albizzia julibrissin (Fabaceae)

In this chapter, quisqualic acid, isolated from the fruits of *Quisqualis indica*, is described. Quisqualic acid is biosynthesized from 2,3-diaminopropionic acid. On the other hand, it can be said that quisqualic acid is biosynthesized from 2-amino-3-ureidopropionic acid, which is a derivative of 2,3-diaminopropionic acid. The former amino acid (2-amino-3-ureidopropionic acid) was isolated from the seeds of the silk tree (*Albizzia julibrissin*) (Fabaceae) and was named albizziin.

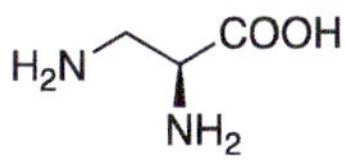

L-2,3-diaminopropionic acid

8.1 *Quisqualis indica* FRUITS AND QUISQUALIC ACID

Quisqualis indica var. *villosa* (Combretaceae) is a vine that grows widely in tropical Asia, and the trunk of this tree may reach 5 m. The fruits are used as a roundworm ascaricide, especially for the infant enterosite illness. The fruits are also applied to maintain the health of the stomach in cases of indigestion, diarrhea, and similar issues.

As an active component with anthelmintic activity, quisqualic acid was isolated and its chemical structure determined [1]. The alkaloid was previously isolated from the fruit of *Quisqualis chinensis* and *Q. indica* and its anthelmintic activity reported [2,3].

Alkaloids
http://dx.doi.org/10.1016/B978-0-12-417302-6.00008-8

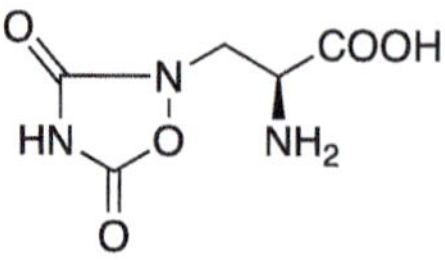

quisqualic acid

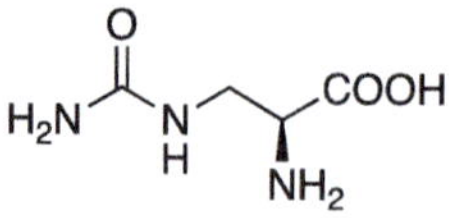

albizziin

Quisqualic acid is an excitatory amino acid, like *N*-methyl-*D*-aspartic acid and kainic acid, and the receptor of this alkaloid is sometimes called the quisqualate receptor [4–6]. X-ray crystallography of quisqualic acid [7] was carried out, and a stereoselective total synthesis of this alkaloid [8] was achieved.

Regarding the biosynthesis of quisqualic acid, it is considered that 2,3-diaminopropionic acid or 2-amino-3-ureidopropionic acid, a derivative of the former amino acid, is a precursor. The latter amino acid structure was attributed to albizziin (2-amino-3-ureidopropionic acid or 3-((aminocarbonyl)-amino-L-alanine)), which was isolated from the seeds of the silk tree (*Albizzia julibrissin*) (Fabaceae) [9,10].

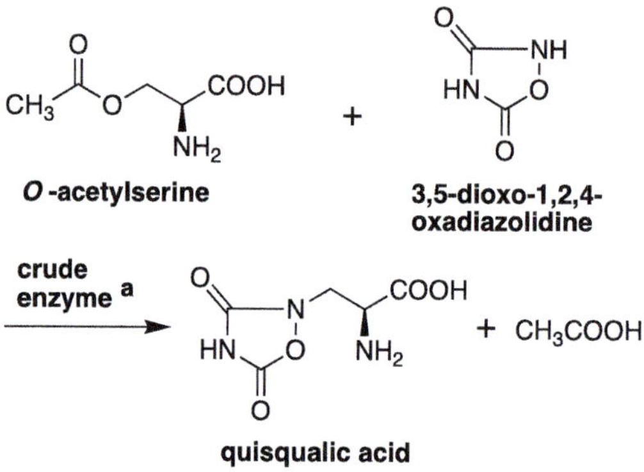

Enzymatic Synthesis of Quisqualic Acid from *O*-Acetylserine and 3,5-Dioxo-1,2,4-oxodiazolidine
[a](Isolated from *P. sativum* or *Q. indica* var. *villosa*)[11]

Quisqualic acid was synthesized by treating a mixture of *O*-acetylserine and 3,5-dioxo-1,2,4-oxadiazolizine with a crude enzyme prepared from an extract of the embryo of *Pisum sativum* or the stems and leaves of *Q. indica* var. *villosa* [11].

LITERATURE CITED

[1] T. Takemoto, S. Takagi, T. Nakajima, K. Koike, Yakugaku Zasshi 95 (1975) 176.
[2] Y.-C. Tuan, C.-H. Li, T.-C. Ch'en, Yao Hsueh Hsueh Pao 5 (1957) 87 [Chem. Abstr., 56, 14896b (1958)].

[3] S.-T. Fang, J.-H. Chu, Hua Hsueh Hsueh Pao 30 (1964) 226 [Chem. Abstr., 61, 7359f (1967)].
[4] H. Shinozaki, I. Shibuya, Neuropharmacology 13 (1974) 665.
[5] K. Koshiya, Life Sci. 37 (1985) 1373.
[6] J.T. Greenamyre, J.M.M. Olson, J.B. Penny Jr., A.B. Young, J. Pharmacol. Exp. Ther. 233 (1985) 254.
[7] J.L. Flippen, R.D. Gilardi, Acta Crystallogr. B32 (1976) 951.
[8] J.E. Baldwin, R.M. Adlington, D.J. Birch, J. Chem. Soc. Chem. Commun. 1985 (1985) 256.
[9] R. Gmelin, G. Strauss, G. Hasenmaier, Z. Naturforsch. 13b (1958) 252.
[10] R. Gmelin, G. Strauss, G. Hasenmaier, Hoppe-Seyler's Zeit. Physiol. Chem. 314 (1959) 28.
[11] I. Murakoshi, F. Kato, J. Haginiwa, T. Takemoto, Chem. Pharm. Bull. 22 (1974) 473.

CHAPTER 9

Compounds Derived from Anthranilic Acid

Hydrangea macrohylla f. *normalis* (Hydrangeaceae)

Anthranilic acid, which is derived from chorismic acid, is a nonessential amino acid.

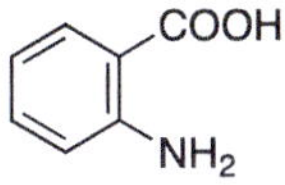

anthranilic acid

On the other hand, chorismic acid, which is derived from shikimic acid, is also a precursor of phenylalanine and tyrosine, which are essential amino acids [1]. Among the alkaloids, there is a group derived specifically from anthranilic acid, and this chapter presents some of these alkaloids.

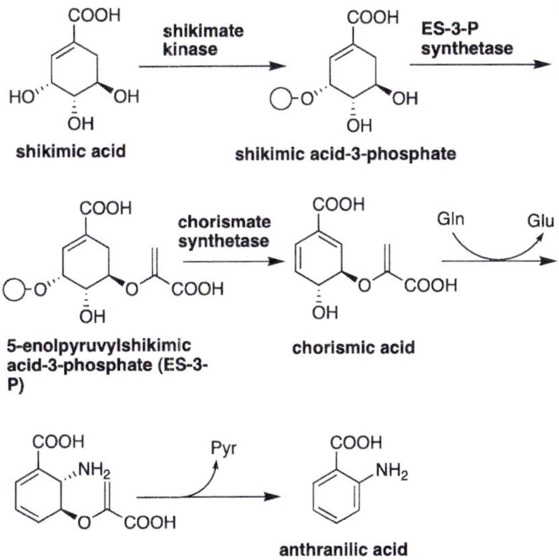

Biosynthetic Pathway of Anthranilic Acid

http://dx.doi.org/10.1016/B978-0-12-417302-6.00009-X

In the biosynthetic pathway of the alkaloids derived from anthranilic acid, the carbon atom corresponding to the carboxyl carbonyl group is included in the resulting alkaloids, except for those alkaloids with the phenazine skeleton. The main groups of alkaloids derived from anthranilic acid are classified as quinoline, acridone, and quinazoline alkaloids. These alkaloids are typically isolated from rutaceous plants, except for febrifugine and related alkaloids, which possess the quinazoline skeleton; these are isolated from plants of the Saxifragaceae family.

In this chapter, the alkaloids of higher plant origin and the alkaloids obtained from microorganisms, such as the pseudan derivatives, as well as pyocyanine and related compounds that possess the phenazine skeleton, are discussed.

LITERATURE CITED

[1] G.A. Dardenne, P.O. Larsen, E. Wieczorkowska, Biochim. Biophys. Acta 381 (1975) 416.

9.1 *Orixa japonica* AND QUINOLINE ALKALOIDS

Orixa japonica is a rutaceous shrub distributed in China, Korea, and Japan. The crude drug prepared from the stems and roots of this plant is called "Jo-zan" in Japanese herbal markets. On the other hand, the crude drug derived from the roots of *Dichroa febrifuga*, which will be described in Section 9.5, is also called Jo-zan. In an ancient Chinese herbal text from the middle of the seventh century, the description of the plant material used as Jo-zan was definitely that of *O. japonica*, and the true material for Jo-zan must have been *O. japonica* at that time. Subsequently *D. febrifuga* was increasingly used as the material for Jo-zan to replace *O. japonica*. This matter will be discussed again in Section 9.5.

As the chemical constituents of the stems and roots of *O. japonica*, quinoline alkaloids, such as kokusagine and orixine, were isolated [1]. On the other hand, dictamnine was obtained from *Dictamnus albus* subsp. *dasycarpus* [2]. The root bark of this plant is used for the symptoms of jaundice in Western medicine, under the name of "Hakusen-pi" in the Eastern system of medicine.

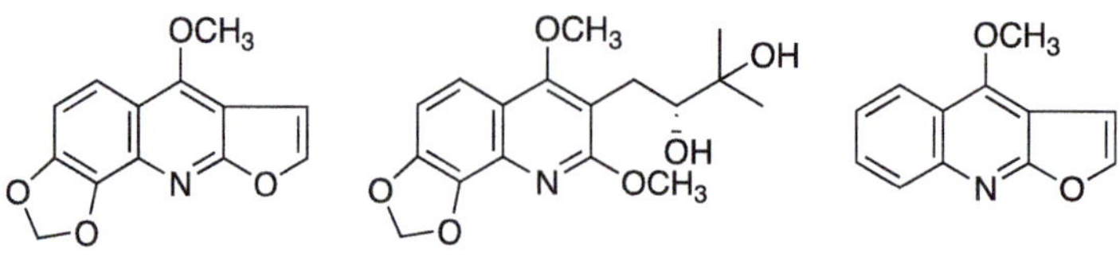

kokusagine **(+)-orixine** **dictamnine**

Kokusagine, orixine, and dictamnine are representative quinoline alkaloids isolated at the beginning of the studies of this type of alkaloid. Among them, the biosynthesis of dictamnine was well studied, and it was suggested that this alkaloid was composed of anthranilic acid, acetic acid, and isoprene moieties [3]. According to this report, the C-4 position of dictamnine was labeled with ^{14}C by feeding [$^{14}COOH$]anthranilic acid, and C-2 or C-3 labeled dictamnine was obtained by feeding ^{14}C-labeled acetate (C-1 or C-2 position, respectively).

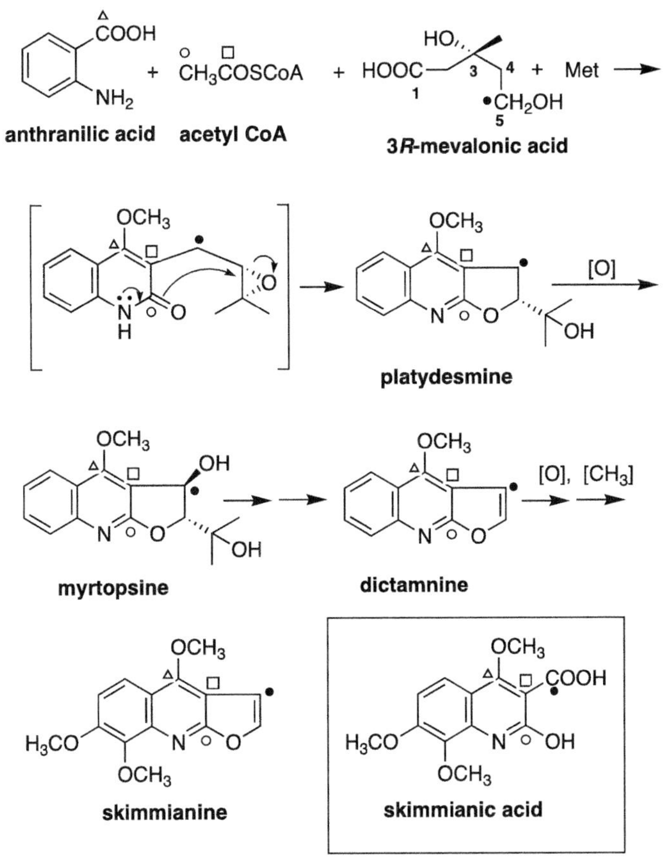

Biosynthetic Pathway of Dictamnine and Skimmianine

Quinoline alkaloids with the furan moiety, such as kokusagine, dictamnine, and skimmianine, are called furoquinoline alkaloids, and it became apparent that a part of the furan moiety of these alkaloids was derived from an isopentenylpyrophosphate unit (IPP) and that a C_3 moiety of the C_5 unit was eliminated during the biosynthetic formation of these alkaloids, as described above. On the other hand, the C_5 unit was retained in the case of orixine and related alkaloids.

The biosynthesis of the furan moiety was investigated for skimmianine, a dictamnine-related alkaloid, using *Fagara coco* (Rutaceae). Thus, when C-4- or C-5-labeled 3*R*-mevalonic acid was used in the biosynthetic pathway, ^{14}C-labeled skimmianine was observed in both cases. When the labeled skimmianine was degraded, radioactive skimmianic acid was obtained from C-4-labeled 3*R*-mevalonic acid, but radioinactive skimmianic acid was obtained when C-5-labeled 3*R*-mevalonic acid was fed. Consequently, it was concluded that the C-1′ and C-2′ carbons (furan moiety) of skimmianine were derived from the C-5 and C-4 carbons of mevalonic acid, respectively [4]. The biosynthetic precursor of dictamnine is therefore a quinoline alkaloid with a prenyl moiety, derived from dimethylallylpyrophosphate (DMAPP), at the C-3 position, such as platydesmine [5]. It was also shown that the methyl carbon of the methoxyl moiety at the C-4 position of skimmianine was derived from methionine.

With the possibility of the 1-deoxyxylulose 2-C-methyl-D-erythritol 4-phosphate (MEP) pathway to the hemiterpene units, IPP and DMAPP, operating in higher plants, it will be of interest to establish whether the five-carbon unit incorporated into the quinoline nucleus is preferentially derived from mevalonic acid or 1-deoxyxylulose.

In the biosynthesis of the furoquinoline alkaloids, in order to form the five-membered ring, a ring closure procedure must occur; and, as the biosynthetic precursor of these alkaloids, quinoline alkaloids with an epoxide moiety were considered [6]. Namely, in the biosynthesis of dictamnine, it was considered that the ring closure occurred on an unidentified precursor with an epoxide moiety to form platydesmine, which was converted into myrtopsine and finally dictamnine. Thus, quinoline alkaloids with an epoxide moiety in the side chain were considered to be important biosynthetic precursors of both the quinoline alkaloids with a C_5 unit at the C-3 position, such as orixine and related alkaloids, and the furoquinoline alkaloids. Such biosynthetic precursors with an epoxide moiety were not isolated previously, probably because they might spontaneously cyclize. Finally, 2′*R*-preorixine, with an epoxide moiety in the side chain, and considered to be the direct biosynthetic precursor of orixine, was isolated from the hexane extract of the stems of *O. japonica*, and the stereochemistry at the C-2′ position was clarified to be *R* [7,8].

(-)-preorixine

The distribution of quinoline alkaloids among higher plants is limited, and the number of quinoline alkaloids is also limited. Thus, *O. japonica* is definitely one of the treasure houses of quinoline alkaloids, and more than 23 quinoline alkaloids have been isolated from this plant material to date [9].

Among some other quinoline alkaloids, *N*-methylflindersine was isolated from rutaceous plants such as *O. japonica* and *Fagara chalybea* and showed antifeedant activity against insects (*Spodoptera exempta*) [10]. Edulinine, on the other hand, isolated from *Zanthoxylum simulans* (Rutaceae), was shown to possess analgesic and anticonvulsive activities [11].

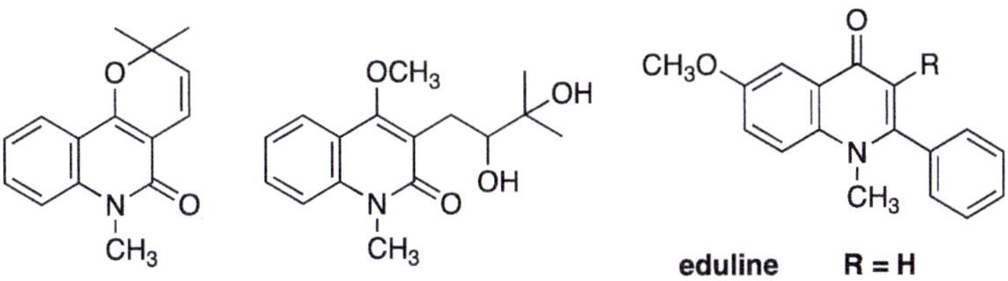

***N*-methylflindersine** **edulinine** **eduline** R = H **japonine** R = OCH_3

LITERATURE CITED

[1] M. Terasaka, Yakugaku Zasshi 53 (1933) 1046.
[2] Y. Asahina, T. Ohta, M. Inubuse, Yakugaku Zasshi 50 (1930) 1117.
[3] I. Monkovic', I.D. Spenser, A.O. Plunkett, Can. J. Chem. 45 (1967) 1935.
[4] A.O. Colonna, E.G. Gros, Chem. Commun. 1970 (1970) 674.
[5] J.F. Collins, W.J. Donnelly, M.F. Grundon, K.J. James, J. Chem. Soc. Perkin Trans. 1974 (1974) 2177.
[6] R.M. Bowman, J.F. Collins, M.F. Grundon, J. Chem. Soc. Perkin Trans. 1973 (1973) 626.
[7] S. Funayama, T. Kageyama, K. Murata, M. Adachi, S. Nozoe, Heterocycles 35 (1993) 607.
[8] S. Funayama, K. Murata, S. Nozoe, Chem. Pharm. Bull. 44 (1996) 1885.
[9] S. Funayama, K. Murata, S. Nozoe, Nat. Med. 50 (1996) 182.
[10] I. Kubo, 153, in: T. Hidaka et al. (Ed.), Konnchu no Seiri to Kagaku, Kitami-Shobo, 1973.
[11] R.-Y. Ji, Kagaku to Kogyo 45 (1992) 419.

9.2 QUINOLINE ALKALOIDS OF MICROBIAL ORIGIN

Pseudans are produced by *Pseudomonas aeruginosa*, a Gram-negative microorganism, and are classified as quinoline alkaloids. There are several kinds of pseudans, and each alkaloid bears a side chain of different length. Among the pseudans, the biosynthetic route to pseudan VII was clarified by feeding experiments, as shown in the following figure.

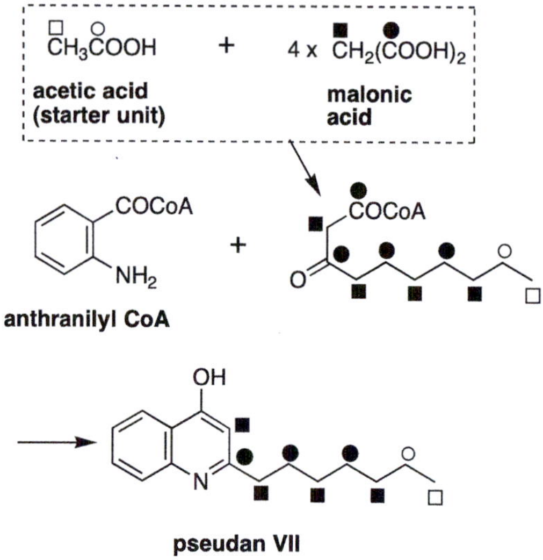

Biosynthetic Pathway of Pseudan VII

Through these experiments, it was confirmed that pseudan VII was biosynthesized from an anthranilyl CoA unit and an unbranched chain that was derived through the polyketide biosynthetic pathway. The remaining part of the quinoline chromophore was derived from malonyl CoA [1]. During the experiments regarding the biosynthesis of this alkaloid, it was found that about twice the level of radioisotope was incorporated when [2-^{14}C]acetate was fed compared with the experiments when [1-^{14}C]acetate was fed. This phenomenon was explained in that the labeled acetate was first incorporated into malonic acid, which was in equilibrium with malonyl CoA, and which was then incorporated into the alkaloid. Namely, when [2-^{14}C]acetate was fed, the C-2 position of the malonyl CoA was labeled and all of the ^{14}C label was incorporated into the resulting alkaloid. On the other hand, when [1-^{14}C]acetate was fed, the C-1 and C-3 positions (carbonyl moiety) of the resulting malonyl CoA would be equally labeled. So, in this case, because only two of the three carbons of the malonyl moiety are incorporated into the alkaloid with the loss of either the C-1 or C-3 carbon, the incorporation rate of the radioisotope is half of the former case.

LITERATURE CITED

[1] C. Ritter, M. Luckner, Eur. J. Biochem. 18 (1971) 391.

9.3 ACRONYCINE AND ACRIDONE ALKALOIDS

Acronycine was isolated from the bark of the Australian scrub ash, *Baurella simplicifolia* (=*Acronychia baueri*) (Rutaceae), together with related alkaloids [1]. Acronycine is classified as an acridone alkaloid, and acridone is an oxidized derivative of acridine. Acridine itself was isolated from coal tar in the nineteenth century, but the isolation of acronycine and related alkaloids was the first occasion that the acridine (acridone) unit was isolated from a higher plant.

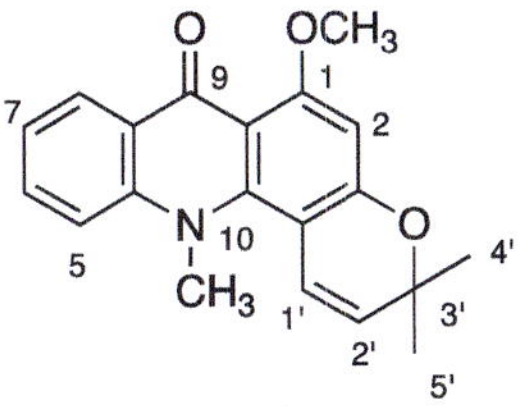

acronycine

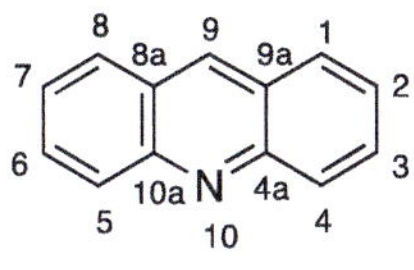

acridine

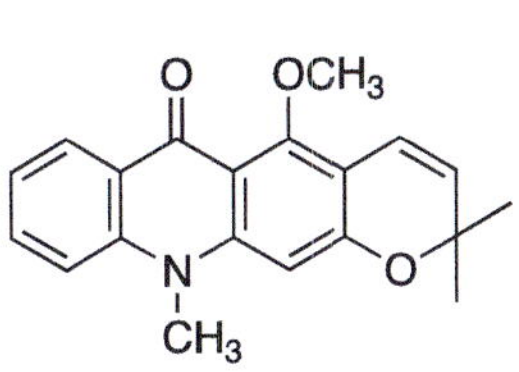

isoacronycine

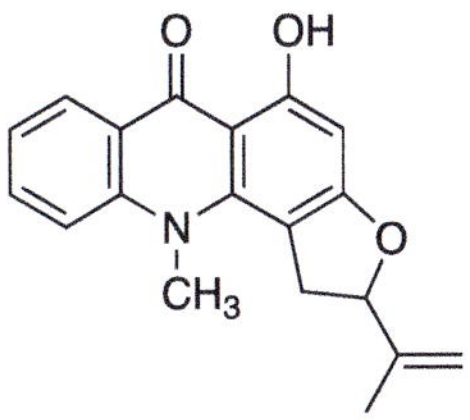

rutacridone

At first, the chemical structure of acronycine was not clarified as to whether the C_5 unit was attached in an angular form or linear form to the acridone moiety. Finally, the chemical structure was determined to be the angular form by x-ray crystallography of the bromo derivative of acronycine [2]. Since then, the isomer in which the C_5 unit is attached to the acridone moiety in the linear form has been called isoacronycine.

Svoboda et al., of Eli Lilly and Co., found that acronycine showed the widest antitumor spectrum against tumors of experimental animals among all the alkaloids tested [3]. Acronycine was also tested clinically, but it did not show the expected activity compared with the results with experimental animals, probably because of the insolubility of this compound in the aqueous solvents used.

The biosynthetic pathway to rutacridone, which was isolated from *Ruta graveolens* (Rutaceae), was studied in detail using the cell culture method [4,5]. According to these results, the biosynthetic precursors of rutacridone are anthranilic acid, acetic acid, and an isopentenyl unit, as in the case of the quinoline alkaloids described in the previous section. However, in the biosynthesis of rutacridone, two additional C_2 units participate in the biosynthesis of the acridone nucleus compared with the biosynthesis of quinoline alkaloids.

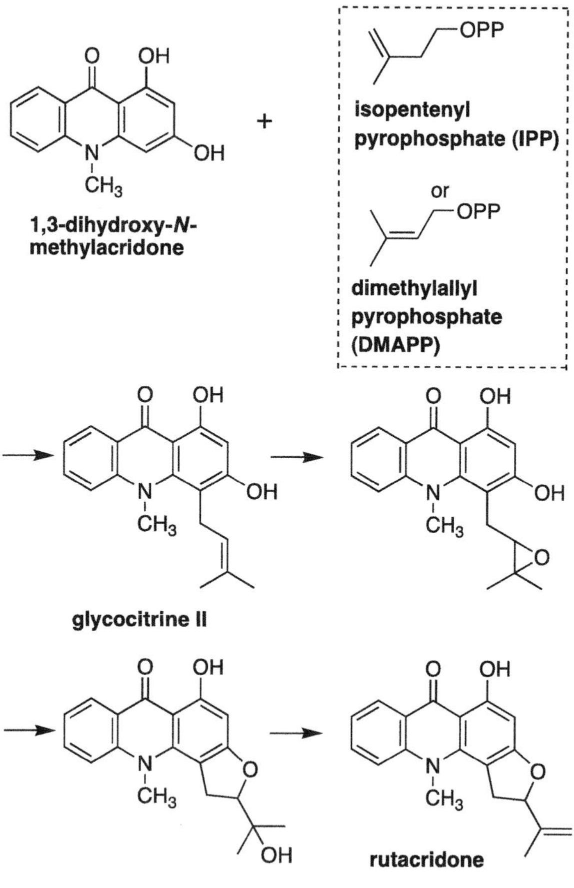

Biogenetic Conversion Route of 1,3-Dihydroxy-*N*-methylacridone into Rutacridone in the Microsomes of *Ruta graveolens*

In the biosynthesis of rutacridone, it is proposed that 1,3-dihydroxyacridone is first formed from anthranilic acid and three C_2 units, then the N-10 nitrogen is methylated to form 1,3-dihydroxy-*N*-methylacridone. Next, a C_5 unit, IPP or DMAPP, is attached to 1,3-dihydroxy-*N*-methylacridone to form glycocitrine II, which is probably oxidized to produce an as-yet-unidentified epoxide. The epoxide is cyclized and dehydrated to give rutacridone [4]. Though rutacridone is a small molecule, as in the case of the quinoline alkaloids, three main biosynthetic precursors are involved in the biosynthesis of this alkaloid. Namely, the shikimic acid, the polyketide, and probably the isoprenoid pathways all provide precursors for the biosynthesis of rutacridone.

The first acridone alkaloid dimers with a C–C junction, the semisynthetic dimers AB-1 and AB-2, were reported in 1985 [6]. AB-1 is a dimer with two noracronycine units and AB-2 was constructed with a noracronycine and an isonoracronycine unit.

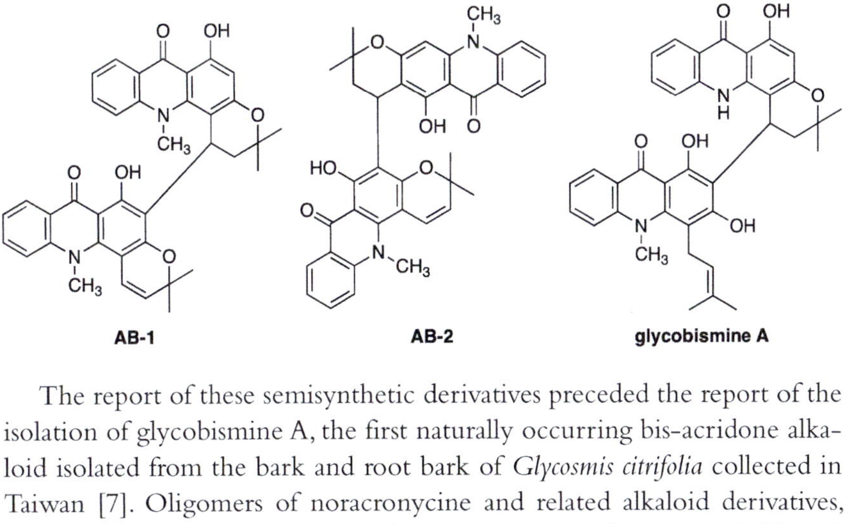

AB-1 **AB-2** **glycobismine A**

The report of these semisynthetic derivatives preceded the report of the isolation of glycobismine A, the first naturally occurring bis-acridone alkaloid isolated from the bark and root bark of *Glycosmis citrifolia* collected in Taiwan [7]. Oligomers of noracronycine and related alkaloid derivatives, dimers and trimers, tetramers, and pentamers were also reported [8], and these oligomers and the chemical reactions concerning acronycine and related compounds were reviewed [9,10].

LITERATURE CITED

[1] G.K. Hughes, F.N. Lahey, J.R. Price, L.J. Webb, Nature 162 (1948) 223.
[2] J.Z. Gougoutas, B.A. Kasuki, Acta Crystallogr. 26B (1970) 853.
[3] G.H. Svoboda, G.A. Poore, P.J. Simpson, G.B. Boder, J. Pharm. Sci. 55 (1966) 758.
[4] W. Maier, B. Schumann, D. Gröger, FEBS Lett. 263 (1990) 289.
[5] W. Maier, A. Baumert, B. Schumann, H. Furukawa, D. Gröger, Phytochemistry 32 (1993) 691.

[6] S. Funayama, G.A. Cordell, H. Wagner, H.L. Lotter, J. Nat. Prod. 47 (1984) 143.
[7] H. Furukawa, T.-S. Wu, C.-S. Kuoh, T. Sato, Y. Nagai, K. Kagei, Chem. Pharm. Bull. 32 (1984) 1647.
[8] S. Funayama, G.A. Cordell, R.D. Macfarlane, C.J. McNeal, J. Org. Chem. 50 (1985) 1737.
[9] S. Funayama, G.A. Cordell, Heterocycles 29 (1989) 815.
[10] S. Funayama, G.A. Cordell, Recent Res. Devel. Phytochem. 1 (1997) 349.

9.4 QUINOLINE ALKALOIDS ISOLATED FROM HIGHER PLANTS OTHER THAN THE RUTACEAE FAMILY

As described in Section 9.1, quinoline alkaloids have been mainly isolated from rutaceous plants, and very few such alkaloids have been isolated from plants in other families.

Quinine and related alkaloids isolated from the bark and other parts of *Cinchona* plants, such as *Cinchona ledgeriana* and *Cinchona succirubra* (Rubiaceae), can be classified as quinoline alkaloids because these alkaloids also possess a quinoline moiety. However, the biosynthetic origin of the chromophore of these alkaloids is tryptophan rather than anthranilic acid. Namely, the quinoline moiety is formed by the oxidative transformation of the indole nucleus during biosynthesis, as described in Section 2.17.

The roots of *Melochia tomentosa* (Sterculiaceae), which grows wild in the tropics, are used as an antiinflammatory for the throat in traditional medicine, and melochinone was isolated from the roots of this plant. The chemical structure was determined by x-ray crystallography and it was found that this alkaloid possesses a 7-membered ring and a benzene ring, as well as a quinoline moiety [1]. The chemical shift (^{1}H NMR) of the methylene moiety of the 7-membered ring attached to the quinoline moiety was shifted to lower field at δ3.56 (^{2}H, t), probably because of the proximate carbonyl moiety.

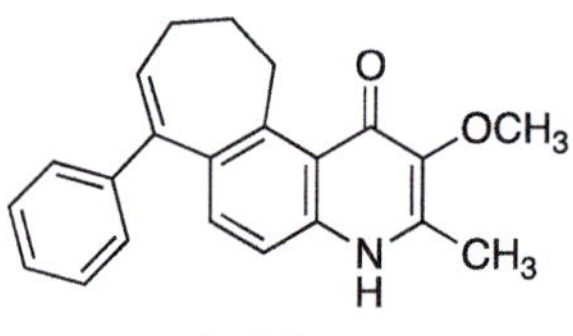

melochinone

The biosynthesis of melochinone has not yet been clarified. Thus, the quinoline skeleton of this alkaloid might be formed by the transformation of tryptophan, as in the case of quinine, or it might be formed by a completely different biosynthetic route. The most plausible explanation is that this chromophore is derived from anthranilic acid and a polyketide with a benzoate and two acetate units attached at C-5, and consequently this alkaloid is described here.

One of the simplest quinoline alkaloids obtained from higher plants is echinopsine, which was isolated from the seeds of *Echinops ritro* (Astraceae) [2]. The chemical structure of this alkaloid was determined in 1922 [3] and it was synthesized by several groups [4–8]. No studies on the biosynthesis of this alkaloid have been conducted, but it is possible that anthranilic acid is the biosynthetic precursor.

echinopsine

LITERATURE CITED

[1] G.J. Kapadia, B.D. Paul, J.V. Silverton, H.M. Fales, E.A. Sokoloski, J. Am. Chem. Soc. 97 (1975) 6814.
[2] M.M. Greshoff, Rec. Trav. Chim. 19 (1900) 360.
[3] E. Späth, A. Kolbe, Monatsh. 43 (1922) 469.
[4] H. Kondo, S. Igawa, Yakugaku Zasshi 51 (1931) 702.
[5] J.A.C. Allison, J.T. Braunholtz, F.G. Mann, J. Chem. Soc. 1954 (1954) 403.
[6] L.C. King, S.V. Abramo, J. Org. Chem. 23 (1958) 1609.
[7] S. Kamiya, Chem. Pharm. Bull. 10 (1962) 669.
[8] J.R. Merchant, V. Shankaranarayan, (London), Chem. Ind. 1979 (1979) 320.

9.5 FEBRIFUGINE AND RELATED ALKALOIDS

Dichroa febrifuga (Saxifragaceae) is an evergreen shrub that grows wild in the southern area of China and the northern area of India. The dried roots of this plant material are now used in Chinese herbal medicine, where it is marketed under the name "chang-shan" (Japanese name "Jo-zan"). However, in an old book concerning herbal medicine published in 659 AD in China, the original description of this crude drug is not that of *D. febrifuga*, but of *O. japonica* (Rutaceae), as described in Section 9.1. Thus, the material used currently for the crude drug and that used in former times seem to be different. The crude drug "Johzan" was used for the symptoms of malaria, and the crude drug originating from *D. febrifuga* was called at that time "Keikotsu Johzan." It was superior, particularly concerning the activity against the symptoms of malaria, to the crude drug originating from

O. japonica, and gradually the former plant material replaced the latter. It was already described that the latter plant is a rich source of various quinoline alkaloids (Section 9.1).

When the extract of the roots of *D. febrifuga* was given to 13 patients with tertian malaria (Section 2.17) two to three times a day for five days (average) orally (corresponding to 0.03–0.06 g of the extract and 7.5–15.0 g of the crude drug), the febrifuge activity of the extract was almost the same as in the control group (152 patients) given quinine, whereas the antiprotozoan activity of the former group appeared one day delayed compared with the group using quinine.

The febrifuge and antiprotozoan activities of the crude extract of *D. febrifuga* were reconfirmed in animal studies. For example, the febrifuge activity was shown by feeding the extract of the roots of this plant to febrile rabbits given the *Escherichia coli* vaccine. Also, it was found that no protozoa were detected after the oral administration of the extract twice a day (equivalent to 1 g of the crude drug/kg) for 7 days to chicks infected with *Plasmodium gallinaceum*, and the survival time of the chicks was extended. However, eventually the sickness caused by the protozoa returned by this method. The extract of the leaves of *D. febrifuga* showed almost the same activity as the dose of one-fifth of the extracts of the roots. From the roots and leaves of this plant, two neutral compounds, dichrins A and B, and two alkaloids, dichroines A and B, were isolated. Among these compounds, only dichroine B was said to be active against the malaria caused by *P. gallinaceum* [1].

On the other hand, the isolation of α-, β-, and γ-dichroines, together with 4-quinazolone and an alkaloid ($C_{18}H_{23}N_3O_3$), from the 90% aq. ethanol extract of the roots of *D. febrifuga* was reported [2]. The molecular formula of the α-, β-, and γ-dichroine was common ($C_{16}H_{21}N_3O_3$), and it was found that these alkaloids were interconvertible. Among these alkaloids, γ-dichroine showed the strongest activity against *P. gallinaceum*, and the weakest was α-dichroine. In this connection, it was reported that the anti–*P. gallinaceum* activity of γ-dichroine was 148 times stronger than that of quinine.

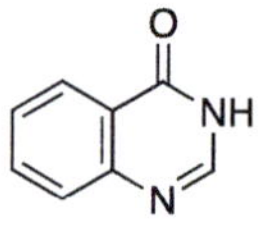

4-quinazolone

Two alkaloids, febrifugine and isofebrifugine, were isolated from the roots of *D. febrifuga* [3]. According to the report, febrifugine was also isolated from the leaves, and isofebrifugine was transformed into febrifugine when the former compound was heated. Febrifugine showed 100 times stronger activity than quinine against *Plophurae lophurae* (protozoa) [4]. In addition, the crude drug named "Jo-zan" prepared from *D. febrifuga* showed anti–*Plophurae falciparum* activity at a concentration of 0.025 μg/ml (EC_{50}) [5]. It is probable that dichroine B and γ-dichroine correspond to febrifugine.

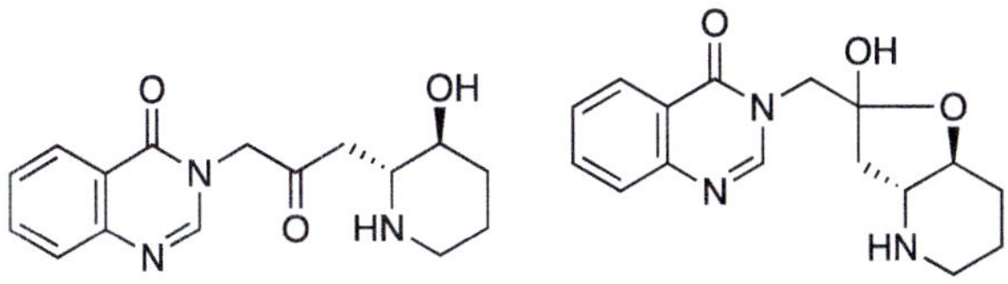

febrifugine **isofebrifugine**

The chemical structures of febrifugine and isofebrifugine were clarified mainly by their chemical reactions [6,7]. Febrifugine, isofebrifugine, and 4-quinazolone all possess the quinazoline nucleus. There are no reports regarding the biosynthesis of these alkaloids, but it is considered that a part of all of these alkaloids is derived from an anthranilic acid unit as a biosynthetic precursor. In addition, febrifugine, as well as isofebrifugine, were isolated from the flowers of the Japanese shrub *Hydrangea macrophylla* subsp. *macrophylla* forma *macrophylla* as the anti–*Eimeria tenella* (coccidium that lives only in the blind gut of chicken) principle. Concerning this anticoccidium activity, febrifugine was shown to be active, whereas isofebrifugine was inactive [8].

LITERATURE CITED

[1] C.S. Jang, F.Y. Fu, C.Y. Wang, K.C. Huang, G. Lu, T.C. Thou, Science 103 (1946) 59.
[2] T.-Q. Chou, F.Y. Fu, Y.S. Kao, J. Am. Chem. Soc. 70 (1948) 1765.
[3] J.B. Koepfli, J.F. Mead, J.A. Brockman Jr, J. Am. Chem. Soc. 69 (1947) 1837.
[4] A.K. Frederick Jr, C.F. Spencer, K. Folkers, J. Am. Chem. Soc. 70 (1948) 2091.
[5] K. Murata, S. Fushiya, T. Ohta, Y. Oshima, Y. Wataya, K.-S. Kim, S. Funayama, 117th Annual Meeting of the Pharmaceutical Society of Japan (Tokyo), Book of Abstracts, 2–140, 1997.
[6] J.B. Koepfli, J.F. Mead, J.A. Brockman Jr, J. Am. Chem. Soc. 71 (1949) 1048.
[7] J.B. Koepfli, J.A. Brockman Jr, J. Moffat, J. Am. Chem. Soc. 72 (1950) 3323.
[8] M. Kato, M. Inaba, H. Itahana, E. Ohara, K. Nakamura, S. Uesato, H. Inoue, T. Fujita, Shoyakugaku Zasshi 44 (1990) 288.

9.6 HARMALA ALKALOIDS AND VASICINE (PEGANINE)

Vasicine, a constituent of the Indian shrub *Adhatoda vasica* (Acanthaceae), possesses the same quinazoline skeleton as febrifugine and related alkaloids described in Section 9.5. *A. vasica* is used as an antiasthma drug in traditional medicine in India, and bronchodilation activity was observed for vasicine [1].

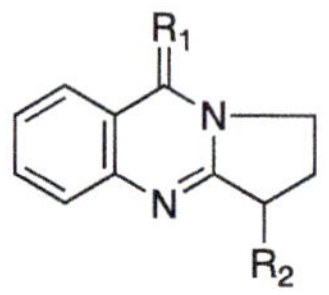

vasicine (peganine)	**$R_1 = H_2$, $R_2 = OH$**
vasicinone	**$R_1 = O$, $R_2 = OH$**
desoxyvasicinone	**$R_1 = O$, $R_2 = H$**

Vasicine was also isolated from the seeds of *Peganum harmala* (Zygophyllaceae) [2]; consequently, the same alkaloid is also known as peganine. Two groups reported the total synthesis of *dl*-vasicine (peganine) [3,4]. The related quinazoline alkaloids, vasicinone and desoxyvasicinone, were also isolated from *P. harmala* [5].

In the biosynthesis of vasicine (peganine), it is well established that anthranilic acid is involved, but different results were reported regarding the biosynthetic origin of the other part of this alkaloid. Namely, in *A. vasica* it was reported that the remaining carbons were derived from aspartic acid and acetic acid [6]. On the other hand, in the seeds of *P. harmala*, the same unit was reported to be derived from ornithine [5]. According to the latter report, it was shown that radioactivity was equally distributed at the C-1 and C-10 positions of the isolated vasicine (peganine) when [2-^{14}C]ornithine was fed by the wick method [7] to *P. harmala*. It was also shown that the C-1 and C-10 positions were labeled in vasicine (peganine) by feeding [5-^{14}C]ornithine or [1,4-$^{14}C_2$]putrescine. On the other hand, a randomly labeled alkaloid was obtained when [5-^{14}C]glutamic acid or [5-^{14}C]proline was used as a precursor. From these observations, it was concluded that in the biosynthesis of vasicine (peganine), ornithine is involved by way of a symmetrical biosynthetic intermediate, probably putrescine.

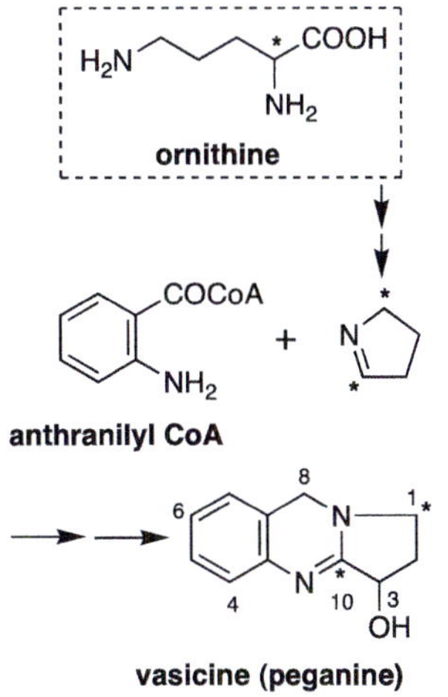

Biogenetic Pathway of Vasicine

It is interesting that harmine, which is derived from tryptophan (Section 2.6), coexists with vasicine (peganine), which is derived from an anthranilic acid precursor in *P. harmala*. Thus, tryptophan in which the benzene moiety was labeled with ^{14}C was fed in the same way as described above to *P. harmala*, and it was found that the ^{14}C was incorporated into vasicine (peganine), although the incorporation rate was low (0.071%). Thus, it was estimated that this plant also possesses the biosynthetic pathway to convert tryptophan into anthranilic acid.

LITERATURE CITED

[1] D.R. Mehta, J.S. Naravane, R.M. Desai, J. Org. Chem. 28 (1963) 445.
[2] E. Späth, E. Nikawitz, Chem. Ber. 67 (1934) 45.
[3] E. Späth, F. Kuffner, N. Platzer, Chem. Ber. 68 (1935) 699.
[4] P. Southwich, J. Casanova Jr, J. Am. Chem. Soc. 80 (1958) 1168.
[5] D.R. Liljegren, Phytochemistry 10 (1971) 2661.
[6] S. Johne, K. Waiblinger, D. Gröger, Pharmazie 28 (1973) 403.
[7] D.R. Liljegren, Phytochemistry 7 (1968) 1299.

9.7 PHENAZINE ALKALOIDS DERIVED FROM MICROORGANISMS

Pyocyanine is a deep blue pigment produced by *P. aeruginosa*, and this alkaloid shows antibiotic activity against Gram-positive microorganisms.

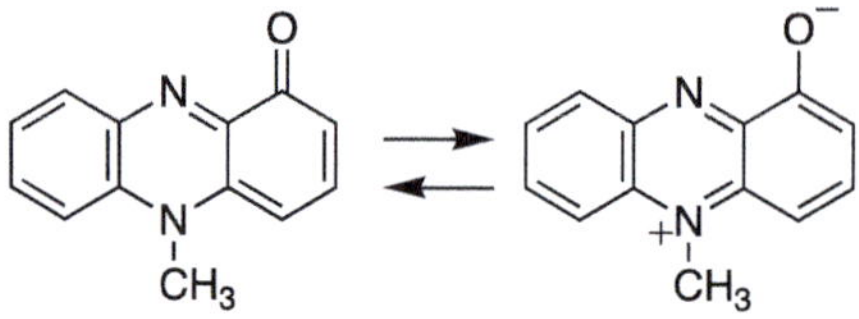

pyocyanine

According to a report concerning the biosynthesis of the pyocyanine-related alkaloid iodinin, the C-4a, C-5a, C-9a, and C-10a positions were labeled in iodinin by feeding [6-^{13}C]shikimic acid to the iodinin-producing microorganism *Brevibacterium iodinum* [1].

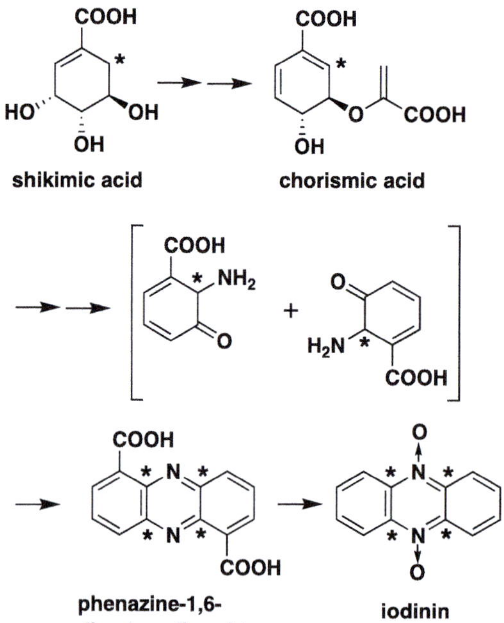

Biosynthetic Pathway of Iodinin

Consequently, it was considered that iodinin was derived from anthranilic acid, or a related intermediate, via phenazine-1,6-dicarboxylic acid. The intermediates for the formation of this alkaloid remain to be clarified.

Phenazinomycin, also having a deep blue color, is an antibiotic produced by *Streptomyces* sp. WK-2057. This antibiotic showed cytotoxicity against HeLa S3 cells (MIC, 0.8 μg/ml) in vitro and antitumor activity against S180 implanted mice (ILS, 130%, 22.2 mg/kg/day x 9, ip) in vivo [2,3].

Phenazinomycin is characterized by a phenazine chromophore with a sesquiterpene moiety. The chromophore (*N*-demethylpyocyanine) of this alkaloid corresponds to that of a pyocyanine derivative and is thought to be derived from anthranilic acid or a related compound, as in the case of pyocyanine.

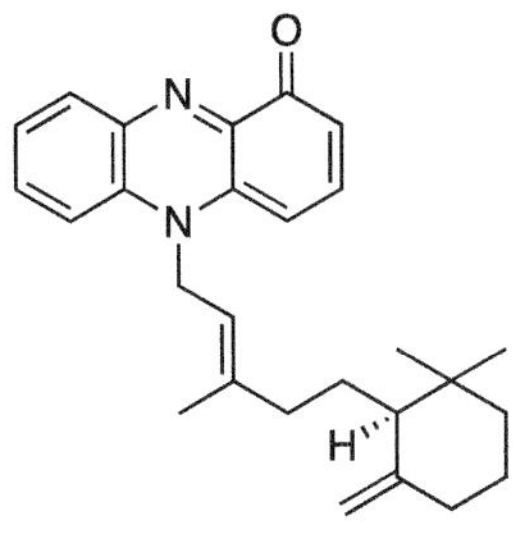

phenazinomycin

The stereochemistry of the sesquiterpene moiety of phenazinomycin was concluded to be *S* by comparing the CD spectra of the compound obtained by ozonolysis of this antibiotic with the reported data [4]. When phenazinomycin was left at room temperature for one week, it was degraded to give two aldehydes and *N*-demethylpyocyanine [3].

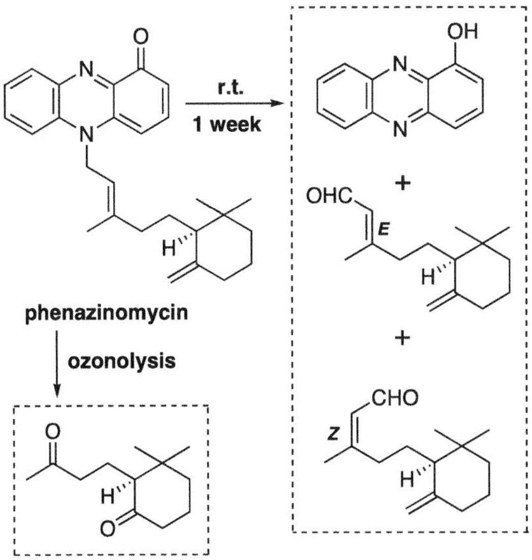

Degradation Reactions of Phenazinomycin

LITERATURE CITED

[1] U. Hollstein, D.L. Mock, R.R. Sibbit, U. Roisch, F. Lingens, Tetrahedron Lett. 19 (1978) 2987.

[2] S. Funayama, S. Eda, K. Komiyama, S. Omura, T. Tokunaga, Tetrahedron Lett. 30 (1989) 3151.

[3] S. Funayama, S. Eda, M. Ishibashi, K. Komiyama, S. Omura, T. Tokunaga, The 1989 International Chemical Congress of Pacific Basin Societies (Honolulu, USA), Book of Abstracts Part 1, BIOS, 1989. 344.

[4] J.D.P. Teresa, E. Caballero, C. Caballero, M. Medarde, A.F. Barrero, H. Grande, Tetrahedron Lett. 19 (1978) 3491.

CHAPTER 10

Alkaloids Derived from Nicotinic Acid

Ricinus communis (Euphorbiaceae)

Nicotinic acid is composed of a carboxylic acid attached at the C-3 position of a pyridine ring, and, in the widest sense, this alkaloid is a type of amino acid.

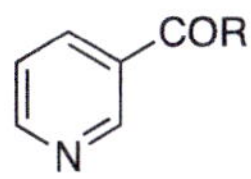

nicotinic acid **R = OH**
nicotinamide **R = NH_2**

There are two alternative biosynthetic pathways to nicotinic acid, depending on the organism. Nicotinic acid and nicotinamide are biosynthesized from tryptophan through a C-2/C-3 cleavage pathway that involves the formation of kynurenine and 3-hydroxyanthranilic acid in animals and in *Neurosporia* sp. (Figure). On the other hand, in higher plants, this alkaloid is derived from aspartic acid by the addition of glycerol or a C_3 unit equivalent.

Alkaloids
http://dx.doi.org/10.1016/B978-0-12-417302-6.00010-6

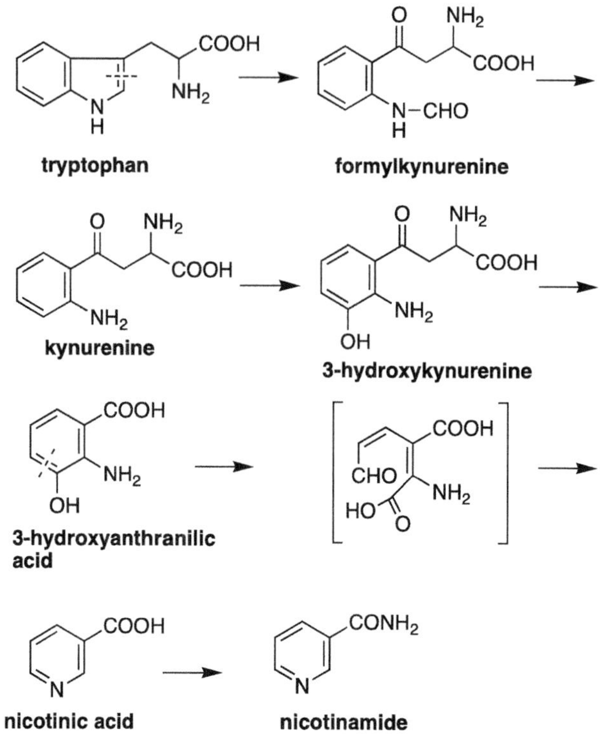

Transformation of Tryptophan into Nicotine and Nicotinamide

Nicotinic acid is also known as niacin, and the deficiency syndrome pellagra occurs when dietary niacin is lacking. It is thought that about 15 mg/day of niacin is needed for an adult. The main sources of niacin in the diet are tryptophan-containing proteins, like meats and foods that also contain nicotinic acid itself. However, a very small amount of tryptophan is included in corn. So, pellagra will occur frequently on a diet derived from corn as a main food.

Simple examples of alkaloids that contain a nicotinic acid residue as a partial structure include nicotine and anabasine. These alkaloids also possess a pyrrolidine and a piperidine ring in their partial structure, respectively, in addition to the pyridine ring derived from nicotinic acid. The origins of the aliphatic portions of these alkaloids are ornithine and lysine, respectively. In this volume, these particular alkaloids are described in the sections on ornithine-derived and lysine-derived alkaloids, respectively.

On the other hand, NAD (nicotinamide adenine dinucleotide), known as coenzyme I and II, and NADP (nicotinamide adenine dinucleotide phosphate) are derivatives of nicotinamides. The chemical structures of NAD, NADP, and the reduced form of these alkaloids, NADH and NADPH (nicotinamide adenine dinucleotide phosphate reduced), are shown. Isonicotinic acid hydrazide (INH or isoniazid) is a synthetic derivative of nicotinic acid and has potent antibacterial activity against *Mycobacterium tuberculosis* (Section 13.2) [1,2].

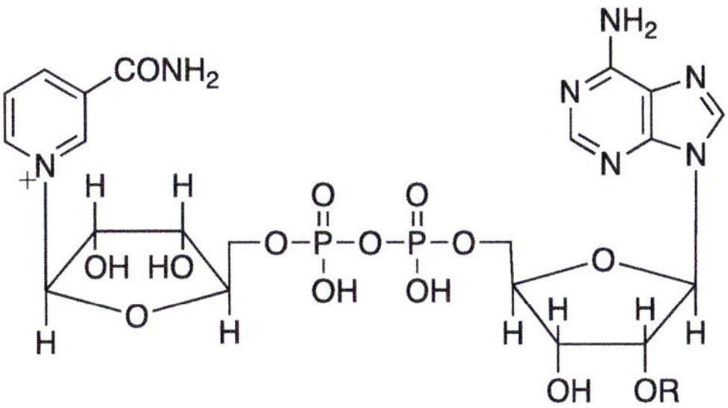

NAD **R = H**
NADP **R = PO_3H_2**

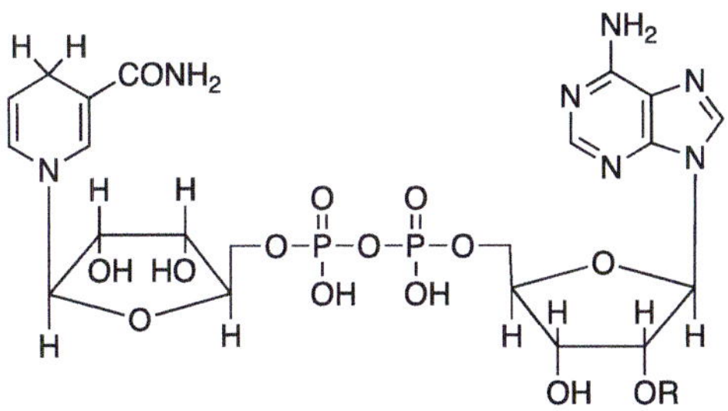

NADH **R = H**
NADPH **R = PO_3H_2**

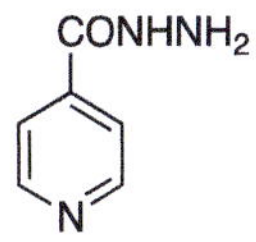

isonicotinic acid hydrazide
(= INAH, isoniazid)

Other simple derivatives of nicotinic acid in nature include arecoline, obtained from the seed of the areca nut (*Areca catechu*, Arecaeae) and ricinine, obtained from the seed of the castor oil plant (*Ricinus communis*, Euphorbiaceae).

LITERATURE CITED

[1] H. H. Fox, US 2596069 (1952) [Chem. Abstr., 46, 43625 (1952)].
[2] K. Johnsson, D.S. King, P.G. Schultz, J. Am. Chem. Soc. 117 (1995) 5009.

10.1 NICOTINE AND ANABASINE

Nicotine is the main alkaloid of tobacco (*Nicotiana tabacum*) of the Solanaceae. Anabasine is the main alkaloid of *Anabasis aphylla* (Chenopodiaceae), although trace amounts are also present in tobacco. Both nicotine and anabasine possess strong insecticidal activity. The structures of these alkaloids also have similarities in that one contains a pyrrolidine ring derived from ornithine and the other a piperidine ring derived from lysine, both of which are joined at C-3 of the pyridine ring, itself derived from nicotinic acid [1]. These alkaloids were described in detail in Chapters 3 and 4 on ornithine- and lysine-derived alkaloids, respectively. The procedure for the formation of nicotine and anabasine by condensation of Δ^1-pyrrolidine and Δ^1-piperidine with a nicotinic acid moiety is shown in the figure [2].

When [2–^{14}C]nicotinic acid is fed to the biosynthetic system of nicotine and anabasine, [2–^{14}C]nicotine and [2–^{14}C]anabasine are obtained.

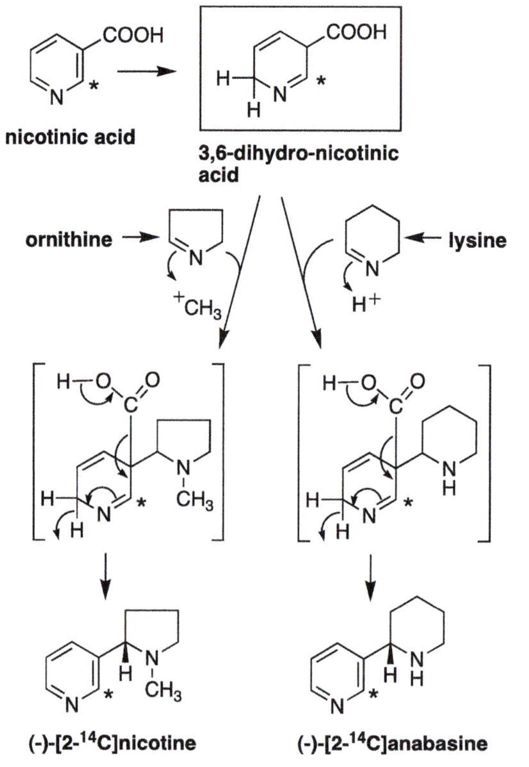

Biosynthetic Routes of (-)-Nicotine and (-)-Anabasine

On the other hand, when [6-^{3}H]nicotinic acid was fed to the above system, the total amount of ^{3}H was considerably decreased compared with the result when [2–^{3}H], [4-^{3}H], or [5-^{3}H]nicotinic acid was used as a precursor. This supports the idea that 3,6-dihydronicotinic acid exists as an intermediate in the biosynthetic pathway. The stage at which decarboxylation occurs has not been finally clarified, but it seems that it is in concert with the condensation reaction of 3,6-dihydronicotinic acid with Δ^1-pyrrolidine or Δ^1-piperidine, because the label from [2–^{14}C]nicotinic acid remains at the C-2 of nicotine and is not randomized between C-2 and C-6 through the involvement of a symmetrical intermediate.

Regarding the biosynthesis of the nicotinic acid moiety, [3–^{14}C] D-glyceraldehyde was incorporated into *Nicotiana rustica* (Solanaceae). According to the result, the uptake of ^{14}C was regioselectively concentrated (55%) at the C-4 position of the pyridine ring [3]. Once again, this also demonstrates that the precursor uniting with aspartic acid to form quinolinic acid is not symmetrical, because the uptake of ^{14}C is not dispersed between C-4, C-5, and C-6. The biogenetic pathway for nicotine is shown in the figure. The ★ label in the figure shows the ^{14}C label, and the numbers in the parentheses on the pyridine ring show the ^{14}C uptake rate as a percentage (%) [4].

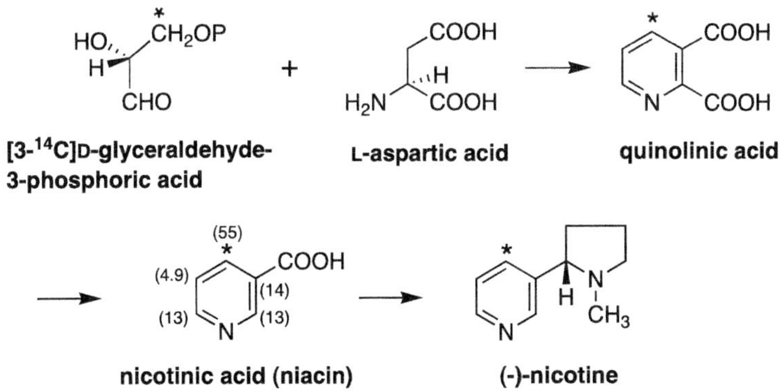

Postulated Biosynthetic Route of Nicotinic Acid and (-)-Nicotine

(−)-Anatabine is one of the principal additional alkaloids of tobacco [5]. The absolute configuration of this alkaloid was deduced to be *S* [5], and the synthesis of the racemate was achieved [6].

Structurally, anatabine is apparently a derivative of (−)-anabasine in which dehydrogenation at the 4′- and 5′-positions of the piperidine ring has occurred. However, it was established that although the piperidine ring of (−)-anabasine is derived from lysine, the piperideine ring of (−)-anatabine is derived from nicotinic acid [7]. Therefore, (−)-[6,6′-^{14}C]anatabine is obtained when [6–^{14}C]nicotinic acid is administered to *Nicotiana glutinosa* (Solanaceae).

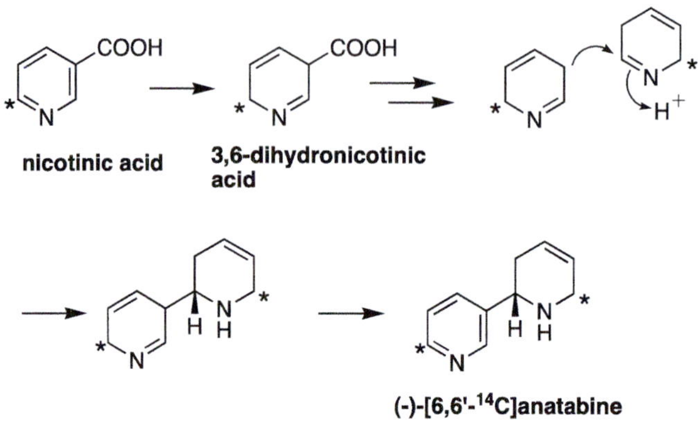

Postulated Biosynthetic Route of (-)-Anatabine

Nicotine is a powerful poison, and it is considered that oral administration of 50–100 mg of nicotine to a person can be lethal [8].

LITERATURE CITED

[1] M.L. Solt, R.F. Dawson, D.R. Christman, Plant Physiol. 35 (1960) 887.
[2] R.F. Dawson, D.R. Christman, A. D'Adamo, M.L. Solt, A.P. Wolf, J. Am. Chem. Soc. 82 (1960) 2628.
[3] J. Fleeker, R.U. Byerrum, J. Biol. Chem. 242 (1967) 3042.
[4] E. Späth, F. Kesztler, Chem. Ber. 70 (1937) 2450.
[5] R. Lukes, A.A. Arojan, J. Kovär, K. Bläha, Colln. Czech. Chem. Commun. 27 (1962) 751.
[6] P.M. Quan, T.K.B. Karns, L.D. Quinn, J. Org. Chem. 30 (1965) 2769.
[7] E. Leete, S.A. Slattery, J. Am. Chem. Soc. 98 (1976) 6326.
[8] M. Hesse, Alkaloids. Nature's Curse or Blessing? Wiley-VCH, 2002, p. 358.

10.2 NIACIN AND VITAMIN B_6

In 1926, Goldberger et al. named a substance as an antipellagra factor, which prevents human pellagra symptoms, and it was included in vitamins [1]. At that time, it was considered that pellagra occurred through the lack of the identical factor that caused black tongue in dogs (Canine black tongue). Subsequently, it was found that nicotinic acid cured black tongue of the dog, and nicotinamide was isolated as an active constituent of the antipellagra factor of the liver [2,3]. The structures of nicotinic acid and nicotinamide were presented earlier in this chapter. Nicotinic acid was named because it was first obtained by nitric acid oxidation of nicotine [4]. Among nicotinic acid and its derivatives, those compounds that show biological activity equal to nicotinamide are known generically as niacin.

It was explained above how an antipellagra factor was included in the vitamin B_2 complex. However, it was demonstrated that vitamin B_2 (riboflavin), isolated from the vitamin B_2 complex, did not show antipellagra activity in rats. As a result, the antipellagra factor separated from the vitamin B_2 complex was called vitamin B_6 [5]. Consequently, the factor that cures human pellagra is clearly different from this vitamin B_6. Thus, as the cure factors for black tongue disease of the dog, which resembles human pellagra, nicotinic acid and nicotinamide were isolated.

Isolation of vitamin B_6 was reported almost simultaneously from five sources in 1938. Three reports were from rice bran [6–8] and two reports of this alkaloid were from yeast [9,10]. Kuhn et al. submitted the structure with the name of adermine, and György suggested this alkaloid be called pyridoxine [11,12].

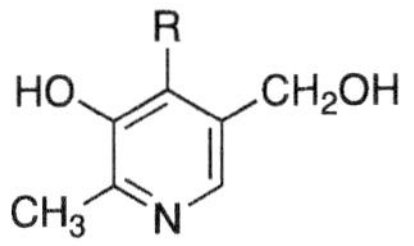

pyridoxine (= pyridoxol)	**R = CH_2OH**
pyridoxal	**R = CHO**
pyridoxamine	**R = CH_2NH_2**

Subsequently, in addition, the aldehyde and amide derivatives of pyridoxine were discovered and were named pyridoxal and pyridoxamine, respectively. The structures of these alkaloids were determined by syntheses [13,14].

It was found that pyridoxine was biosynthesized from three units of triose by using a blocked mutant of *Escherichia coli* in which the conversion of pyridoxine to pyridoxal was prevented [15]. One of these three units is introduced as C_2 (acetaldehyde) by way of pyruvic acid, and the other two units are incorporated intact (Figure). The portions shown in the figure in the bold lines show the triose unit, and the incorporation results using [1,3–^{14}C]glycerol and [2–^{14}C]glycerol are displayed. Nitrogen is incorporated in the NH_3 form during a later stage in the biosynthesis. Sometimes, both nicotinamide and nicotinic acid (niacin) are referred to as vitamin B_3.

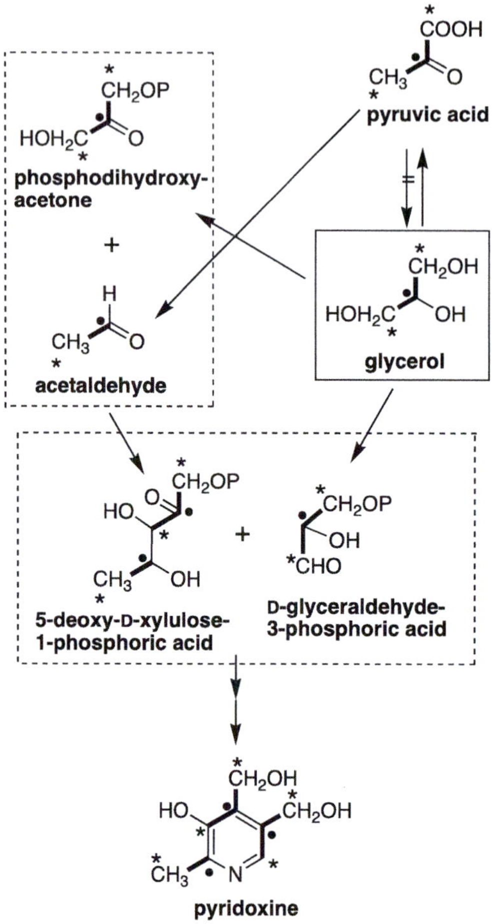

Biosynthetic Route of Pyridoxine

LITERATURE CITED

[1] J. Goldberger, G.A. Wheller, Public Health Rep. 41 (1926) 1025.
[2] C.A. Elvehjem, R.J. Madden, F.M. Strong, D.W. Woolley, J. Am. Chem. Soc. 59 (1937) 1767.
[3] C.A. Elvehjem, R.J. Madden, F.M. Strong, D.W. Woolley, J. Biol. Chem. 123 (1938) 137.
[4] H. Weidel, Liebigs Ann. 165 (1873) 328.
[5] P. György, Biochem. J. 29 (1935) 741.
[6] S. Lepkovsky, J. Biol. Chem. 124 (1938) 125.
[7] J.C. Keresztesy, J.R. Stevens, J. Am. Chem. Soc. 60 (1938) 1267.
[8] A. Ichiba, K. Michi, Sci. Pap. Inst. Phys. Chem. Res. 34 (1938) 623.
[9] R. Kuhn, G. Wendt, Chem. Ber. 71 (1938) 1118.

[10] P. György, J. Am. Chem. Soc. 60 (1938) 983.
[11] R. Kuhn, K. Westphal, G. Wendt, O. Westphal, Naturwiss. 27 (1939) 469.
[12] P. György, R.E. Eckhardt, Nature 144 (1939) 512.
[13] S.A. Harris, D. Heyl, K. Folkers, J. Am. Chem. Soc. 66 (1944) 2088.
[14] S.A. Harris, D. Heyl, K. Folkers, J. Biol. Chem. 154 (1944) 315.
[15] R.E. Hill, F.J. Rowell, R.N. Gupta, I.D. Spenser, J. Biol. Chem. 247 (1972) 1869.

10.3 *Areca* NUT AND ARECOLINE

Areca catechu (Arecaceae) is an evergreen tree from the Malay Peninsula, and it is widely cultivated in the tropical zone. The dried matured seed of *A. catechu* is known in Japan as "Binro-ji" and used for diuresis, as a purgative, and as an antiparasitic in Kampo medicine. This substance was used as a teniacide in Europe in the nineteenth century. In Southeast Asia, there is a widespread custom to chew the mixture of the chopped seed of *A. catechu*, lime, and Gambir wrapped with betel (the leaf of *Piper betle* (Piperaceae)) as a favorite food and stimulant. When the material is chewed, the inside of the mouth becomes vivid red.

Areca catechu seeds contain arecoline as the main alkaloid, and arecaidine, the *N*-demethylated derivative of arecoline, is present as a minor constituent [1]. This crude drug is used as a raw material for the manufacture of arecoline hydrobromic acid salt.

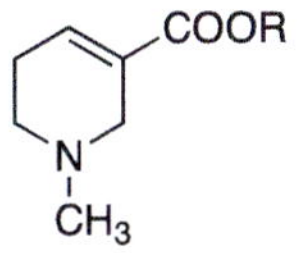

arecoline **R = CH_3**
arecaidine **R = H**

Arecoline possesses a parasympathetic nerve system excitatory effect and a central nervous system depressant effect, similar to, and stronger than, pilocarpine. Nicotinic activity is also shown. Therefore, pupil reduction and gland secretion improvement activity are observed when taking this crude drug. The secretion of sweat, saliva, digestive juice, etc. are improved when *A. catechu* seeds are chewed. Arecoline is used for the treatment of glaucoma to reduce ocular pressure. Such activity is not demonstrated by arecaidine.

Arecoline was isolated by the pharmacist E. Jahns of Göttingen at the end of the nineteenth century, and a postulated structure of this alkaloid was also presented [2]. The proposed structure of arecoline suggested in 1891 was correct, except for the position of a double bond. Subsequently, syntheses were conducted by several routes and the structure was determined [3,4].

Arecoline is thought to be derived with nicotinic acid (niacin) as a precursor (Section 10.1). On the other hand, coniine, a toxic alkaloid of *Conium maculatum* (Apiaceae) and also possessing the piperidine skeleton, is biosynthesized through the polyketide pathway, which is completely different from that of arecoline. Coniine will be discussed in Chapter 15.

LITERATURE CITED

[1] K. Hess, F. Leibbrandt, Chem. Ber. 51 (1918) 806.
[2] E. Jahns, Arch. Pharm. 229 (1891) 673.
[3] C. Mannich, Chem. Ber. 75B (1942) 1480.
[4] I.A. Kozello, A.Y. Gasheva, V.I. Khmelevskii, Khim.-Farm. Zh. 10 (1976) 90 [Chem. Abstr., 86, 171205a (1977)].

10.4 CASTOR AND RICININE

Castor (*R. communis* (Euphorbiaceae)) is an annual grass from India, and castor oil is expressed from the seeds. A very toxic protein, ricin, is also present in the seed. Ricin is constituted of two protein chains, one of M.Wt. 30,000 Daltons (A chain), and the other M.Wt. 33,000 Daltons (B chain). Ricin is regarded as one of the most toxic substances known [1].

Other than the toxic protein described above, an alkaloid ricinine was reported by Tuson in 1846 [2]. Ricinine causes nausea, convulsions, and hypotension. Further, a person may fall into a comatose state and die. Ricinine was synthesized [3] and its biosynthesis was studied [4,5].

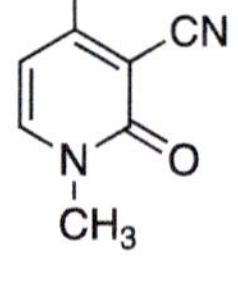

ricinine

According to Waller and Henderson, ricinine is present in every part of the young plants of *R. communis*, and the quantity of this alkaloid reaches 1 mg/1 g of undried plant [4]. When nicotinamide labeled with ^{3}H on the aromatic ring was fed to cultivated seedlings of *R. communis* hydroponically, labeled nicotinamide was well incorporated into ricinine. It was also established that the nitrogen atoms of the pyridine ring and of the CN group originate from the corresponding nitrogens of nicotinamide by feeding ^{3}H and ^{15}N double-labeled nicotinamide. Essery et al. examined the ricinine obtained by administration of [1–^{14}C]glycerol or [2–^{14}C]glycerol into the culture medium of *R. communis* [5]. As a result, in the case in which the former glycerol was administered, the C-4 and C-6 positions were labeled, and when the latter glycerol was administered the C-5 position was labeled. Consequently, it was concluded that carbons 4–6 of ricinine were derived specifically from glycerol or its related derivatives.

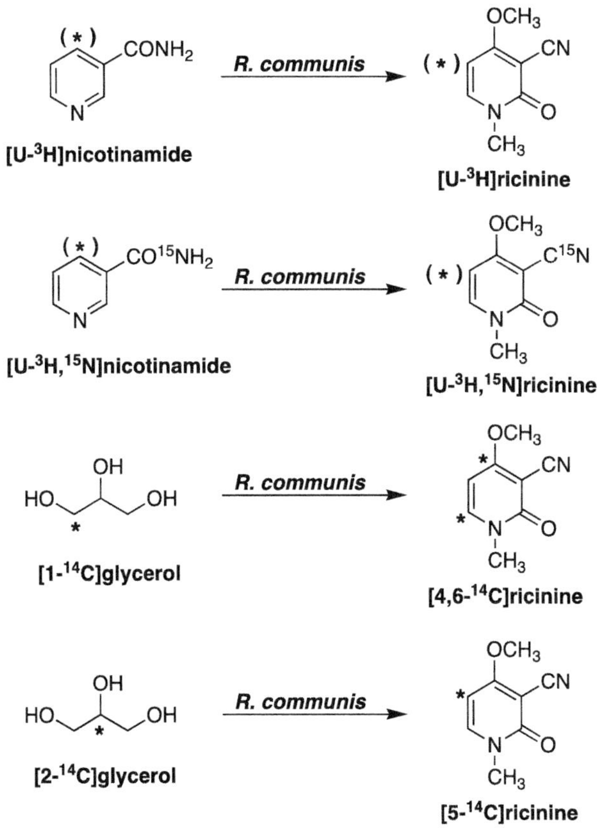

Biosynthesis of Ricinine in *R. communis*

It was shown that nicotinamide was derived from nicotinic acid, and that nicotinic acid was biosynthesized from aspartic acid and glycerol, as described at the beginning of this chapter. Therefore, ricinine, as in the case of arecoline and nicotinic acid, is also an alkaloid derived from aspartic acid.

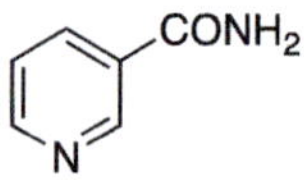

nicotinamide

When ricinine-3,5–^{14}C was administered to castor by way of the stem, it was found that the alkaloid was translocated to the seed [6]. It was also shown that ricinine-3,5–^{14}C was converted to respiratory $^{14}CO_2$ both in the dark and in the light [6].

LITERATURE CITED

[1] P.E. Thorpe, D.W. Mason, A.N.F. Brown, S.J. Simmonds, W.C.J. Ross, A.J. Cumber, J.A. Forrester, Nature 297 (1982) 594.
[2] B. Böttcher, Chem. Ber. 51 (1918R) 673.
[3] G. Schroeter, C. Seidler, M. Sulzbacher, R. Kanitz, Chem. Ber. 65 (1932) 432.
[4] G.R. Waller, L.M. Henderson, J. Biol. Chem. 236 (1961) 1186.
[5] J.M. Essery, P.F. Juby, L. Marion, E. Trumbull, Can. J. Chem. 41 (1963) 1142.
[6] G.R. Waller, J.L.-C. Lee, Plant Physiol. 44 (1969) 522.

CHAPTER 11

Alkaloids Derived from Nucleic Acids and Related Compounds

Coffea arabica (Rubiaceae)

Generally, nucleic acids and related compounds are not described as alkaloids. However, in this volume because the definition of alkaloids is expanded according to recent trends, these compounds are also considered.

As a base part of nucleic acids and related alkaloids, a purine or pyrimidine nucleus is the principal structural element. These base elements comprising purine and pyrimidine systems in nucleotides are biosynthetically derived from amino acids. However, since the bases themselves are regarded as the fundamental chemical compound classes in the organism, these alkaloids are described according to those bases rather than according to the original amino acids.

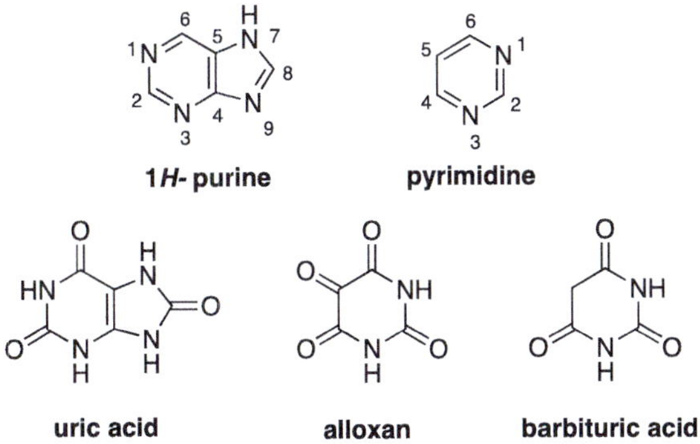

Because of the vast scope of this topic, if nucleic acids and related alkaloids such as genes were shown in detail, in this chapter only a limited number of generally known alkaloids of this group will be presented.

Alkaloids
http://dx.doi.org/10.1016/B978-0-12-417302-6.00011-8

Uric acid is an alkaloid possessing a purine skeleton, and through oxidation, alloxan with a pyrimidine skeleton is obtained, as first reported in 1818. Subsequently it was established that alloxan possessed the ability to cause hyperglycemia experimentally in mice, and this alkaloid is used for this very important purpose in experimental models for diabetes research.

As with other examples, among the alkaloids possessing a pyrimidine skeleton, barbituric acid, which became the synthetic building block of various sleeping aids, is known.

11.1 PURINE BASES AND CAFFEINE

Nucleosides are alkaloids constructed with an adenine (abbreviated as A) or a guanine (G) base as a purine derivative, or a cytosine (C) or thymine (T) base of the pyrimidine type. D-Ribose or D-2-deoxyribose is attached as an *N*-glycoside. For example, adenosine is an alkaloid constructed with an adenine base and a D-ribose unit attached at *N*-9, and guanosine is an alkaloid constructed with a guanine base and a D-ribose moiety [1,2].

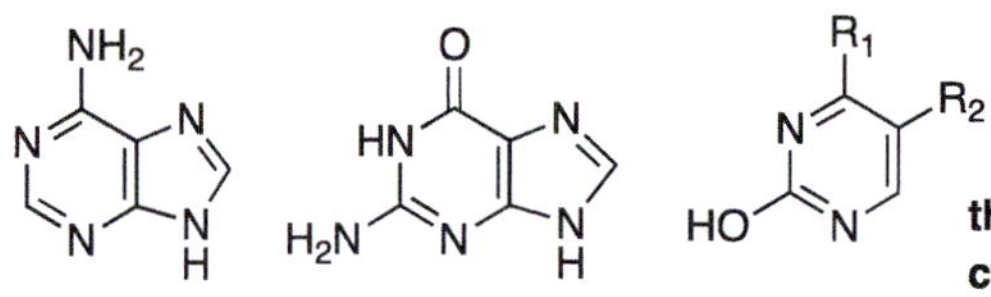

adenine **guanine**

thymine	$R_1 = OH, R_2 = CH_3$
cytosine	$R_1 = NH_2, R_2 = H$
uracil	$R_1 = OH, R_2 = H$

On the other hand, nucleic acids are high-molecular-weight compounds with long chains composed of many nucleosides combined through esterification of phosphoric acid at the 5′-position with the 3′-position of the other sugar moiety.

A nucleoside in which a hydroxyl group of one of the sugar moieties is esterified with a phosphoric acid is called a nucleotide. For example, placing a phosphate at the 5′-position of adenosine and guanosine, each possessing purine base, generates nucleotides, and these are named 5′-adenylic acid (AMP) and 5′-guanylic acid (GMP), respectively, and are utilized biochemically in the formation of ribonucleic acid (RNA).

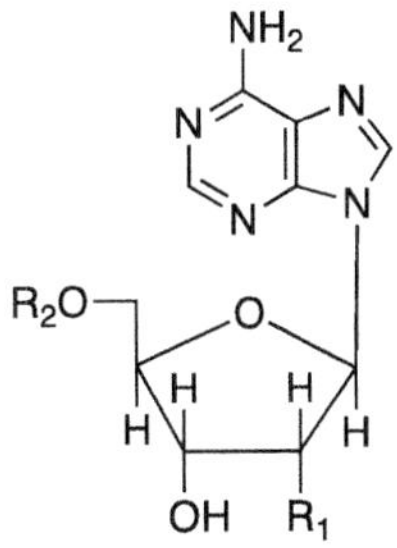

adenosine $R_1 = OH, R_2 = H$
2'-deoxyadenosine $R_1 = R_2 = H$
5'-adenylic acid (AMP) $R_1 = OH, R_2 = PO_3H_2$

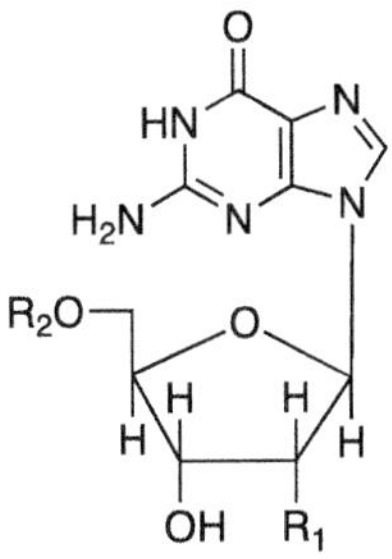

guanosine $R_1 = OH, R_2 = H$
2'-deoxyguanosine $R_1 = R_2 = H$
5'-guanylic acid (GMP) $R_1 = OH, R_2 = PO_3H_2$

In the meantime, in the formation of deoxyribonucleic acid (DNA), the nucleosides 2′-deoxyadenosine and 2′-deoxyguanosine, which lack the oxygen atom at the 2′ position in the sugar moieties, are utilized. These compounds are referred to as 2′-deoxyadenylic acid (dAMP) and 2′-deoxyadenylic acid (dGMP), respectively. A nucleic acid is therefore a polymer of nucleotides.

The most important nucleic acids are DNA, which localizes in the chromosome of the nucleus of the cell of flora and fauna, and RNA. Among the forms of RNA, messenger RNA (mRNA) reads the information of DNA and carries it to the ribosome where a protein is synthesized. Transfer RNA (tRNA) promotes protein synthesis by arranging the amino acids according to the mold of mRNA and then polymerizes them.

As described previously, the combined sugars are different between DNA and RNA; namely, in the former D-2-deoxyribose is combined, and in the latter D-ribose is combined. On the other hand, as for the bases in DNA, four types of bases, A, G, C, and T, are used, and in RNA, A, G, C, and U are used. Thus U is used in the case of RNA instead of T in the case of DNA. An example of the coupling scheme of DNA is shown in the figure, where a partial structure of GGA (dGMP–dGMP–dAMP as nucleotide), from the 5′ position to the 3′ position, is shown.

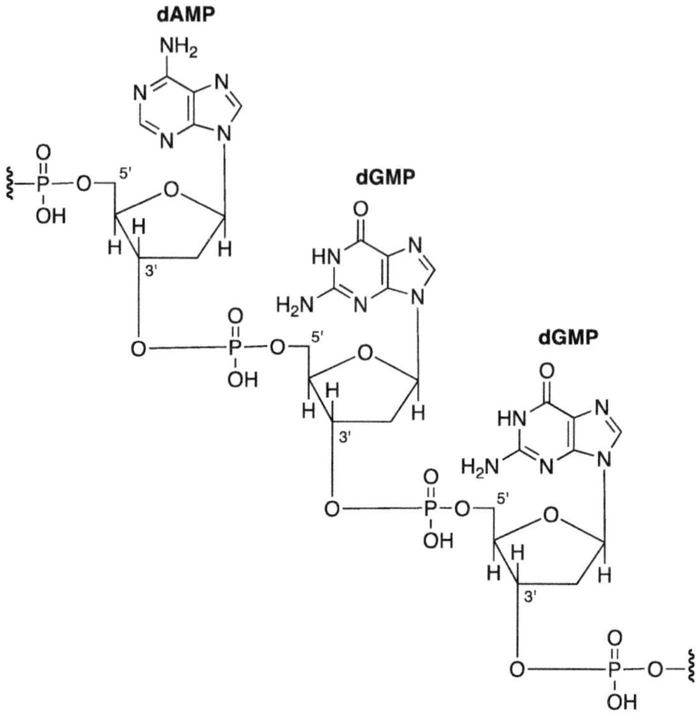

Example of A Partial Structure of DNA (G-G-A)

Purines also occur in common beverages. For example, it is customary to drink tea, coffee, cocoa, mate, guarana, and other related stimulating drinks in many places in the world. Coffee, red tea, green tea, and cocoa are representative drinks that are widely appreciated all over the world. Among these, coffee is prepared from the seeds of *Coffea arabica* or *Coffea robusta* (Rubiaceae), and red tea and green tea are prepared from the leaves of *Camellia sinensis* (Theaceae). Cocoa is prepared from the seeds of *Theobroma cacao* (Sterculiaceae). These drinks all contain purine derivatives, i.e., caffeine, theobromine, and theophylline. The history of the research on caffeine and theophylline is old, and these alkaloids were isolated as long ago as 1820 by Pelletier and Caventou. Total syntheses of these alkaloids were attempted at the end of the nineteenth century [3,4].

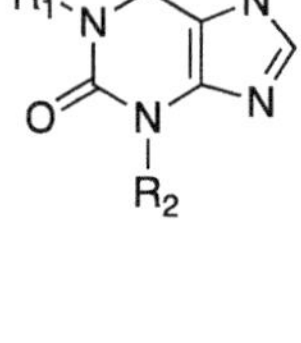

xanthine $R_1 = R_2 = R_3 = H$
caffeine $R_1 = R_2 = R_3 = CH_3$
theobromine $R_1 = H, R_2 = R_3 = CH_3$
theophylline $R_1 = R_2 = CH_3, R_3 = H$

Caffeine shows slight central neuron excitatory action, accelerates the depressed center function, increases activity generally, and improves the depressive state. The basic purine skeleton in which the 2 and 6 positions became carbonyl groups, as occurs in caffeine, theobromine, and theophylline, is called a xanthine. Therefore, these alkaloids are also known as xanthine derivatives.

It has been established that the xanthine nucleus of caffeine and theophylline is biosynthesized from glycine, whereas the nitrogen atoms originate from glutamine and aspartic acid, a carbonyl group from carbon dioxide; and two C_1 units originate from another source, as shown in the figure.

Adenosine, in which D-ribose is combined with adenine as an *N*-glycoside, has already been mentioned. Adenosine triphosphate (ATP) is an alkaloid in which a unit of three phosphoric acid moieties combine at the C5′ position of the sugar unit of an adenosine unit. ATP is known as an alkaloid that is rich in energy. When ATP is hydrolyzed, it releases a phosphoric acid moiety to form adenosine diphosphate, and energy is generated accordingly.

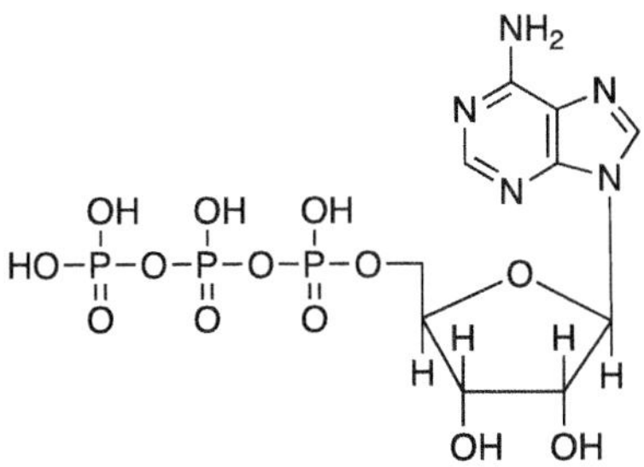

adenosine triphosphate (ATP)

ATP was isolated from the muscle of rabbits [5], and its chemical structure was reported initially by a Japanese scientist [6]. Among the adenosine derivatives, cyclic 3′,5′-adenosine mono phosphate (cyclic AMP/cAMP) is an important alkaloid contained in animal cells as an energy in various chemical and biological activities.

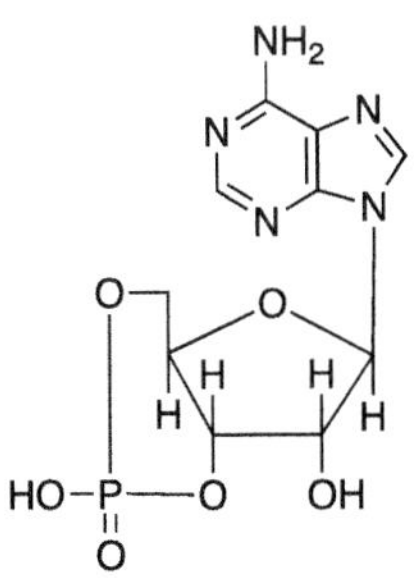

cyclic AMP

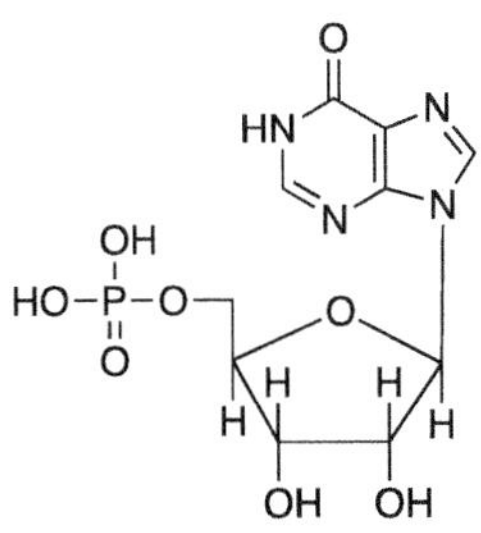

5'-inosinic acid (IMP)

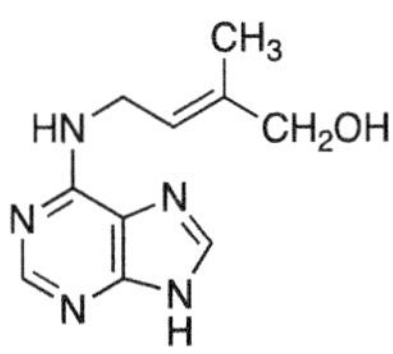

zeatin

cAMP is derived from ATP, and the formation of this alkaloid is catalyzed by adenylate cyclase. cAMP is transformed into AMP in tissues by a reaction catalyzed by cAMP phosphodiesterase. The intracellular cAMP concentration is usually about 1 mM. cAMP was isolated in 1957 [7] and was identified subsequently as one of the resulting compounds when ATP was treated with an aqueous solution of barium hydroxide.

The chemical structure of cAMP was deduced to be adenosine-3′,5′-phosphoric acid [8]. This was finally confirmed by x-ray crystallographic analysis, including the configuration [9], and its total synthesis has been achieved [10,11].

On the other hand, there is also an example of the isolation of cAMP from a higher plant. The dried fruit of the jujube (*Zizyphus jujuba* var. *inermis*) of the family Rhamnaceae is called zizyphi fructus (Japanese name, Taiso), and it is used in Kampo medicine. It was reported that this material contains 0.03–0.16 mg/g of cAMP [12].

Other known purines in addition to those mentioned above include 5′-inosinic acid, which is known as the "umami-taste" substance of Katsuobushi (dried bonito preparation), and zeatin, which promotes cell division (such a compound is referred to as a cytokinin).

Inosinic acid was found many years ago by Liebig, and it was pointed out that this compound possesses "umami-taste" in 1847. On the other hand, zeatin was isolated from the immature seed of corn of *Zea mays* (Poaceae), and it was the first naturally occurring cytokinin to be isolated [13,14]. Subsequently, various cytokinins were isolated from natural sources, and all of them are purine derivatives [15].

In the biosynthesis of nucleosides, ribose combines with an intermediate from the initial stage of purine ring construction. It was found that the base portion was constructed from glycine and glutamine, and that a nitrogen atom originated from aspartic acid, and of the two C_1 units one is derived from carbon dioxide and the other from a molecule of formic acid.

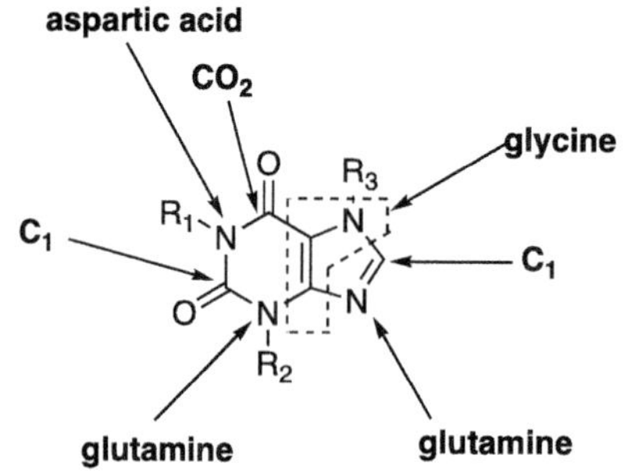

Biosynthetic Origins of Xanthine Skeleton of Caffeine and Related Alkaloids

Initially, the C_1 position of the ribose-5-phosphoric acid (ribose-5-phosphate) unit is activated by phosphorylation with ATP, and it becomes 5-phospho-D-ribosyl-1-pyrophosphoric acid (PRPP). This pyrophosphoric acid unit is replaced with an activated ammonia derived from a glutamine with inversion of the stereoconfiguration at the C_1 position. It is this nitrogen atom that subsequently becomes N-9 of the purine base.

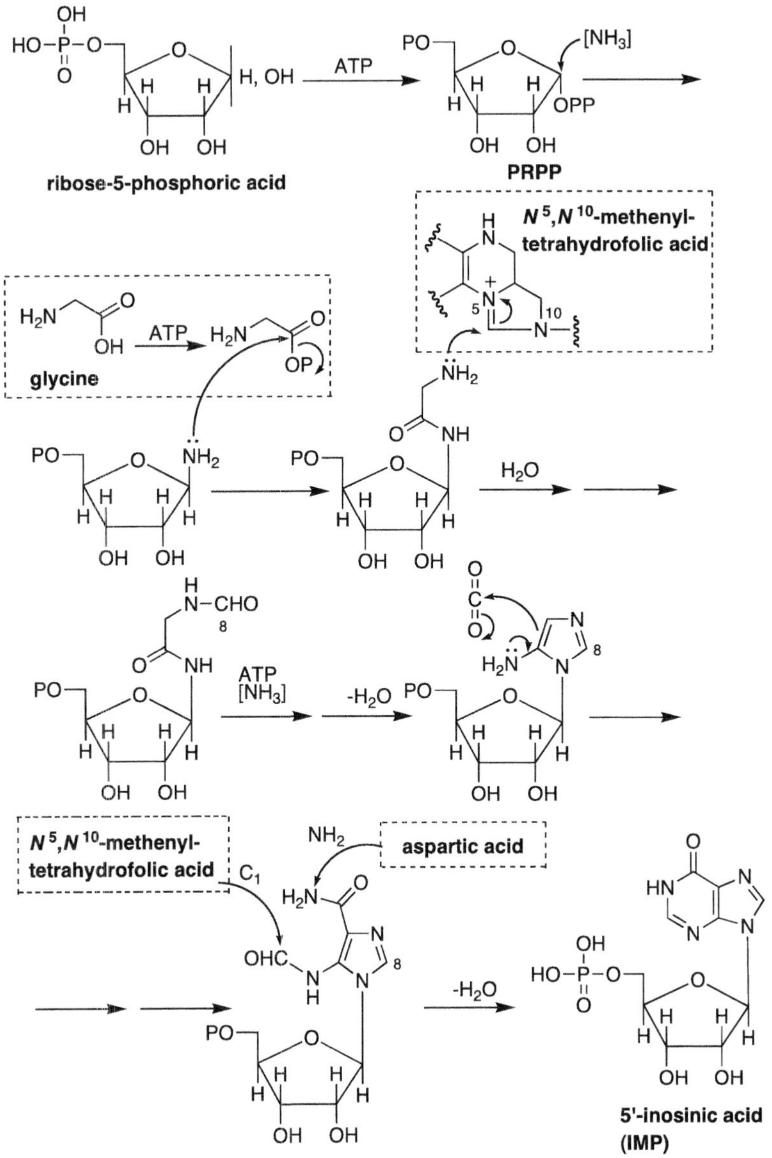

Biosynthetic Pathway of 5'-Inosinic Acid

Next, a glycine moiety activated by ATP is introduced as an amide, and an ammonia moiety derived from glutamine is introduced to form an imidazole ring. In addition, carbon dioxide and a nitrogen atom derived from an aspartic acid moiety are introduced, and the biosynthesis of inosinic acid (IMP) is completed.

IMP is also an important biosynthetic intermediate in the biosynthesis of XMP and GMP from AMP. The conversion from AMP into XMP is carried out by the further oxidation of IMP at the C_6 position. GMP is formed by amination of the C_2 position of XMP with ammonia derived from an aspartic acid. These biosynthetic procedures are presented in the scheme [16].

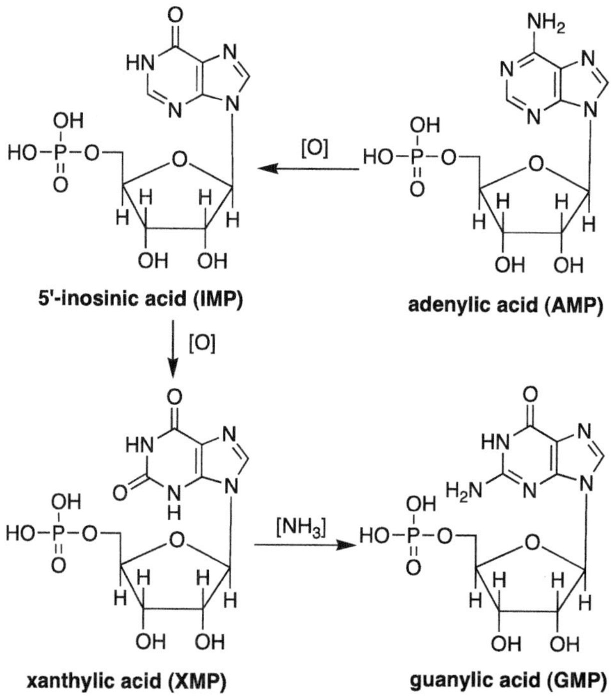

Biosynththic Pathway from Adenylic Acid to 5'-Inosinic Acid and Guanylic Acid

The biosynthesis of caffeine was investigated through incorporation experiments using the stems of *Coffea* and *Camellia* plants [17]. As a result, it became clear that methylation of the N-7 position occurred on xanthosine derived from XMP, which originated from AMP via IMP. Caffeine is then biosynthesized through 7-methyl xanthosine, 7-methyl xanthine, and theobromine, as shown in the figure.

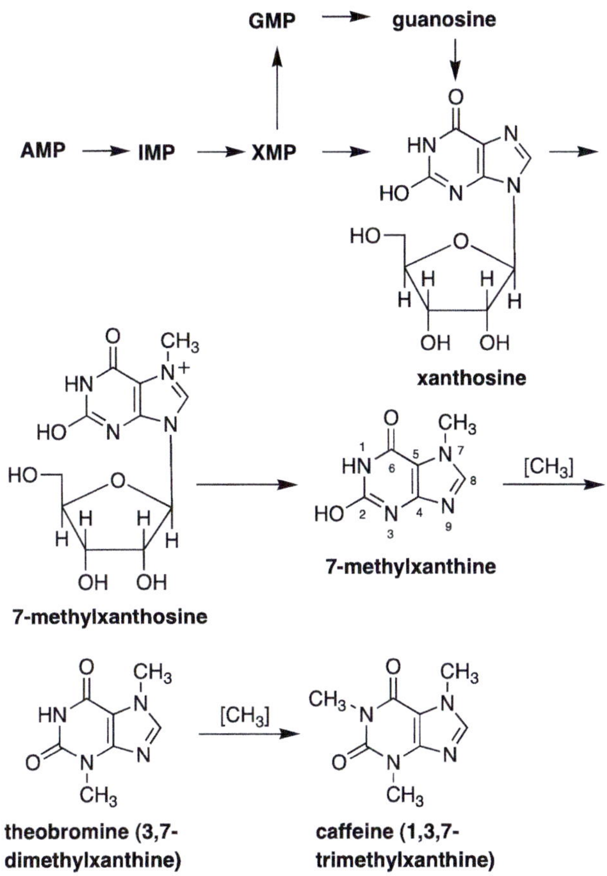

Biosynthetic Pathway of Caffeine *via* Adenosine and Theobromine

LITERATURE CITED

[1] P.A. Levene, R.S. Tipson, J. Biol. Chem. 94 (1932) 809.
[2] P.A. Levene, R.S. Tipson, J. Biol. Chem. 97 (1932) 491.
[3] E. Fischer, L. Ach, Chem. Ber. 28 (1895) 2473.
[4] E. Fischer, L. Ach, Chem. Ber. 28 (1895) 3135.
[5] C.H. Fiske, Y. Subbarow, Science 70 (1929) 381.
[6] K. Makino, Biochem. Zeitschr. 278 (161) (1935), [*Chem. Abstr.*, 29, 8020 (1935).
[7] T.W. Rall, E.W. Sutherland, J. Berthet, J. Biol. Chem. 224 (1957) 463.
[8] D. Lipkin, W.H. Cook, R. Markham, J. Amer. Chem. Soc. 81 (1959) 6198.
[9] K. Watenpaugh, J. Dow, L.H. Jensen, S. Furberg, Science 159 (1968) 206.
[10] M. Smith, G.I. Drummond, H.G. Khorana, J. Amer. Chem. Soc. 83 (1961) 698.
[11] R.K. Borden, M. Smith, J. Org. Chem. 31 (1966) 3247.
[12] K. Hanabusa, J. Cyong, M. Takahashi, Planta Med. 42 (1981) 380.
[13] D.S. Letham, Life Sci. 2 (1963) 569.

[14] D.S. Letham, J.S. Shannon, I.R. McDonald, Proc. Chem. Soc. 1964 (1964) 230.
[15] B.A. McGaw, in: P.J. Davies (Ed.), Plant Hormones and Their Role in Plant Growth and Development, Kluwer Academic Publishers, Dordrecht, 1990R, p. 76.
[16] J.M. Buchanan, S.C. Hartman, Adv. Enzymol. 21 (1959R) 199.
[17] T. Suzuki, H. Ashihara, G.R. Waller, Phytochemistry 31 (1992) 2575.

11.2 PYRIMIDINE BASES, 5-FLUOROURACIL, AND VITAMIN B_1

As described above, the bases of DNA are composed of four structures, namely adenine and guanine (purine bases) and cytosine and thymine (pyrimidine bases). On the other hand, in the case of RNA, one of the four bases, uracil, is utilized instead of thymine of DNA within the four kinds of bases. Uracil possesses a structure in which the fifth methyl group of thymine is replaced with a hydrogen atom.

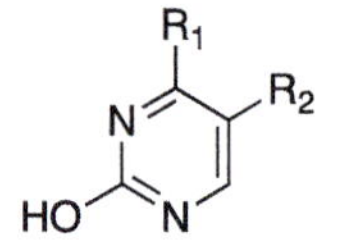

thymine $R_1 = OH, R_2 = CH_3$
cytocine $R_1 = NH_2, R_2 = H$
uracil $R_1 = OH, R_2 = H$

In the case of DNA, a D-2-deoxyribose molecule is combined to each of the bases to form a nucleoside, and the nucleosides are then combined with each other with a phosphoric acid to form a polymer (DNA). On the other hand, in the case of RNA, D-ribose, instead of D-2-deoxyribose, is combined to each of the bases to form a nucleoside, and as in the case of DNA, these nucleosides are combined with each other to form a polymer (RNA). Among the bases within DNA and RNA, adenine and guanine have been described in the preceding section. In this section, cytosine, thymine, and uracil, which are pyrimidine bases, will be described. Purine derivatives exist as a constituent unit of nucleic acids and as many kinds of monomers, and these are also present in natural products, such as caffeine, inosinic acid, and cytokinin. On the other hand, as natural products, pyrimidine derivatives are rather rare. Nucleosides composed of pyrimidine bases cytosine, thymine, and uracil coupled with D-ribose are known as cytidine, thymidine, and uridine, respectively. Among these alkaloids, cytidine was first isolated from the nucleic acid of yeast [1,2], and thymidine was isolated from thymonucleic acid [3,4]. In the meantime, uridine was obtained by the weak alkali hydrolysis [5] of the nucleic acids originating from yeast.

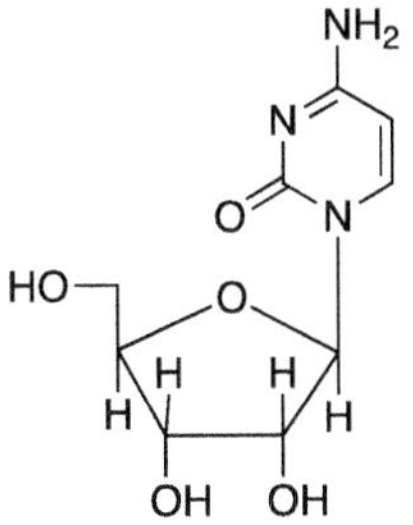

cytidine

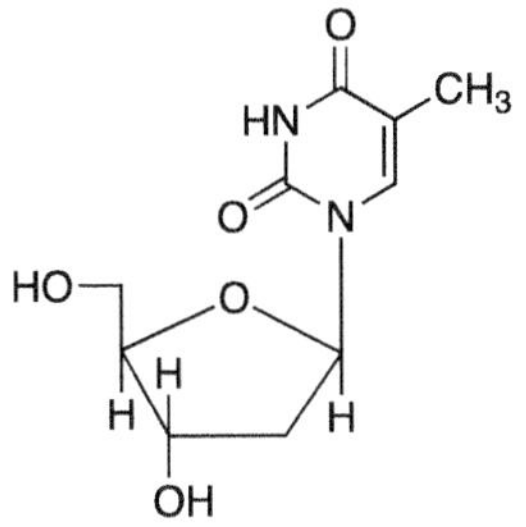

thymidine

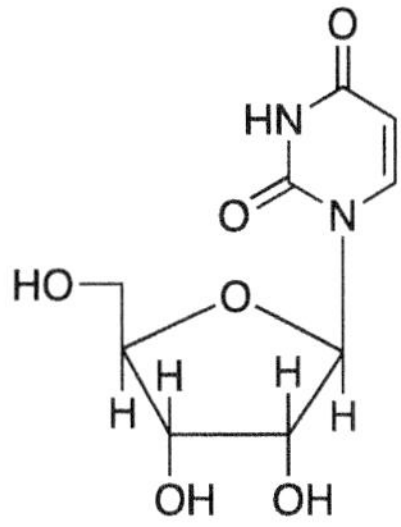

uridine

Cytidylic acid, thymidylic acid, and uridylic acid (UMP) are compounds in which the sugar moiety of each of the related nucleosides described above is phosphorylated at the 5′ position. The sugar moiety combined with the uracil base of the uridylic acid used for RNA formation is D-ribose, whereas the sugar combined with the thymine of thymidylic acid used in DNA formation is D-2-deoxyribose. 2′-Deoxycytidylic acid is used in RNA synthesis, whereas cytidylic acid is used in the formation of DNA.

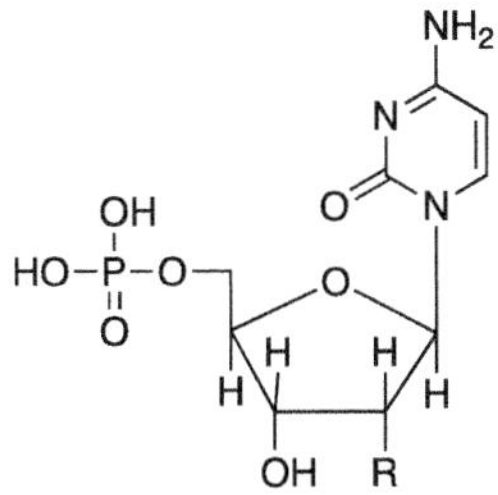

5'-cytidylic acid (CMP) **R = OH**
2'-deoxy-5'-cytidylic acid (dCMP) **R = H**

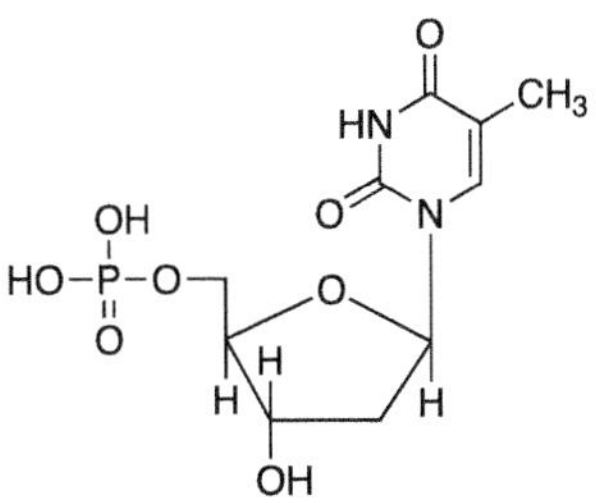

5'-tymidylic acid (TMP)

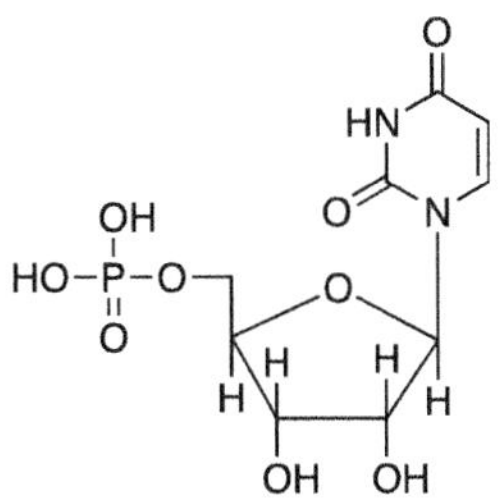

5'-uridylic acid (UMP)

As described in the previous section, in the biosynthesis of inosinic acid, ribose combines with the purine base in the initial stages of purine skeleton synthesis. On the other hand, in the biosynthesis of the pyrimidine nucleosides, ribose is introduced after the completion of the synthesis of the pyrimidine base. In the biosynthesis of uridylic acid, carbamylation of aspartic acid occurs initially with carbamoyl phosphate to form dihydroorotic acid, and this becomes orotic acid through dehydration via the action of the flavin oxidation–reduction system. Next, ribose is introduced at the N-1 position by 5-phospho-α-D-ribosyl-1-pyrophosphoric acid (PRPP), and the orotic acid becomes orotidylic acid and is transformed into UMP by decarboxylation.

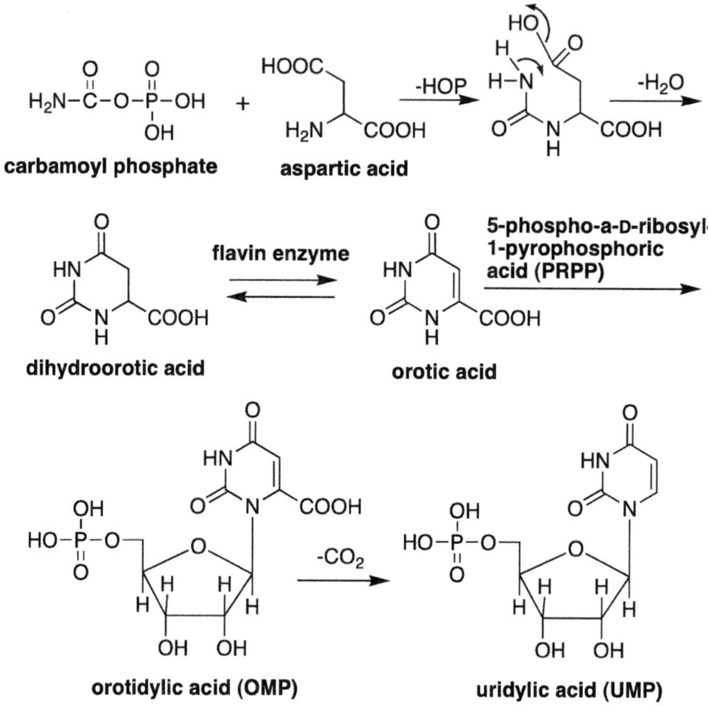

Biosynthetic Pathway of Uridylic Acid

5-Fluorouracil (5-FU) is an anticancer agent possessing a pyrimidine skeleton. 5-FU was synthesized in 1956 and its anticancer activity was reported subsequently [6,7]. As a result of the incorporation of 5-FU in the assimilation pathway as a pyrimidine base, 5-FU shows anticancer activity by inhibiting the biosynthesis of nucleic acids.

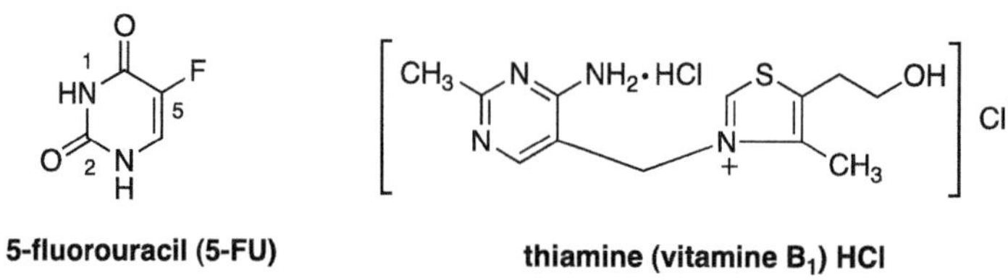

Beriberi is a nutritional deficiency that is generated by the lack of vitamin B_1, a pyrimidine derivative. Beriberi was a disease of unknown origin in former times and was recognized as a particularly horrible disease, because it sometimes caused death by cardiopathy. In 1910, at the conference of Tokyo Chemical Association, Dr U. Suzuki reported obtaining a substance from rice bran named "aberic acid," which meant that it acted against beriberi.

This substance was also reported under the name "oryzanine" in 1912 by the same investigator. The name oryzanine is derived from the scientific name of the rice plant (*Oryza sativa*) [8,9]. This material was effective for the treatment of multiple neuritis in chickens. At that time, although no chemical description of the active component was given, it was reported that unmilled rice was effective in treating the symptoms of beriberi [10]. In the meantime, a substance effective for the disease known as "birds' polished rice syndrome" was obtained by C. Funk et al. at the Lister Laboratory in London in 1911–1912, and it was discovered that a substance that showed the same effect was also present in brewer's yeast. It was reported that the active substance contained nitrogen atom(s), was shown to be basic, and was regarded as a kind of amine. From these points of view, this amine was named vitamine (<vital amine) [11,12].

Because oryzanine was isolated from the same material and, as described above, it was effective for the treatment of chickens with multiple neuritis, it was apparent that oryzanine was the same substance as the vitamine reported by Funk et al.

The announcement of the discovery of oryzanine by Suzuki et al. in Tokyo (in 1910) was earlier than that by Funk et al. However, only the achievement of Funk et al. and the name of "vitamin(e)" have been retained internationally. The active substances isolated by Suzuki et al. and Funk et al. were probably not pure compounds. Subsequently, the active component was purified and crystallized as its hydrochloride. The yield of the active component was 100 mg from 300 kg of material [13].

With respect to the chemical structure, it was clarified that a sulfur atom was present [14]. It was Dr Katashi Makino, a physician of the South Manchuria Railway Hospital (Dairen, Manchuria, the Great Empire of Japan), who suggested that the pyrimidine ring and the thiazole ring were combined through a methylene moiety [15]. The chemical structure of vitamin B_1 hydrochloride (thiamin hydrochloride) was also supported by other researchers [16], and the structure was established by chemical synthesis [17]. As described in Chapter 11 (Section 11.1), Dr Makino also provided the first correct chemical structure of ATP.

As the existence of various small and indispensable chemical factors was clarified, it was advocated that these substances should be called vitamines A, B, C, etc. As mentioned above, the common name "vitamine" was proposed for those basic substances (amine) that are necessary (vital) for a life. However, gradually it was demonstrated that these small and indispensable factors are not always amines, and it was decided that these compounds should be referred to as "vitamin" (which excludes the meaning of amine), and not as "vitamine" [18]. This is the origin of the name "vitamin B_1."

LITERATURE CITED

[1] P.A. Levene, W.A. Jacobs, Chem. Ber. 43 (1910) 3150.
[2] P.A. Levene, F.B. La Forge, Chem. Ber. 45 (1912) 608.
[3] P.A. Lenene, E.S. London, J. Biol. Chem. 83 (1929) 793.
[4] P.A. Levene, S. Tipson, J. Biol. Chem. 109 (1935) 623.
[5] R.J.C. Harris, J.F. Thomas, J. Chem. Soc. 1948 (1948) 1936.
[6] R. Duschinsky, W.G. Farkas, C. Heidelberger, U.S. Patent 2,970,139, 1961 [Chem. Abstr., 55, P14490f (1961)].
[7] C. Heidelberger, F.J. Ansfield, Cancer Res. 23 (1963) 1226.
[8] U. Suzuki, T. Shimamura, J. Tokyo Chem. Soc. 32, 4, 144, 335, 874. (1911), [*Chem. Abstr.,* 6, 251 (1912)].
[9] U. Suzuki, T. Shimamura, S. Odake, Biochem. Z. 43 (1912) 89 [*Chem. Abstr.,* 6, 2774 (1912)].
[10] W.P. Chamberlain, Philipp. J. Sci. 6 (1911) 133 [*Chem. Abstr.,* 5, 3484 (1911)].
[11] C. Funk, Brit. Med. J. II (1912) 787 [*Chem. Abstr.,* 7, 492 (1913)].
[12] C. Funk, Pharm. J. 89 (1913) 351 [*Chem. Abstr.,* 7, 680 (1913)].
[13] B.C.P. Jansen, W.F. Donath, Verslag Akad. Wetenschappen Amsterdam 35 (1927) 923 [*Chem. Abstr.,* 21, 2150 (1927)].
[14] A. Windaus, R. Tschesche, H. Ruhkopf, F. Laquer, F. Schultz, Hoppe-Seyler's Z. Physiol. Chem. 204 (1932) 123.
[15] K. Makino, T. Imai, Hoppe-Seiler's Z. Physiol. Chem. 239 (1936) I.
[16] R.R. Williams, J. Am. Chem. Soc. 58 (1936) 1063.
[17] R.R. Williams, J.K. Cline, J. Am. Chem. Soc. 58 (1936) 1504.
[18] J.C. Drummond, Biochem. J. 14 (1920) 660.

11.3 PTERIDINE SKELETON AND FOLIC ACID

The pteridine skeleton is constructed between a pyrimidine and a pyrazine moiety, which has been described in the preceding section. Some compounds that possess this skeleton include folic acid, which is known as an anti–pernicious anemia agent and as a growth promotion factor for various microorganisms; tetrahydrofolic acid, which is a biologically important coenzyme; and riboflavin (vitamin B_2).

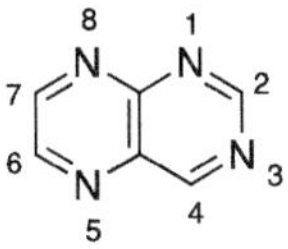

pteridine

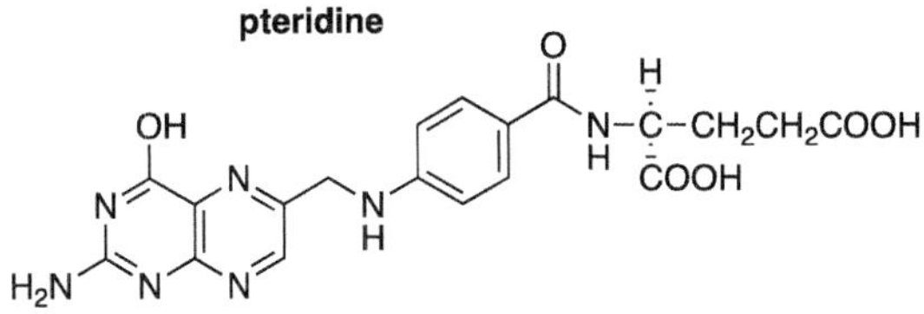

folic acid

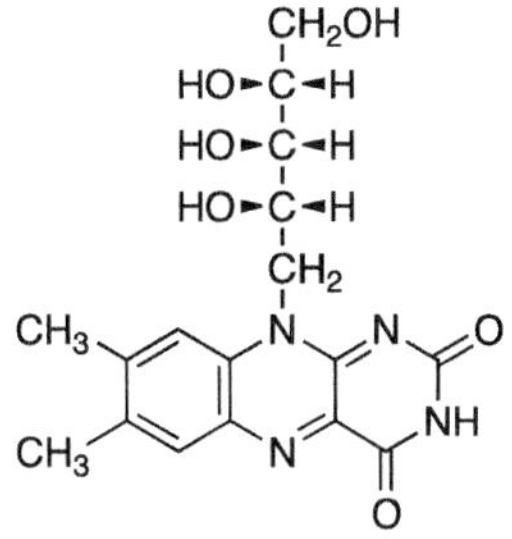

riboflavin

Folic acid was purified in the middle of the twentieth century, and research on the chemical structure of this compound and on its synthesis have been continuously conducted [1,2]. The first total synthesis of riboflavin was achieved in 1935 [3,4]. Riboflavin is a yellow compound, and its solutions emit a yellow-green fluorescence. Riboflavin is sensitive to light, and it gave lumiflavin when exposed to a light under basic conditions. Under neutral conditions it gave lumichrome by releasing the D-ribose moiety.

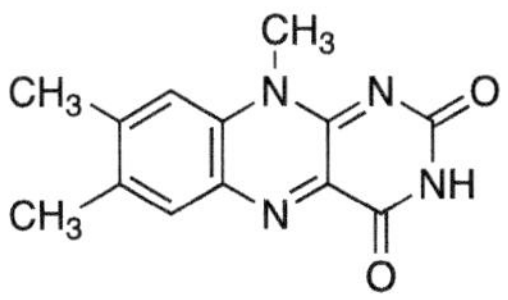

lumiflavine

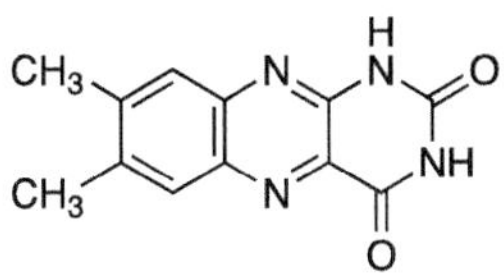

lumichrome

Xanthopterin, leucopterin, chrysopterin, and erythropterin were isolated from butterfly wings and are 2-amino-4-hydroxy pteridine derivatives [5–7]. Among these alkaloids, xanthopterin is a yellow substance and is widely distributed in insects and other animals, and it was also isolated from crabs in the Crustaceae. On the other hand, leucopterin is a colorless material, and it seems that this compound is derived from xanthopterin. In addition, erythropterin is an alkaloid that is responsible for the red and orange colors of the butterfly.

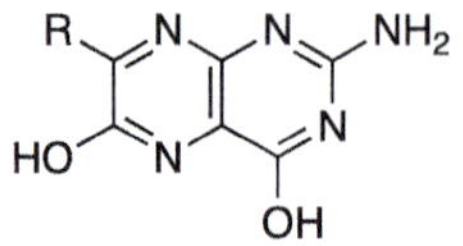

leucopterin **R = OH**
xanthopterin **R = H**
chrysopterin **R = CH_3**

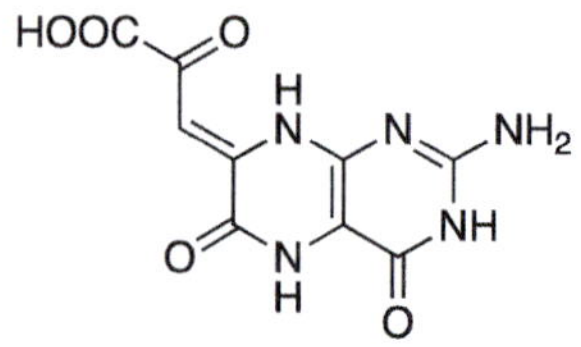

erythropterin

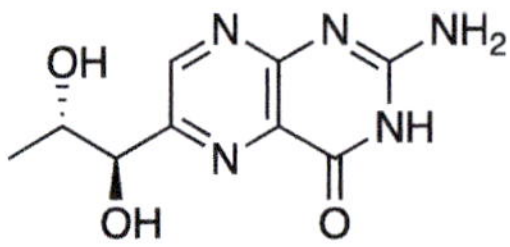

biopterin

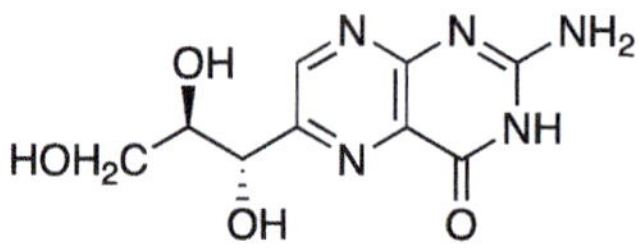

D-erythroneopterin

Various pterin derivatives are detected in human urine, and the main components are biopterin and D-erythroneopterin. These are excreted every day, typically in the amount of 980 mg and 380 mg, respectively. It is said that the level of excretion is not changed, even if a large amount of folic acid is administered orally [8].

LITERATURE CITED

[1] J.J. Pfiffner, S.B. Binkley, E.S. Bloom, B.L. O'Dell, J. Am. Chem. Soc. 69 (1947) 1476.
[2] M. Gates, Chem. Rev. 41 (1947R) 63.
[3] P. Karrer, K. Schöpp, F. Benz, Helv. Chim. Acta 18 (1935) 426.
[4] H. von Euler, P. Karrer, M. Malmberg, K. Schöpp, F. Benz, B. Becker, P. Frei, Helv. Chim. Acta 18 (1935) 522.
[5] H. Wieland, R. Purrmann, Liebigs Ann. der Chem. 544 (1940) 163.
[6] C. Schöpf, E. Becker, Liebigs Ann. der Chem. 507 (1933) 266.
[7] C. Schöpf, E. Becker, Liebigs Ann. der Chem. 524 (1936) 49.
[8] T. Fukushima, T. Shiota, J. Biol. Chem. 247 (1972) 4549.

CHAPTER 12

Alkaloids Possessing the Porphine Skeleton

Camellia japonica (Theaceae)

Pigments of heme in erythrocytes; chlorophyll, which plays an important role in photosynthesis in plants; and vitamin B_{12} are also referred to as porphyrins or corrins, depending on their skeleton.

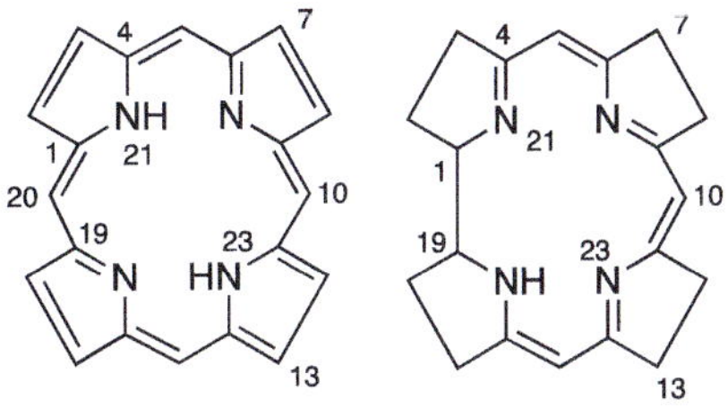

Until now, these corrinoid derivatives were not usually described in alkaloid terms, and were often classified and discussed with purines and pyrimidines as primary metabolites related to life. However, at the beginning of the biosynthesis of porphyrin in the mitochondrion, a C_4 unit is attached to an activated glycine, and during this biosynthetic step decarboxylation is accomplished.

Corrinoids are formed from a porphyrin precursor biosynthesized as described above, and from which one ring carbon (C-20 in the porphine nucleus) has been extruded. On the other hand, in the biosynthesis of the porphyrin skeleton in the chloroplast of a plant, glutamic acid is introduced instead of glycine [1]. Therefore, it can be said that these alkaloids are derived from glycine and glutamic acid, and contain four pyrrole rings.

Alkaloids
http://dx.doi.org/10.1016/B978-0-12-417302-6.00012-X

Here, these alkaloids will be described in this chapter as alkaloids based on a porphine skeleton. Their biosynthesis is completely different from that of prodigiosin described in the chapter on alkaloids derived from proline (Chapter 5). Total synthesis of chlorophyll a was achieved [2,3], and the assignment of all resonances in the ^{13}C nuclear magnetic resonance (NMR) spectrum of chlorophyll b has been accomplished [4]. Also the biosynthesis and chemistry of the chlorophylls have been reviewed [5–8].

LITERATURE CITED

[1] S.I. Beale, Ann. Rev. Plant Physiol. 29 (1978R) 95.
[2] R.B. Woodward, G.L. Closs, E. Le Goff, W.A. Ayer, H. Dutler, W. Leimgruber, J.M. Beaton, J. Hannah, W. Lwowski, F. Bickelhaupt, F.P. Hauck, J. Sauer, R. Bonnett, S. Ito, Z. Valenta, P. Buchschacher, A. Langemann, H. Volz, J. Am. Chem. Soc. 82 (1960) 3800.
[3] R.B. Woodward, Pure Appl. Chem. 2 (1963) 383.
[4] N. Risch, H. Brockmann, Tet. Lett. 24 (1983) 173.
[5] M.O. Senge, Photochem. Photobiol. 57 (1993) 189.
[6] H.M. Rissler, E. Collakova, D.D. Penna, J. Whelan, B.J. Pogson, Plant Physiol. 128 (2002) 770.
[7] O.B. Belyaeva, Photosynth. Res. 76 (2003) 405.
[8] R.D. Willows, Nat. Prod. Rep. 20 (2003) 327.

12.1 HEME AND CHLOROPHYLL

The red pigment hemoglobin, which exists in blood erythrocytes and which plays an important role in oxygen transport, consists of heme and globin. Heme has a chemical structure based on a porphyrin skeleton and is comprised of four pyrrole moieties. As alkaloids containing a porphyrin skeleton, chlorophyll a and vitamin B_{12} are very important for life. In order to elucidate the origin of the carbons contained in the pyrrole rings of the porphyrin skeleton, an uptake experiment using ^{14}C-labeled precursors was conducted [1,2]. As a result, it was established that a nitrogen atom and the carbon atoms attached to the nitrogen atom, namely an alpha carbon atom of the pyrrole ring and a methine carbon that connects the pyrrole rings, were derived from glycine. All of the other carbon atoms were derived from succinyl CoA, which was formed from acetic acid via the citric acid cycle. In the initial steps on the pathway, 5-aminolevulinic acid (=δ-aminolevulinic acid), which is formed from an activated glycine and succinic acid CoA, is dimerized to form porphobilinogen (PBG) with a pyrrole nucleus. Next, uroporphyrinogen I, a colorless alkaloid, is formed as a tetramer of PBG by an enzymatic reaction catalyzed by a deaminase and is composed of alternating acetic acid and propionic acid units. This alkaloid is transformed into uroporphyrin I by oxidation with oxygen.

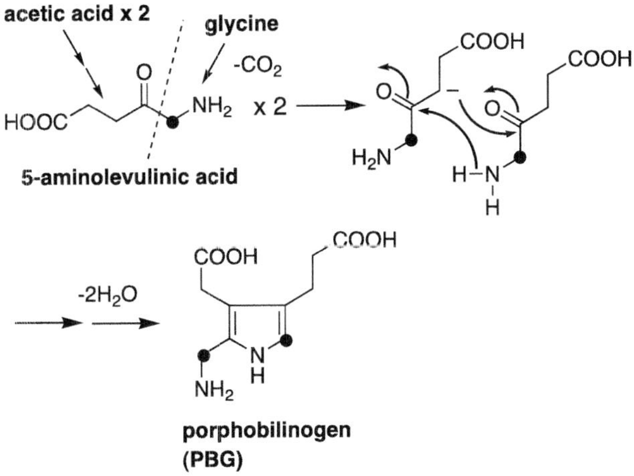

Biosynthesis of Porphobilinogen (PBG)

Uroporphyrin I is an alkaloid containing an aromatic 18π system. On the other hand, in uroporphyrinogen III, which is an intermediate in the biosynthesis of heme, the A–C rings are the same as uroporphyrinogen I; namely, the acetic acid and propionic acid units on the pyrrole ring are alternate, whereas in the D ring the arrangement of the acetic acid and propionic acid moieties on the pyrrole ring are reversed.

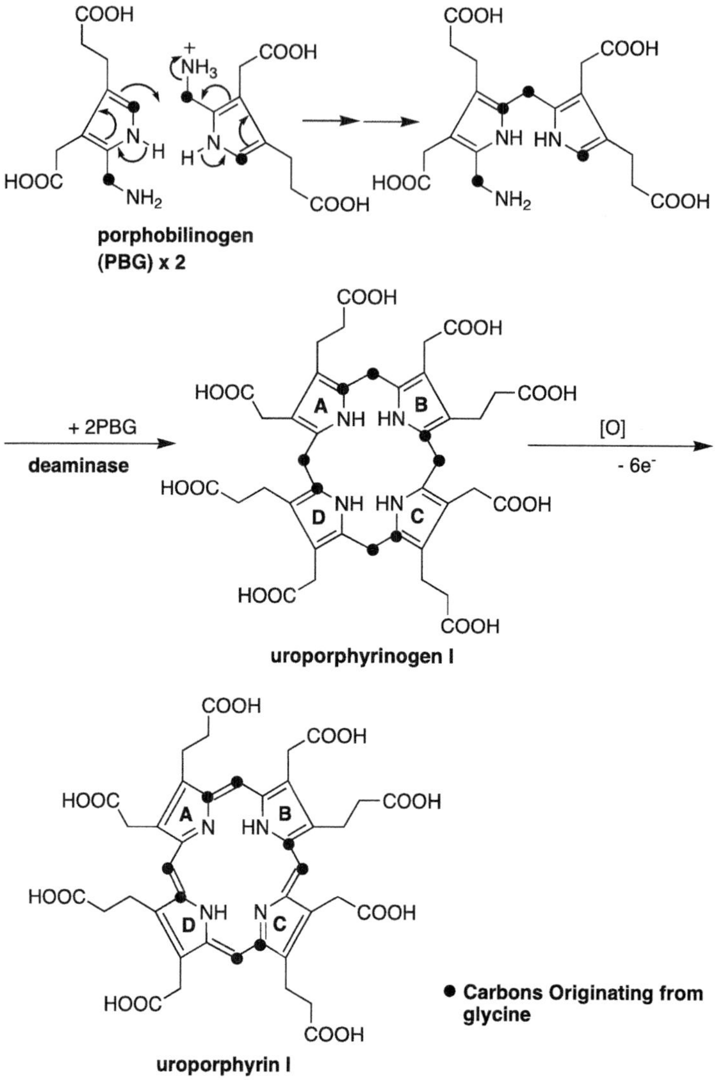

Biosynthetic Route of Uroporphyrin I

Uroporphyrinogen III, a biosynthetic intermediate of heme, is also biosynthesized through PBG via 5-aminolevulinic acid, as in the case of uroporphyrinogen I. However, in the biosynthesis of uroporphyrinogen III, after the tetramerization of PBG, a second enzyme, a cosynthetase, is involved in the reaction, and uroporphyrinogen III is formed by the transformation of an intermediate. Uroporphyrinogen III is then oxidatively transformed into uroporphyrin III.

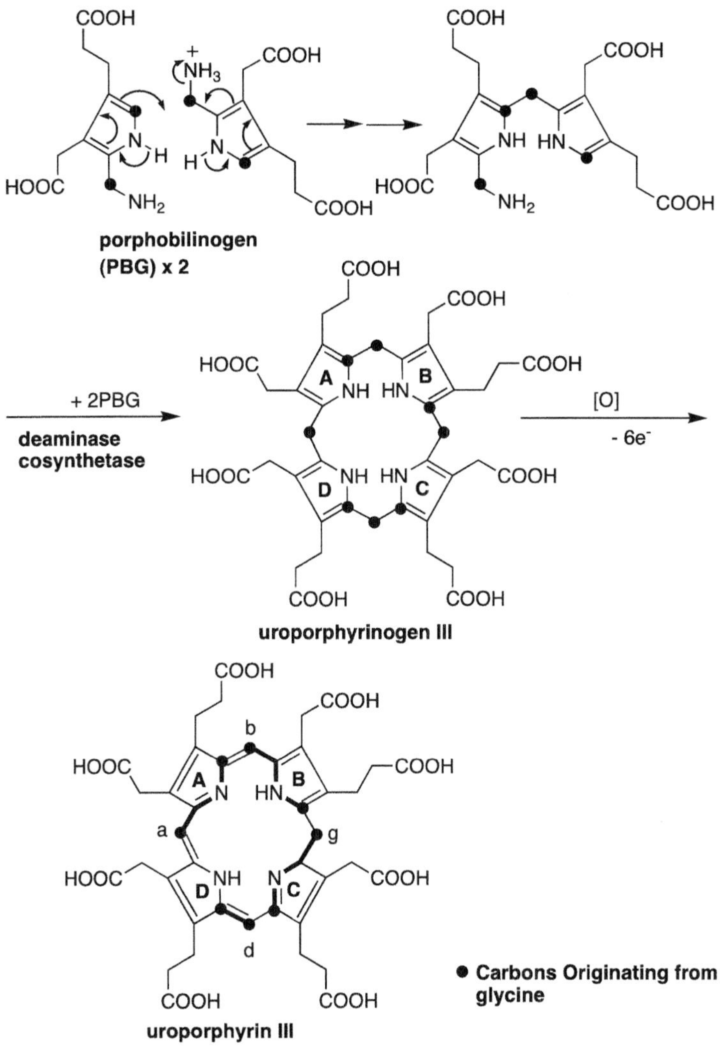

Biosynthetic Route of Uroporphyrin III

The cosynthetase itself cannot transform uroporphyrinogen I into uroporphyrinogen III, and it is possible to separate the two enzymes, the deaminase and the cosynthetase. It was also reported to be possible to identify the intermediate (tetramer), which changes slowly into uroporphyrinogen I by a nonenzymatic chemical reaction if the action of the deaminase is removed. If the intermediate is treated with the cosynthetase, it is immediately transformed into uroporphyrinogen III. It was also shown that the biosynthesis of the porphyrin ring is initiated from the A ring and advances in the sequence of the B–C–D rings [3-5].

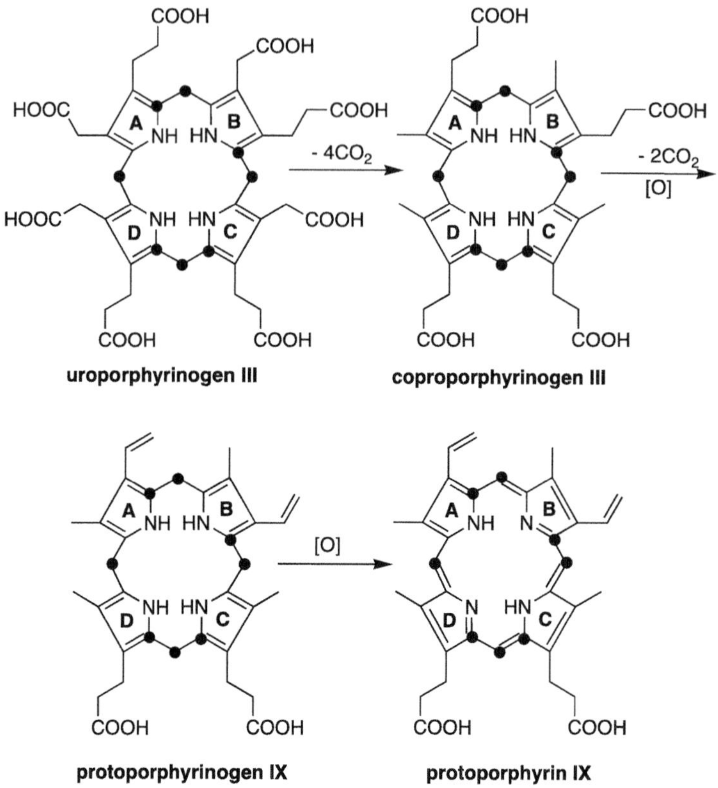

Biosynthetic Route from Uroporphyrinogen I to Protoporhyrin IX (• Carbons Originating from Glycine)

The ^{13}C labeling pattern of uroporphyrin III shown in the figure was reported using the uroporphyrin III formed by a multienzyme complex obtained from chicken blood or euglena (*Euglena gracilis*) of algae. That is to say, PBG doubly labeled with ^{13}C (90%) is diluted with PBG (4-fold) without the label, and is administered to the enzyme reaction system.

In the ^{13}C NMR spectrum of uroporphyrin III, at the α, β, and γ carbons only a ^{3}J long-range coupling (J=5.5 Hz) between the ^{13}C–^{13}C, as shown (bold line) in the figure, was observed. On the other hand, at the δ carbon a ^{1}J–^{13}C–^{13}C coupling (J=72 Hz) was observed [4]. This shows that in the formation of uroporphyrin III, a bond between a pyrrole ring and a linkage derived from a glycine is broken, and an intramolecular rearrangement has occurred in the D unit. In subsequent steps, the four COOH moieties of the four acetic acid moieties attached to the pyrrole rings of uroporphyrinogen III are decarboxylated to form four methyl groups and thus copropophyrinogen III (figure).

Subsequently, decarboxylation of the COOH moieties of the propionic acid residues in the A and B rings produces protoporphyrinogen IX, which is transformed into protoporphyrin IX by enzymatic oxidation. The order of the decarboxylation of the propionic acid moieties of copropophyrinogen III was established using a synthetic sample with a propionic acid unit in the A ring and a vinyl group in the B ring. This model compound, in which decarboxylation of the B ring occurred first, was not metabolized by this enzyme [6].

Protoporphyrin IX is a precursor for the biosynthesis of heme, chlorophyll, and cytochrome. Bile pigment is also derived from heme. The structure of bilirubin, the main component of this bile pigment, is shown [7]. Bilirubin is also a main component of the crude drug "Go-Oh" (bezoar, concave) prepared from the unhealthy calculus in the gallbladder or bile duct of cattle. Bezoar is an animal preparation used for detoxification, as an antipyretic, and as a cardiotonic. The biosynthesis of chlorophylls and bacteriochlorophylls from protoporphyrin IX has been reviewed [8].

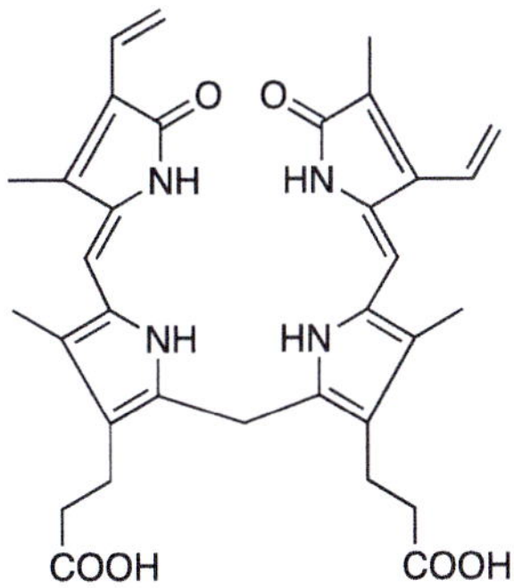

Bilirubin

LITERATURE CITED

[1] D. Shemin, D. Rittenberg, J. Biol. Chem. 166 (1946) 621.
[2] D. Shemin, J. Wittenberg, J. Biol. Chem. 192 (1951) 315.
[3] A.R. Battersby, E. McDonald, Acc. Chem. Res. 12 (1979) 14.
[4] A.R. Battersby, C.J.R. Fookes, G.W.J. Matcham, E. McDonald, Nature 285 (1980) 17.
[5] J.A.S. Cavaleiro, G.W. Kenner, K.M. Smith, J. Chem. Soc. Chem. Commun. (1973) 183.
[6] G. Städeler, Liebigs Ann. 132 (1964) 323.
[7] J. Fog, E. Jellum, Nature 198 (1963) 88.
[8] R.D. Willows, Nat. Prod. Rep. 20 (2003) 327.

12.2 VITAMIN B_{12}

Uroporphyrinogen III, described in the previous Section (12.1), is a precursor of vitamin B_{12}. Vitamin B_{12} is an alkaloid that contains a cobalt atom in the molecule, and was isolated from a marketed liver extract as cyanide-containing crystals [1]. Subsequently, it was found that this alkaloid is highly effective against pernicious anemia [2]. The same alkaloid was also isolated from the cultured broth of an *Actinomyces*, *Streptomyces griseus* [3]. The chemical structure of the cobalt complex was finally clarified by x-ray crystallography [4]. The remarkable total synthesis of this alkaloid was achieved by Woodward and Eschenmoser and their coworkers [5].

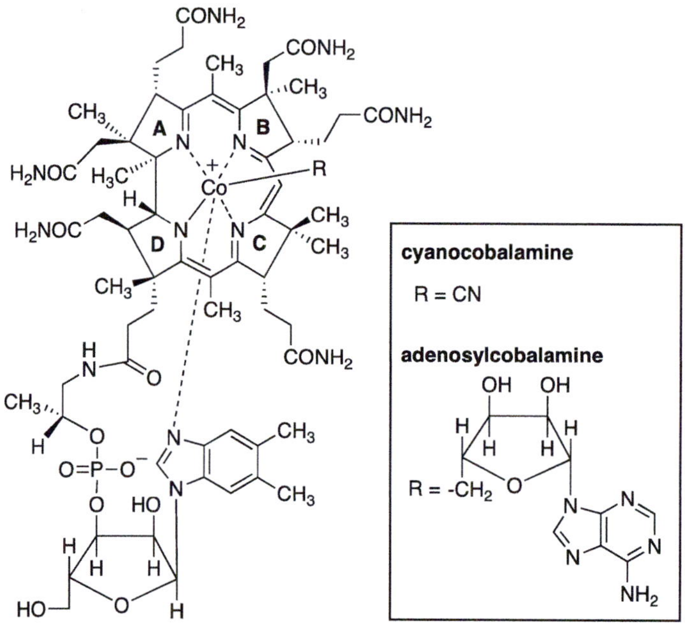

Vitamin B_{12}

The cyanide-containing metabolite is now known as cyanocobalamine, and the unit except for the nitrile (CN) moiety, which attaches to the cobalt atom of the cyanocobalamine, is known as cobalamine. The skeleton of this alkaloid system is known as a corrin.

In the narrow sense, the name vitamin B_{12} indicates cyanocobalamine itself, but this name is also used as a generic name of forms of vitamin B_{12} in which various moieties are combined with the cobalt atom of the cobalamine skeleton. When cyanocobalamine is incorporated into the human body, the nitrile ligand is changed into a 5′-deoxyadenosyl base. This metabolite corresponds to the substance that was formerly regarded as the coenzyme of vitamin B_{12} [6]. This alkaloid was also isolated from the cultured broth of a microorganism, *Propionibacterium shermanii* [7,8].

The biosynthetic pathway of this alkaloid, starting with 5-aminolevulinic acid, has been well studied by incorporation studies using radioisotope-labeled (^{14}C and ^{3}H) and stable isotope-labeled (^{13}C, ^{15}N, and ^{2}H (D)) precursors [9–13]. The biosynthesis of vitamin B_{12} is reviewed [14].

LITERATURE CITED

[1] E.L. Ricks, N.G. Brink, F.R. Koniuszy, T.R. Wood, K. Folkers, Science 107 (1948) 396.
[2] R. West, Science 107 (1948) 398.
[3] E.L. Ricks, N.G. Brink, F.R. Koniuszy, T.R. Wood, K. Folkers, Science 108 (1948) 634.
[4] D.C. Hodgkin, J. Kamper, J. Lindsey, M. Mackay, J. Pickworth, J.H. Robertson, C.B. Shoemaker, J.G. White, R.J. Prosen, K.N. Trueblood, Proc. R. Soc. 242A (1957) 228.
[5] R.B. Woodward, Pure Appl. Chem. 33 (1973) 145.
[6] H.A. Barker, H. Weissbach, R.D. Smyth, Proc. Natl. Acad. Sci. U.S.A. 44 (1958) 1093.
[7] H.A. Barker, R.D. Smyth, H. Weissbach, J.I. Toohey, J.N. Ladd, B.E. Volcani, J. Biol. Chem. 235 (1960) 480.
[8] Y. Sugiyama, et al., A Handbook of the Japanese Pharmacopoeia Fifteenth Edition, C-1575, Hirokawa-Shoten, Tokyo, 2006.
[9] A.R. Battersby, C. Edington, C.J.R. Fookes, J. Chem. Soc. Chem. Commun. (1984) 527.
[10] A.I. Scott, Tetrahedron 50 (1994) 13315.
[11] M. Kodera, F.J. Leeper, A.R. Battersby, J. Chem. Soc. Chem. Commun. (1992) 835.
[12] D. Thibaut, F. Kiuchi, L. Debussche, F. Blanche, M. Kodera, F.J. Leeper, A.R. Battersby, J. Chem. Soc. Chem. Commun. (1992) 982.
[13] D. Thibaut, L. Debussché, D. Fréchet, F. Herman, M. Vuilhorgne, F. Blanche, J. Chem. Soc. Chem. Commun. (1993) 513.
[14] A.R. Battersby, Bioorg. Mecicinal Chem. 4 (1996) 937.

CHAPTER 13

Alkaloids Derived from an *m*-C_7N Unit

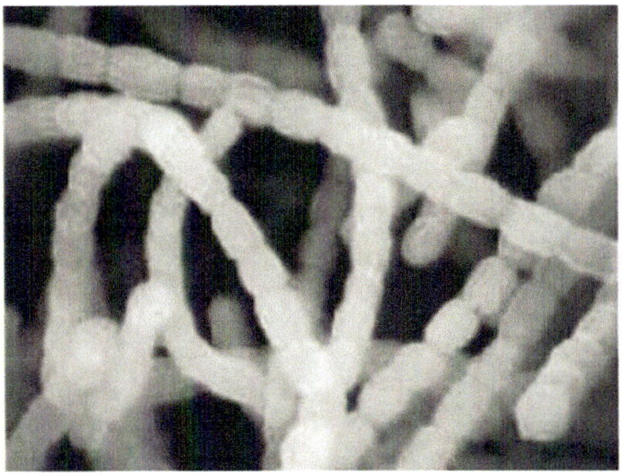

Streptomyces sp. 83-16

There is a group of alkaloids originating from microorganisms that are biosynthesized based on an *m*-C_7N unit as a fundamental unit.

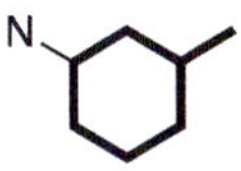

***m*-C_7N Unit**

The *m*-C_7N unit is derived from the shikimate pathway and possesses a C_1 group and an amino group *meta*-disposed on a benzene ring. For example, some of the alkaloids described in this chapter are established to be derived from 3-amino-5-hydroxybenzoic acid, and other isomers are also involved in the biosynthesis of various alkaloids.

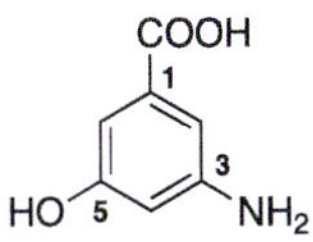

3-amino-5-hydroxy-benzoic acid

There are many alkaloids included in this group that are useful as antibiotics. Mitomycin C is an example of such an alkaloid, and is one of the principally used anticancer antibiotics in Japan as of this writing. Maytansine, which was first isolated from higher plants in the Celastraceae, and which was later shown to be produced by microorganisms (e.g., *Nocardia* sp.), and rifamycin S, which is a material used for the semisynthesis of the antitubercular agent rifampicin, and related alkaloids are also in this category.

Alkaloids
http://dx.doi.org/10.1016/B978-0-12-417302-6.00013-1

13.1 MITOMYCIN C

It was found by Hata et al. of the Kitasato Institute that a culture broth of *Streptomyces caepitosis* exhibited strong antimicrobial activity against Gram-positive bacteria, and the active components were isolated and named as mitomycins A and B [1].

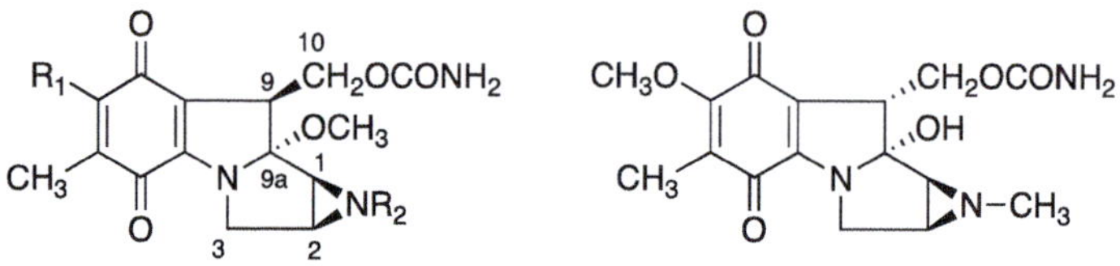

mitomycin A $R_1 = OCH_3$, $R_2 = H$
mitomycin C $R_1 = NH_2$, $R_2 = H$
porfiromycin $R_1 = NH_2$, $R_2 = CH_3$

mitomycin B

It was also found that these alkaloids showed strong antitumor effects in experimental animals [2], although severe toxicity was also observed. However, the third alkaloid isolated, named as mitomycin C, showed excellent antitumor activity, and the toxicity of this alkaloid was low [3]. Mitomycin C is thus utilized clinically as an antitumor antibiotic, especially for stomach cancer [4].The chemical structure of mitomycin C, including its absolute stereochemistry, and that of the related alkaloid porfiromycin have been reported [5,6]. It was established that ^{13}C- and ^{15}N-labeled L-glucosamine or L-mannosamine were not incorporated into mitomycins, whereas, D-glucosamine was incorporated. Consequently, six carbons and a nitrogen of D-glucosamine were incorporated as C_1, C_2, C_3, C_{9a}, C_9, and C_{10} and a nitrogen in the aziridine ring of the mitomycins [7,8].

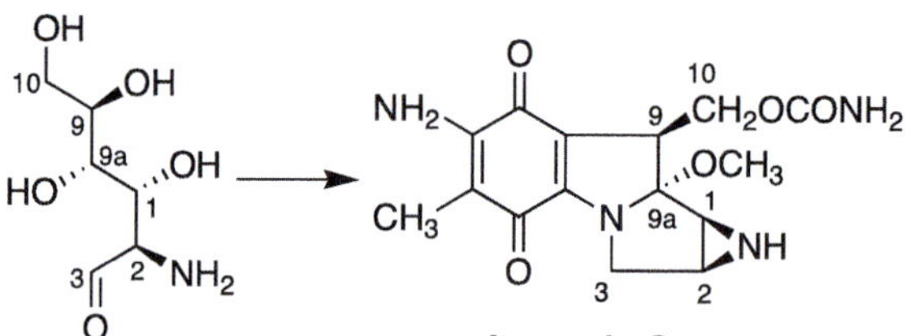

mitomycin C

D-glucosamine

Incorporation of D-Glucosamine into Mitomycins (Indicated for the Case of Mitomycin C; Carbon Numbers of D-Glucosamine Attributed to Those of Mitomycin C)

Thus, a question arose as to whether the absolute structure of the mitomycins that had been reported previously should be reversed. Reinvestigation of the x-ray crystallography of a mitomycin C derivative established that the previous report on the stereochemistry of mitomycins was erroneous [9]. The other unit in the molecule, except for the D-glucosamine-incorporated carbon atoms of the mitomycins, is derived from an m-C_7N unit.

When [carboxy-^{13}C]-3-amino-5-hydroxybenzoic acid was fed to the culture of *Streptomyces verticillatus*, porfiromycin labeled with ^{13}C at the C_6-methyl position was obtained [10].

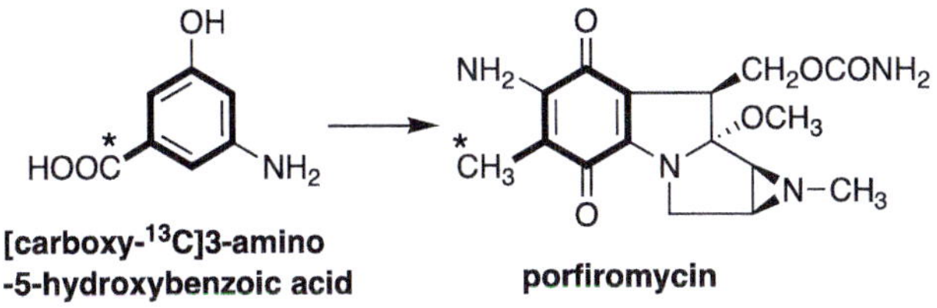

Incorporation of [carboxy-^{13}C]3-amino-5-hydroxybenzoic acid into Porfiromycin

From this result, it was confirmed that 3-amino-5-hydroxybenzoic acid was the direct biosynthetic precursor of the m-C_7N unit of porfiromycin. Because of its chemical structure, it might be thought that tryptophan could be involved in the biosynthesis of the mitomycins, like those alkaloids possessing an indole nucleus, whereas in fact the mitomycins were biosynthesized from an m-C_7N unit and a D-glucosamine moiety.

Mitomycin C showed antimicrobial activities against Gram-positive bacteria, and also against Gram-negative bacteria and acid-fast microorganisms. The antitumor activity of mitomycin C is thought to occur because of the cross-linking of two DNA strands. It was established, using ultraviolet and Fourier transform infrared spectroscopy, that mitomycin C reacted with guanine moieties in DNA at the N_7 position [11,12].

LITERATURE CITED

[1] H. Kanamori, T. Shima, C. Morita, T. Hata, J. Antibiot. 10A (1957) 120.
[2] I. Usubuchi, S. Oboshi, R. Tsuchida, H. Tanabe, N. Narita, Chemotherapy 6 (1958) 378.
[3] S. Wakaki, H. Marumo, K. Tomioka, G. Shimizu, E. Kato, H. Kamada, S. Kudo, Y. Fujimoto, Antibiot. Chemother. 8 (1958) 228.

[4] K. Sukigara, T. Takeishi, T. Noguchi, Chemotherapy 5 (1975) 223.
[5] J.S. Webb, D.B. Cosulich, J.H. Mowat, J.B. Patrick, R.W. Broschard, W.E. Meyer, R.P. Williams, C.F. Wolf, W. Fulmor, C. Pidacks, J.E. Lancaster, J. Am. Chem. Soc. 84 (1962) 3185.
[6] J.S. Webb, D.B. Cosulich, J.H. Mowat, J.B. Patrick, R.W. Broschard, W.E. Meyer, R.P. Williams, C.F. Wolf, W. Fulmor, C. Pidacks, J.E. Lancaster, J. Am. Chem. Soc. 84 (1962) 3187.
[7] U. Hornemann, M.J. Aikman, J. Chem. Soc. Chem. Commun. 88 (1973).
[8] U. Hornemann, J.P. Kehrer, C.S. Nunez, R.L. Ranieri, J. Am. Chem. Soc. 96 (1974) 320.
[9] K. Shirahata, N. Hirayama, J. Am. Chem. Soc. 105 (1983) 7199.
[10] M.G. Anderson, J.J. Kibby, R.W. Rickards, J.M. Rothschild, J. Chem. Soc. Chem. Commun. 1277 (1980).
[11] V.N. Iver, W. Szybalski, Proc. Natl. Acad. Sci. U.S.A. 50 (1963) 355.
[12] G.L. Verdine, K. Nakanishi, J. Am. Chem. Soc. 107 (1985) 6118.

13.2 NAPHTHALENOID ANSAMYCINS AND RIFAMPICIN

The term ansamycins is the name given to a group of antibiotics. "Ansa" means the grip of a bucket and/or flower basket, and the name ansamycin was derived from the shape of the chemical structures of the alkaloids included in this group. Namely, the structures of this group of alkaloids possess a long chain from an aromatic ring, part of a benzene or a naphthalene moiety, and the chain is extended to the other part of the aromatic ring not adjacent to the starting part of the chain. Thus, the chain looks like a grip of the aromatic moiety.

As implied above, there are two kinds of ansamycins, one that possesses a benzene ring and another that possesses a naphthalene ring in the structure: these are known as the benzenoid and naphthalenoid ansamycins, respectively. The *m*-C_7N unit is involved in the biosyntheses of the chromophores of both these ansamycins, and the polyketide biosynthetic pathway is also concerned in the biosynthesis of the ansa chains of both types of ansamycin and a part of the chromophore of the naphthalenoid ansamycins.

In this section, a semisynthetic antibiotic rifampicin, which plays an important role as a chemotherapeutic agent against tuberculosis, and its starting material rifamycin SV and related compounds are described.

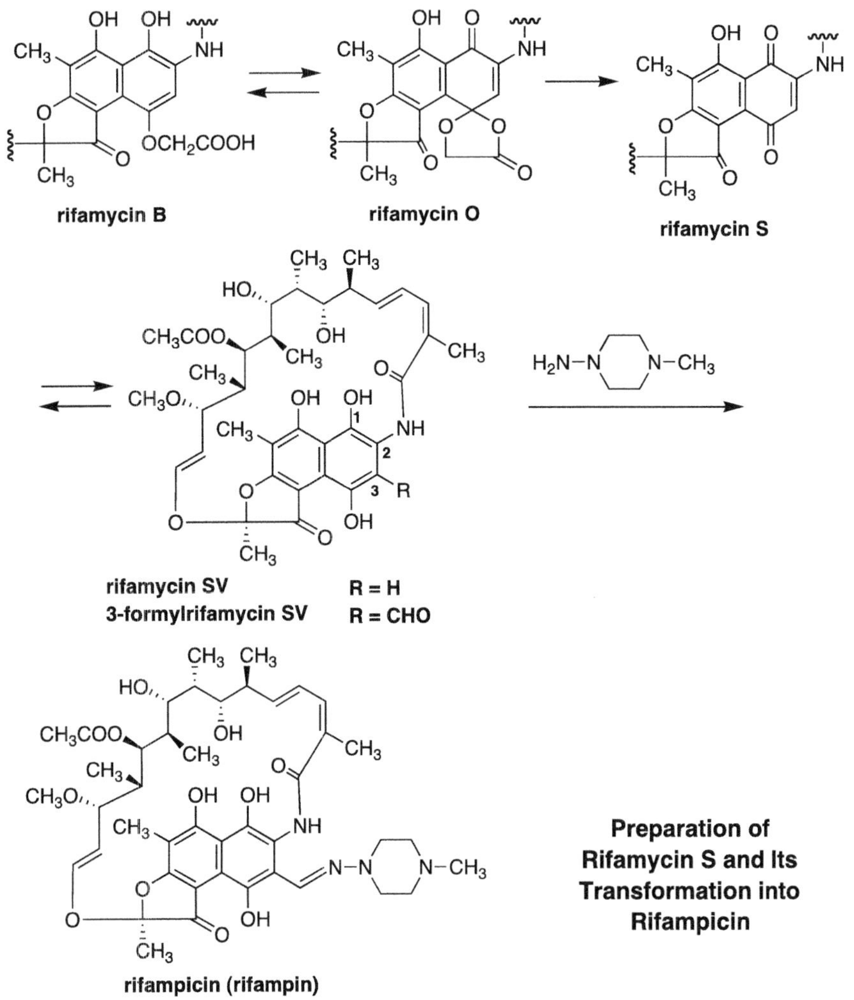

Preparation of Rifamycin S and Its Transformation into Rifampicin

Rifamycin B is an ansamycin antibiotic produced by *Streptomyces mediterranei* (subsequently reclassified as *Nocardia mediterranei*) [1–4]. This antibiotic shows antimicrobial activities against Gram-positive bacteria and acid-fast bacteria, and the toxicity for mammalian systems is low. Rifamycin

O is an oxidized form of rifamycin B, which was obtained by oxidizing rifamycin B with hydrogen peroxide. Conversely, rifamycin O can be transformed into rifamycin B by treating rifamycin O with ascorbic acid.

Rifamycins B and O can be transformed into rifamycin S, and rifamycin SV is obtained by treating rifamycin S with ascorbic acid [5]. Formylation of the C_3 position of rifamycin SV, followed by the combination of this alkaloid with 1-amino-4-methylpiperazine, gave the semisynthetic ansamycin rifampicin. Rifampicin was prepared in 1966, and it is one of the most effective antituberculosis agents at present. The chemical structures of the rifamycins were finally clarified in 1973–1974 [6,7].

Tuberculosis is a disease that has been a scourge to humanity over a very long period. Hippocrates (460?–377? BC) had provided a detailed description of a disease that is attributed to be the present tuberculosis, and in the eighteenth century 20–30% of all deaths in Europe were from tuberculosis.

In Japan, the first cause of death attributed to tuberculosis only occurred in 1950. However, at the first tuberculosis field study in 1953, there were 2.92 million tuberculosis patients and 5.53 million people who were regarded as susceptible to this disease.

The pathogenic bacteria *Mycobacterium tuberculosis* was found in 1882 by Robert Koch (1843–1910), and the discovery of rifampicin occurred 84 years after this event. Now, by treating with the combination of rifampicin and isonicotinic acid hydrazide (also known as isoniazid or INH, and described in Chapter 10), it became possible to completely control internal *M. tuberculosis* by chemotherapy in most cases since the 1970s. However, serious attention is now being paid to the appearance of drug-resistant strains of the microbe.

Regarding the biosynthesis of the ansa chain and part of the chromophore of rifamycin S, it was shown that two molecules of acetic acid and eight molecules of propionic acid were incorporated, as shown in the figure. During the biosynthesis, the methyl moiety (C-34) at the sixth propionic acid residue (C-27, C-28, and C-34) was eliminated, and the bond between C-29 and C-12 at the seventh propionic acid (C-12, C-13, and C-29) was cleaved and an oxygen was inserted [8,9]. The other part of the naphthalene nucleus was shown to be derived from an *m*-C_7N unit. The more detailed biosynthesis of various naphthalenoid ansamycins will be described later (Section 13.4).

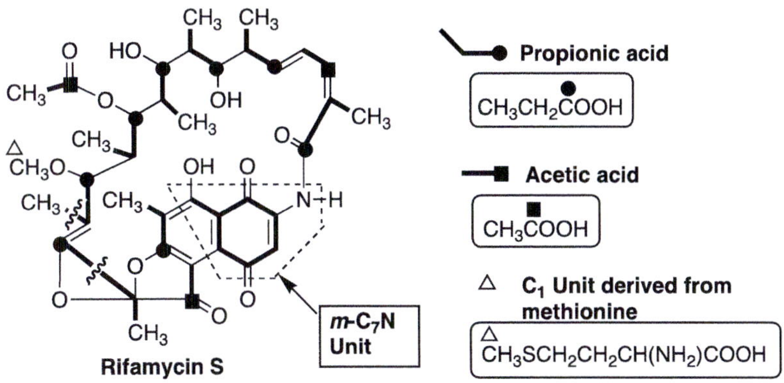

Biosynthetic Units of Rifamycin S

The discovery, classification, biosynthesis, and biological activities of various ansamycin antibiotics are reviewed by Funayama and Cordell [10].

LITERATURE CITED

[1] P. Sensi, A.M. Greco, R. Ballotta, Antibiot. Annu. 262 (1959–1960).
[2] M.T. Timbal, Antibiot. Annu. 271 (1959–1960).
[3] G. Maffii, M.T. Timbal, Antibiot. Annu. 277 (1959–1960).
[4] S. Füresz, R. Scotti, Antibiot. Annu. 285 (1959–1960).
[5] W. Oppolzer, V. Prelog, P. Sensi, Experientia 20 (1964) 336.
[6] W. Oppolzer, V. Prelog, Helv. Chim. Acta 56 (1973) 2287.
[7] M. Brufani, S. Cerrini, W. Fedeli, A. Vaciago, J. Mol. Biol. 87 (1974) 409.
[8] R.J. White, E. Martinelli, G.G. Gallo, G. Lancini, P. Beynon, Nature 243 (1973) 273.
[9] M. Brufani, D. Kluepfel, G.C. Lancini, J. Leitich, A.S. Mesentsev, V. Prelog, F.P. Schmook, P. Sensi, Helv. Chim. Acta 56 (1973) 2315.
[10] S. Funayama, G.A. Cordell, Stud. Nat. Prod. Chem. 23 (2000) 51.

13.3 BENZENOID ANSAMYCINS AND MAYTANSINE

Maytansine is an alkaloid first isolated from *Maytenus ovatus* (Celastraceae) [1], and maytansine is classified as a benzenoid ansamycin. The structure, including absolute stereochemistry, was determined by x-ray crystallography of its 3-bromopropyl ether derivative [2], and a total synthesis has been achieved [3].

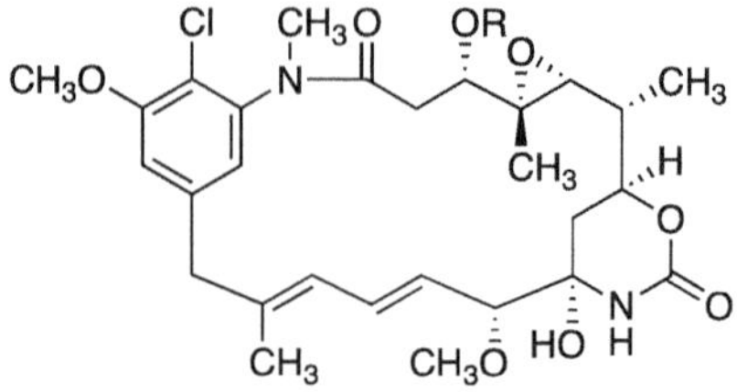

	(S)
maytansine	**R = $COCH(CH_3)N(CH_3)COCH_3$**
ansamitocin P-0	**R = H**
ansamitocin P-1	**R = $COCH_3$**
ansamitocin P-2	**R = $COCH_2CH_3$**
ansamitocin P-3	**R = $COCH(CH_3)_2$**
ansamitocin P-3'	**R = $COCH_2CH_2CH_3$**
ansamitocin P-4	**R = $COCH_2CH(CH_3)_2$**

Maytansine showed activity against many experimental tumor systems both in vitro and in vivo at a very low concentration, and it was reported that this alkaloid was effective clinically against acute lymphatic leukemia, non-Hodgkin lymphoma, melanomata, head and neck cancer, mammary cancer, and ovarian cancer. However, from the results of phase II tests reported until this writing, bioavailability of this alkaloid is low, and the clinical applicability seems to be limited [4].

The yield of maytansine from the plant material was very low, although it was found later, by the same research group who had isolated maytansine, that various alkaloids possessing the same skeleton as maytansine were obtained from the culture broth of *Nocardia* sp. No. C-15,003 (N-1), which was separated from a leaf of *Maytenus* sp. [5,6].

These alkaloids were named as ansamitocins, and the chemical structures of ansamitocins P-1, 2, 3, 3′, and 4 are shown in the figure. These alkaloids are based on ansamitocin P-0, a common parent skeleton, by a reductive hydrolysis of ansamitocins P-1, 2, 3, 3′, and 4 using $LiAlH_4$. Ansamitocin P-0 corresponds to maytansinol, which is the parent skeleton of the above-mentioned maytansine. On the other hand, it was found that ansamitocins P-1 and P-2 isolated from *Maytenus* plants corresponded to maytanacine and maytansinol propionate, respectively [7]. The alkaloids possessing the maytansinol (ansamitocin P-0) skeleton are also referred to as maytansinoids. The synthesis of maytansinol (ansamitocin P-0) as a racemic mixture has been reported [8].

Other benzenoid ansamycins, geldanamycin [9,10] and herbimycin [11,12], are also known. Herbimycin was first isolated as a herbicidal antibiotic and was subsequently found to normalize tumor cells [13].

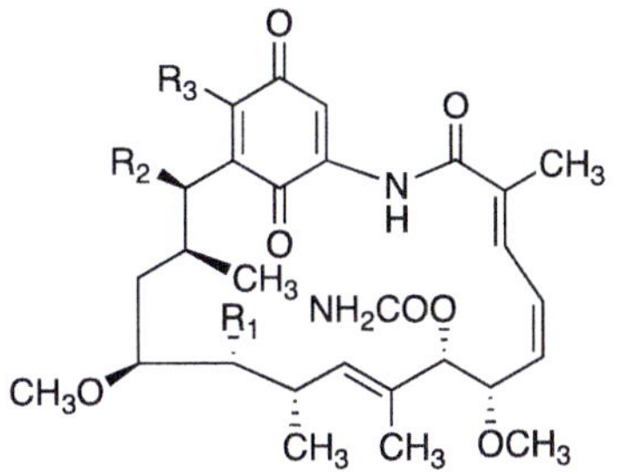

geldanamycin $R_1 = OH, R_2 = H, R_3 = OCH_3$
herbimycin A $R_1 = R_2 = OCH_3, R_3 = H$

The origin of the chromophore moiety of these benzenoid ansamycins is the m-C_7N unit, and no unit(s) derived from the polyketide pathway are concerned in the biosynthesis of the chromophore, unlike the naphthalenoid ansamycins. Therefore, in the benzenoid ansamycins, it can be more clearly seen how the m-C_7N unit is incorporated into the chromophore. This matter will be discussed in Section 13.4.

Geldanamycin and herbimycin possess a *p*-benzoquinone unit and its dihydro derivative as the chromophores, respectively. On the other hand, the trienomycins possess chromophores in which the *ortho*-position of the nitrogen and the C_1 unit of the m-C_7N unit are not oxidized, as in the case of maytansinoids [14–17]. The trienomycins A–E possess a 1,3,5-trisubstituted aromatic moiety; trienomycin A showed strong cytotoxic activity (IC_{50} 5 ng/ml) against HeLa S3 cells in vitro and antitumor activity (151% increase in life span) against sarcoma S180-bearing mice.

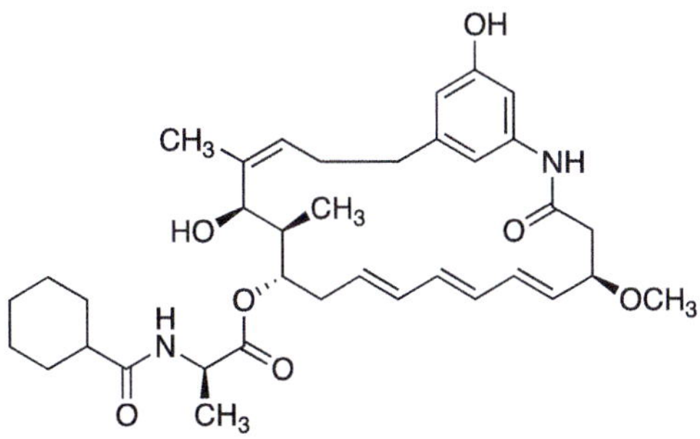

trienomycin A

The acute toxicity of trienomycin A was very low at >400 mg/kg [14,18]. The cytotoxicity of trienomycin A was lowered by reducing the triene moiety and/or by omitting the acyl alanyl moiety of the side chain [19].

LITERATURE CITED

[1] S.M. Kupchan, Y. Komoda, W.A. Court, G.J. Thomas, R.M. Smith, A. Karim, C.J. Gilmore, R.C. Haltiwanger, R.F. Bryan, J. Am. Chem. Soc. 94 (1972) 1354.
[2] R.F. Bryan, C.J. Gilmore, R.C. Haltiwanger, J. Chem. Soc. Perkin Trans. 897 (1973).
[3] E.J. Corey, L.O. Weigel, A.R. Chamberlin, H. Cho, D.H. Hua, J. Am. Chem. Soc. 102 (1980) 6613.
[4] T. Furue, ("*Kogan-busshitu Dehta Shu,*" in Japanese), in: K. Nitta, S. Tsukagoshi (Eds.), Data Book of Antitumor Drugs, Realize Inc., Tokyo, 1984, p. 172.
[5] E. Higashide, M. Asai, K. Ootsu, S. Tanida, Y. Kozai, T. Hasegawa, T. Kishi, Y. Sugino, M. Yoneda, Nature 270 (1977) 721.
[6] S. Tanida, T. Hasegawa, K. Hatano, E. Higashide, M. Yoneda, J. Antibiot. 33 (1980) 192.
[7] S.M. Kupchan, A.R. Branfman, A.T. Sneden, A.K. Verma, R.G. Dailey Jr., Y. Komoda, Y. Nagao, J. Am. Chem. Soc. 97 (1975) 5294.
[8] A.I. Meyers, P.J. Reider, A.L. Campbell, J. Am. Chem. Soc. 102 (1980) 6597.
[9] C. DeBoer, P.A. Meulman, R.J. Wnuk, D.H. Peterson, J. Antibiot. 23 (1970) 442.
[10] K. Sasaki, K.L. Rinehart Jr., G. Slomp, M.F. Grostic, E.C. Olson, J. Am. Chem. Soc. 92 (1970) 7591.
[11] S. Omura, Y. Iwai, Y. Takahashi, N. Sadakane, A. Nakagawa, H. Oiwa, Y. Hasegawa, T. Ikai, J. Antibiot. 32 (1979) 255.
[12] S. Omura, A. Nakagawa, N. Sadakane, Tetrahedron Lett. 20 (1979) 4323.
[13] Y. Uehara, M. Hori, T. Takeuchi, H. Umezawa, Jpn. J. Cancer Res. (Gann) 76 (1985) 672.
[14] I. Umezawa, S. Funayama, K. Okada, K. Iwasaki, J. Satoh, K. Masuda, K. Komiyama, J. Antibiot. 38 (1985) 699.
[15] S. Funayama, K. Okada, K. Iwasaki, K. Komiyama, I. Umezawa, J. Antibiot. 38 (1985) 1677.
[16] H. Nomoto, S. Katsumata, K. Takahashi, S. Funayama, K. Komiyama, I. Umezawa, S. Omura, J. Antibiot. 42 (1989) 479.
[17] A.B. Smith III, J.L. Wood, W. Wong, A.E. Gould, C.J. Rizzo, S. Funayama, S. Omura, J. Am. Chem. Soc. 112 (1990) 7425.
[18] K. Komiyama, Y. Hirokawa, H. Yamaguchi, S. Funayama, K. Masuda, Y. Anraku, I. Umezawa, S. Omura, J. Antibiot. 40 (1987) 1768.
[19] S. Funayama, Y. Anraku, A. Mita, Z.-B. Yang, K. Shibata, K. Komiyama, I. Umezawa, S. Omura, J. Antibiot. 41 (1988) 1233.

13.4 THE BIOSYNTHESIS–STEREOCHEMISTRY MODEL (CELMER'S MODEL) OF THE ANSAMYCINS

About 70 naphthalenoid ansamycins have been reported until now. In most of these ansamycins, the ansa part is composed of 17 carbons, like the rifamycins, and in other instances the ansa moiety is composed of 23 carbons, like the naphthomycins [1]. Models for the biosynthesis and stereochemistry (Celmer's model), as applied to macrolide antibiotics [2], have been developed for these two types of ansamycins [3].

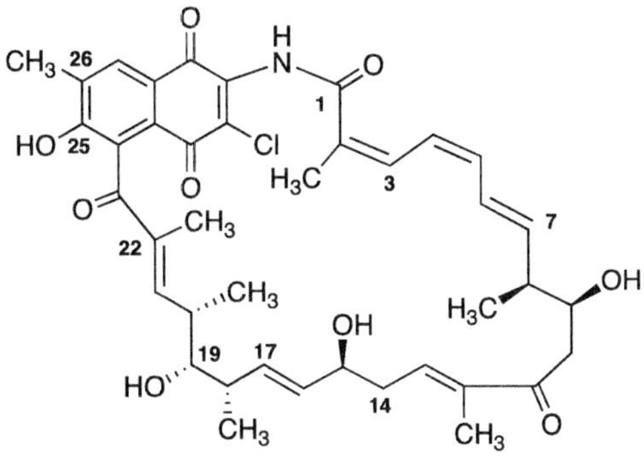

naphthomycin A

With respect to those ansamycins consisting of 17 carbon ansa chains, the moiety except for the part derived from the m-C_7N unit of each alkaloid has been considered [3].Thus, the biogenetic units of these moieties were P-A-P-P-P-P-P-P-A-P (P is propionic acid, and A is acetic acid), with no exception. For example, although it appears that the chemical structures of rifamycin B and streptovaricin C are quite different, the biogenetic units of these two alkaloids, and their sequence, are the same as P-A-P-P-P-P-P-P-A-P.

It was also shown that the absolute configuration between C-8 and C-16 of rifamycin B [4] and streptovaricin C [5] are the same. In addition, a methyl moiety derived from a propionic acid unit attached to the C_1 moiety of the m-C_7N unit corresponds to the methyl moiety at the C-20 position of the naphthalenoid chromophore. In all of these alkaloids, an oxygen atom derived from propionic acid remains at the C-19 position. This information can be applied to the elucidation of the chemical structure and stereochemistry of various naphthalenoid ansamycins, for example awamycin, which belongs to naphthalenoid ansamycins [6].

In the meantime, the biogenetic units between C-11 and C-20 of the naphthalenoid ansamycins with a C_{17} ansa chain, and those between C-17 and C-26 of the naphthalenoid ansamycins with a C23 ansa chain, are the same (P-P-P-A-P).

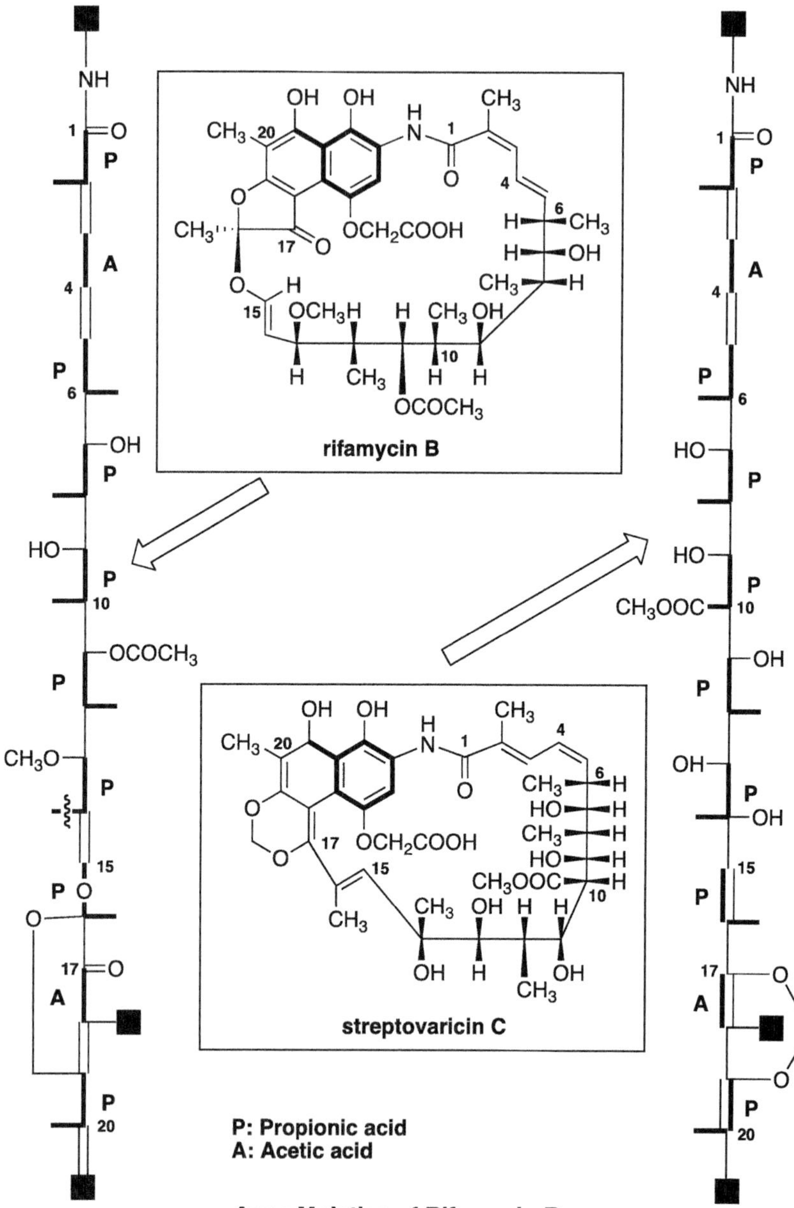

Ansa Moieties of Rifamycin B and Streptovaricin C Indicated by Celmer's Model

In the benzenoid ansamycin antibiotics, the ansa chains of herbimycin A, geldanamycin, and macbecin I [7] are similar. It was shown that the ansa moiety of geldanamycin consisted of four molecules of propionic acid (P), two molecules of glycolic acid (G), and one molecule of acetic acid (A) [8]. Based on this, it is thought that the ansa moiety of herbimycin A is derived from the same units as those of geldanamycin. For macbecin I, the ansa moiety is also estimated to be derived from the same units as those of geldanamycin, except for the C-5 and C-6 positions. Because the relative stereochemistry of herbimycin A and geldanamycin have been reported [9,10], the absolute stereochemistry of these alkaloids can be estimated to be as shown in the figure by comparison with the absolute stereochemistry reported for macbecin I [7].

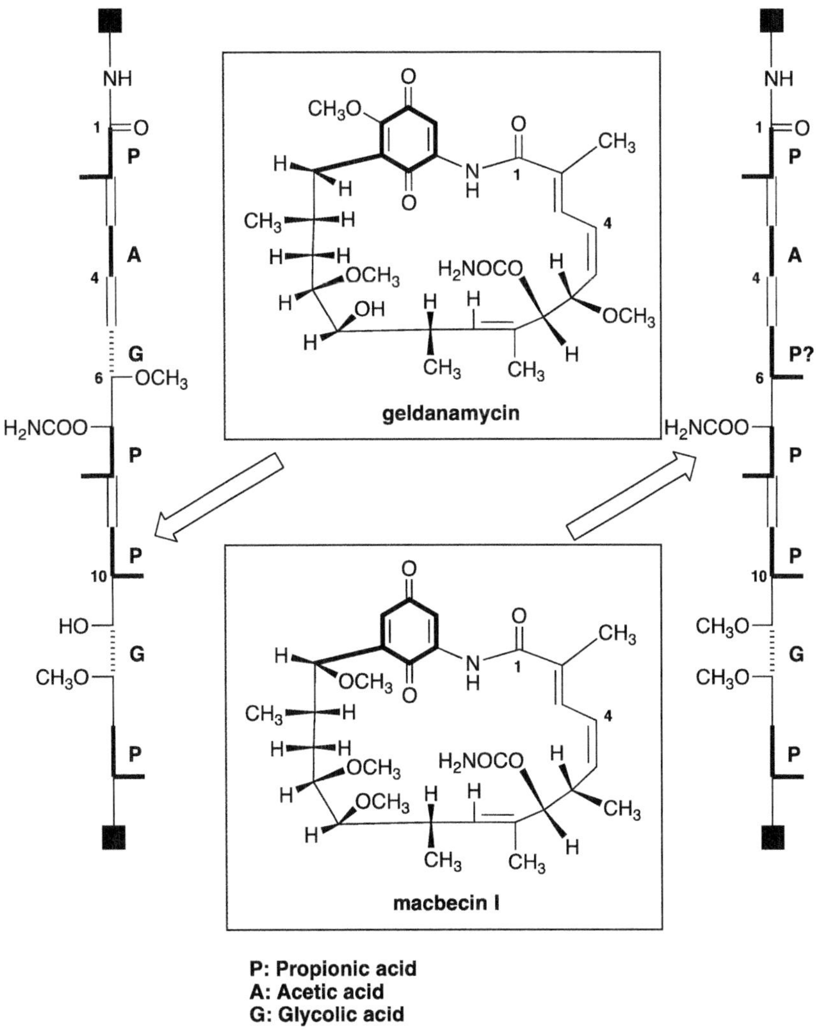

Ansa Moieties of Geldanamycin and Macbecin I Indicated by Celmer's Model

Biosynthesis and biological activities of various ansamycin antibiotics are reviewed by Funayama and Cordell [11].

LITERATURE CITED

[1] W. Keller-Schierlein, M. Meyer, L. Cellai, S. Cerrini, D. Lanba, A. Segre, W. Fedeli, M. Brufani, J. Antibiot. 37 (1984) 1357.
[2] W.D. Celmer, Pure Appl. Chem. 28 (1971) 413.
[3] S. Funayama, A. Nakagawa, S. Omura, 28th Symposium on the Chemistry of Natural Products, Symposium Papers, p. 73 (1986).
[4] M. Brufani, S. Cerrini, W. Fedeli, A. Vaciago, J. Mol. Biol. 87 (1974) 409.
[5] K.L. Rinehart Jr., W.M.J. Knöll, K. Kakinuma, F.J. Antosz, I.C. Paul, A.H.-J. Wang, F. Reusser, L.H. Li, W.C. Krueger, J. Am. Chem. Soc. 97 (1975) 196.
[6] S. Funayama, K. Okada, H. Oka, S. Tomisaka, T. Miyano, K. Komiyama, I. Umezawa, J. Antibiot. 38 (1985) 1284.
[7] M. Muroi, Y. Haibara, M. Asai, K. Kamiya, T. Kishi, 22nd Symposium on the Chemistry of Natural Products, Symposium Papers, p. 386 (1979).
[8] R.D. Johnson, A. Haber, K.L. Rinehart Jr., J. Am. Chem. Soc. 96 (1974) 3316.
[9] A. Furusaki, T. Matsumoto, A. Nakagawa, S. Omura, J. Antibiot. 33 (1980) 781.
[10] K.L. Rinehart Jr., L.S. Shield, Fortschr. Chem. Org. Naturst. 33 (1976) 231.
[11] S. Funayama, G.A. Cordell, Stud. Nat. Prod. Chem. 23 (2000) 51.

CHAPTER 14

Alkaloids Derived from Terpenoids

Aconitum japonicum subsp. *subcuneatum* (Ranunculaceae)

3*R*-Mevalonic acid is formed by the reduction of (3*S*)-3-hydroxy-3-methylglutaryl CoA((3S)HMG-CoA) with NADPH (nicotinamide adenine dinucleotide phosphate/reduced form; Chapter 10). (3*S*)-HMG-CoA is formed by the addition of an acetyl CoA unit to aceto-acetyl CoA, which itself is formed by the aldol condensation of two acetyl CoA units [1].

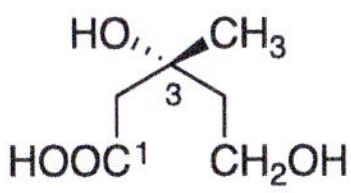

(3*R*)-mevalonic acid

Mevalonic acid (MVA) is transformed into isopentenyl pyrophosphate (IPP) by decarboxylation, and is further transformed into dimethylallyl pyrophosphorate (DMAPP). IPP and DMAPP are the fundamental building blocks that are incorporated into terpenoids. Terpenoids are therefore regarded as compounds derived from the combination of multiple C_5 units. These biosynthetic pathways are demonstrated by specific incorporation studies with $T(^3H)$- and ^{13}C-labeled compounds [2–5].

http://dx.doi.org/10.1016/B978-0-12-417302-6.00014-3

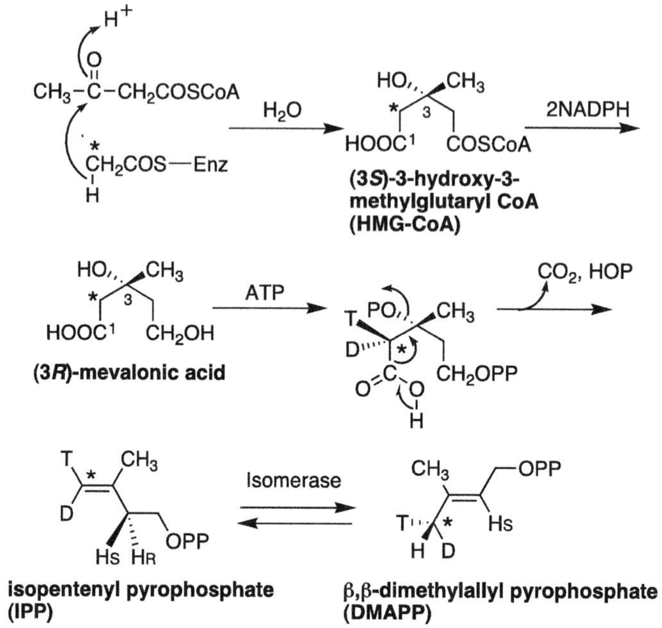

Biosynthetic Route for β,β-dimethylallyl pyrophosphate (DMAPP) and isopenteny Pyrophosphate

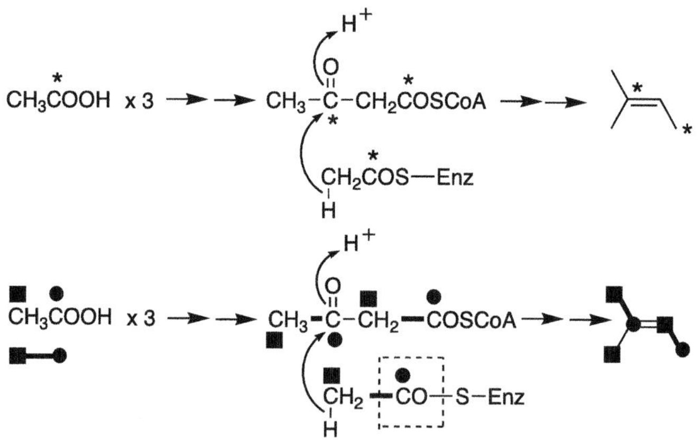

Incorporation Pattern of ^{13}C Labeled Acetic Acid into a C_5 Unit

For many years, it was thought that this pathway was the only one available for the formation of C_5 units. Recently, it was demonstrated that IPP and/or DMAPP were also biosynthesized through 1-deoxy-D-xylulose, and that this biosynthetic route was more common than the biosynthetic route via MVA. This route is also known as the mevalonate-independent pathway or methylerythritol phosphate pathway [6,7].

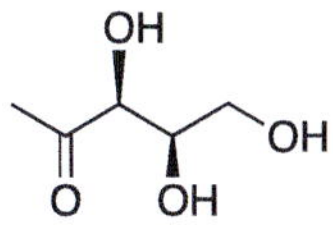

1-deoxy-D-xylulose

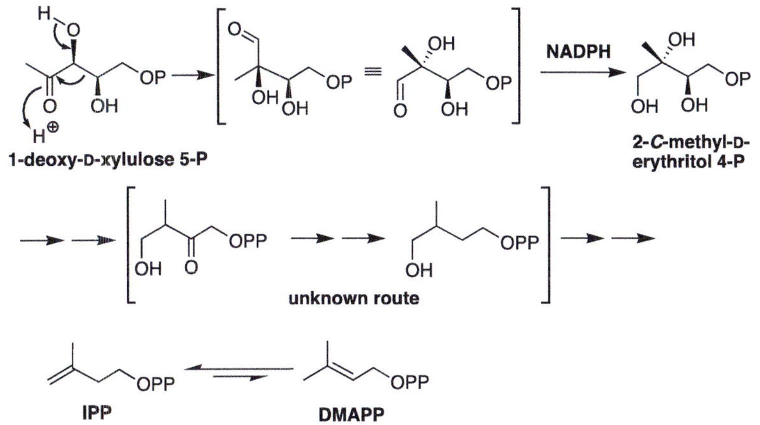

Biosyntheses of IPP and DMAPP *via* Mevalonate-independent Pathway

Those alkaloids formed by introducing nitrogen (in the form of an amine or ammonia, etc.) into preformed terpenoid moieties are known as terpenoid alkaloids. In these alkaloids, the origin of the nitrogen in the molecule is not through the incorporation of the carbon skeleton of an amino acid. Similarly, alkaloids derived through the polyketide pathway, and the alkaloids derived from a C_6-C_1 unit, which are described in the following chapters, are also not directly amino acid derived.

Several individual groups of alkaloids incorporate terpenoid units. For example, acronycine (Section 9.3) and reserpine (Section 2.10) incorporate C_5- or C_{10}-based terpenoid units, respectively. However, the origin of the nitrogen of these alkaloids is amino acids (anthranilic acid and tryptophan, respectively), so these alkaloids are not included in this chapter.

The alkaloids of this chapter are classified according to the classification of terpenoids, namely hemiterpenoid alkaloids (alkaloids with a C_5 unit), monoterpenoid alkaloids (alkaloids with a C_{10} unit), sesquiterpenoid alkaloids (alkaloids with a C_{15} unit), diterpenoid alkaloids (alkaloids with a C_{20}

unit), sesterterpenoid alkaloids (alkaloids with a C_{25} unit), and triterpenoid alkaloids (alkaloids with a C_{30} unit).

According to this classification, actinidine described in this chapter is classified as a monoterpenoid alkaloid, whereas nupharidine and aconitine are classified as sesquiterpenoid and diterpenoid alkaloids, respectively. Batrachotoxin A, which possesses a steroidal skeleton, is classified into steroidal alkaloid. Steroids are derived through the degradation of a triterpenoid skeleton (C_{30} unit) [2,4].

LITERATURE CITED

[1] F. Lynen, U. Henning, Angew.Chem. 72 (1960) 820.
[2] K.J. Stone, W.R. Roeske, R.B. Clayton, E.E. van Tamelen, Chem.Commun. 1969 (1969) 530.
[3] J.W. Cornforth, R.H. Cornforth, C. Donninger, G. Popjak, Proc. Roy. Soc. London Ser. B 163 (1965) 492 [Chem.Abstr., 64, 8549b (1966)].
[4] J.W. Cornforth, R.H. Cornforth, G. Popjak, L.Yengoyan, J. Biol. Chem. 241 (1966) 3970.
[5] S. Funayama, M. Ishibashi, K. Komiyama, S. Omura, J. Org. Chem. 55 (1990) 1132.
[6] H.K. Lichtenhaler, Ann. Rev. Plant Physiol. Plant Mol. Biol. 50 (1999) 47.
[7] M. Rohmer, Nat. Prod. Rep. 16 (1999) 565.

14.1 *Actinidia polygama* AND ACTINIDINE

When the unripe fruits of *Actinidia polygama* (Actinidiaceae) are stabbed by the insect *Pseudosphondylia matatabi*, a gall is formed. The gall is collected, treated with boiling water, and dried to yield "mokutenryou," which is used in Kampo medicine for analgesia and to prevent diarrhea.

The fruits of plants of the genus *Actinidia* cause abnormal excitement for animals of the cat genus, such as cats and lions. Matatabilactone, obtained from "mokutenryou," possesses the same effect [1]. It was found that matatabilactone was a mixture, and that it contained dihydronepetalactone, isodihydronepetalactone, iridomyrmecin, and isoiridomyrmecin [2–4].

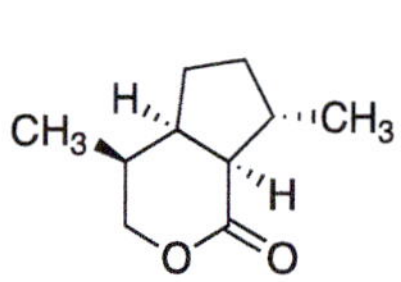

dihydronepetalactone

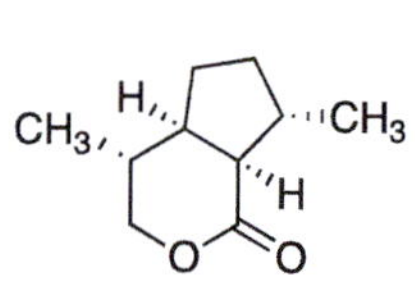

isodihydronepetalactone

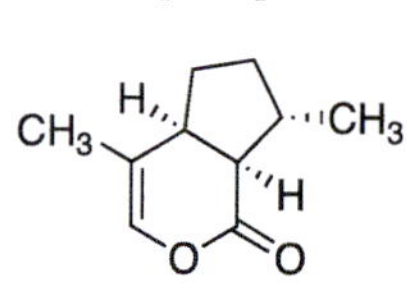

nepetalactone

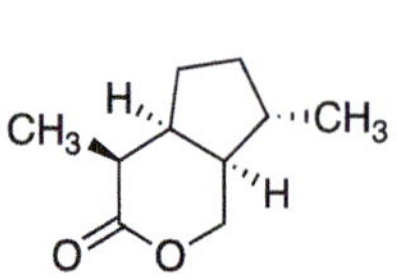

iridomyrmecin

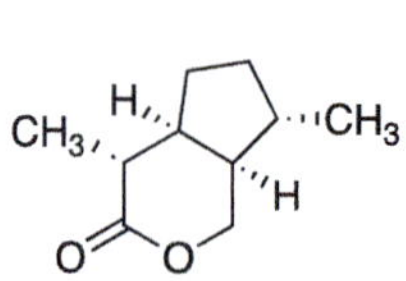

isoiridomyrmecin

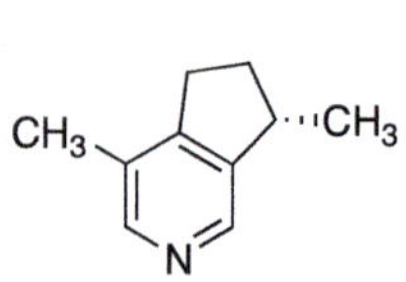

actinidine

All of these components possess the molecular formula $C_{10}H_{16}O_2$ (mw 168 amu). Among these, the former two compounds are the reduced form nepetalactones [5], which is the main component of catnip, *Nepeta cataria* (Labiatae), known as a favorite plant of cats in the United States. The latter two components are also present in the postbinary anal gland secretions of the Argentina ant (*Iridomyrmex humilis*) and *Iridomyrmex nitidus* and work as defensive substances [6–8]. Therefore, these C_{10} compounds with such skeleton are sometimes called iridoids.

In addition to the compounds described above, actinidine is also isolated from mokutenryou [1,2,9]. As estimated by comparing the chemical structures of actinidine and the iridoids described above, these compounds are derived from a common biosynthetic pathway.

The biosynthetic pathway from *l*-citronellal to an enol half acetal is shown in the figure. Actinidine is formed from this enol half acetal type by the incorporation of nitrogen. On the other hand, gentianine (described in the next section) and its related secoiridoids are alkaloids in which introduction of nitrogen occurs into an iridoid that has been cleaved between C-7 and C-8.

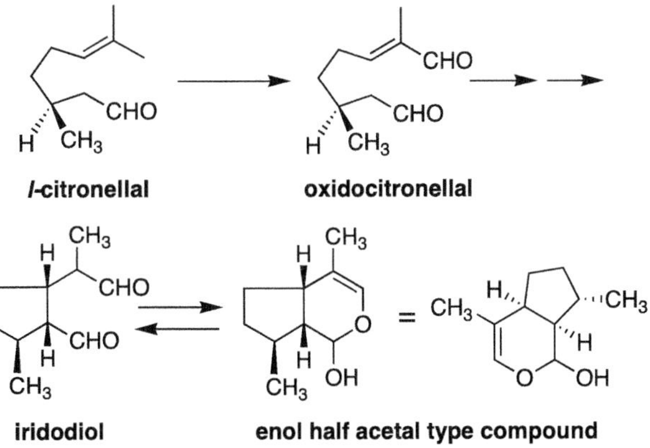

Biosynthetic Route from *l*-Citronellal to an Enol Half Acetal Type Compound

LITERATURE CITED

[1] T. Sakan, A. Fujino, F. Murai, Y. Butsugan, A. Suzui, Bull. Chem. Soc. Jpn 32 (1959) 315.
[2] T. Sakan, A. Funino, F. Murai, Nihon Kagaku Kaishi 81 (1960) 1320.
[3] F. Murai, Nihon Kagaku Kaishi 81 (1960) 1324.
[4] T. Sakan, S. Isoe, S.B. Hyeon, R. Katsumura, T. Maeda, J. Wolinsky, D. Dickerson, M. Slabaugh, D. Nelson, Tetrahedron Lett. 1965 (1965) 4097.
[5] R.B. Bates, E.J. Eisenbraun, S. M. McElvain, J. Am. Chem. Soc. 80 (1958) 3420.
[6] G.W.K. Cavill, D.L. Ford, H.D. Locksley, Aust. J. Chem. 9 (1956) 288.
[7] J. Weatherston, Q. Rev. 21 (1967R) 287.
[8] P. Callant, R. Ongena, M. Vandewalle, Tetrahedron 37 (1981) 2085.
[9] A. Fujino, Nihon Kagaku Kaishi 81 (1960) 1327.

14.2 *Gentiana scabra* AND GENTIANINE

The dried rhizome of *Gentiana scabra* (Gentianaceae) and its variety *G. scabra* var. *buergeri* are known as "Ryutan" and are used as bitters in Japan. *Gentiana lutea* is used for the same purpose in Europe.

The monoterpene gentiopicroside (=gentiopicrin) was obtained as a bitter glycoside from Ryutan [1,2], and the same compound was isolated from *Gentiana kirilowi* [3–5]. Gentianine is an alkaloid derived from swertiamarin and gentiopicroside, and was also isolated from *Swertia japonica* (Gentianaceae). Gentianine is regarded as an artifact of swertiamarin and gentiopicroside during alkaloid extraction using ammonia. Gentianine was obtained when ammonia was used during the processing of *Enicostema littorale* (Gentianaceae). The basic principle was detected at trace levels, but gentianine was not isolated, when ammonia was not used [6]. As bitter principles, gentiopicroside, sweroside, and swertiamarin were isolated from *S. japonica* [7,8].

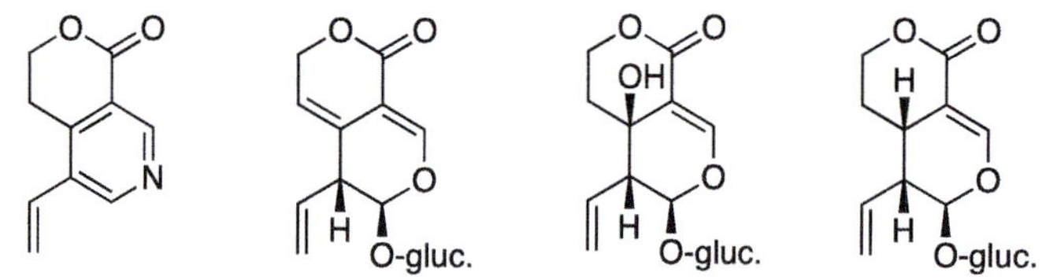

gentianine　gentiopicroside (= gentiopicrin)　swertiamarin　sweroside

These chemical compounds are classified within the monoterpenoids, and are also known as secoiridoids. In the figure, the biosynthetic pathway from *l*-citronellal to sweroside, swertiamarin, and gentiopicroside is shown [9]. The study of the biosynthetic pathway was carried out using *S. japonica* or *Gentiana triflora,* and it was shown that gentiopicroside was derived from sweroside via swertiamarin. The structure elucidation, synthesis, and biosynthesis of the monoterpenoid alkaloids were reviewed by Cordell [10,11].

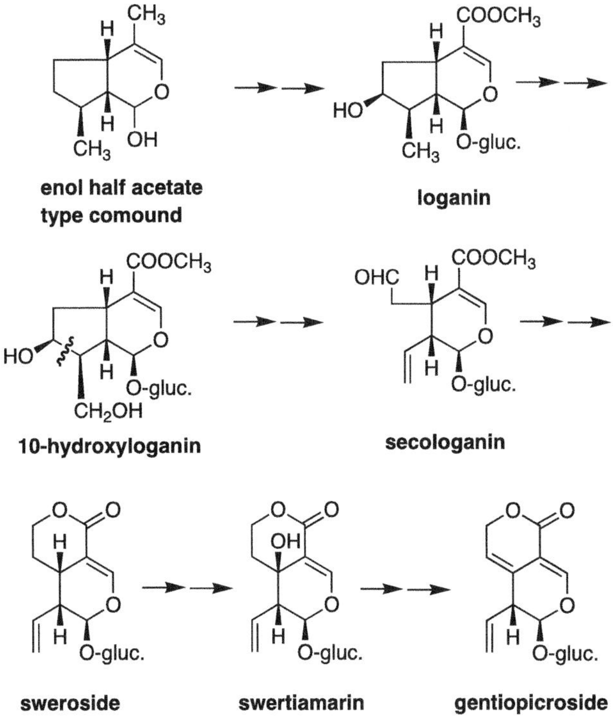

Biosynthetic Route from Enol Half Acetate Type Compound to Sweroside, Swertiamarin and Gentiopicroside

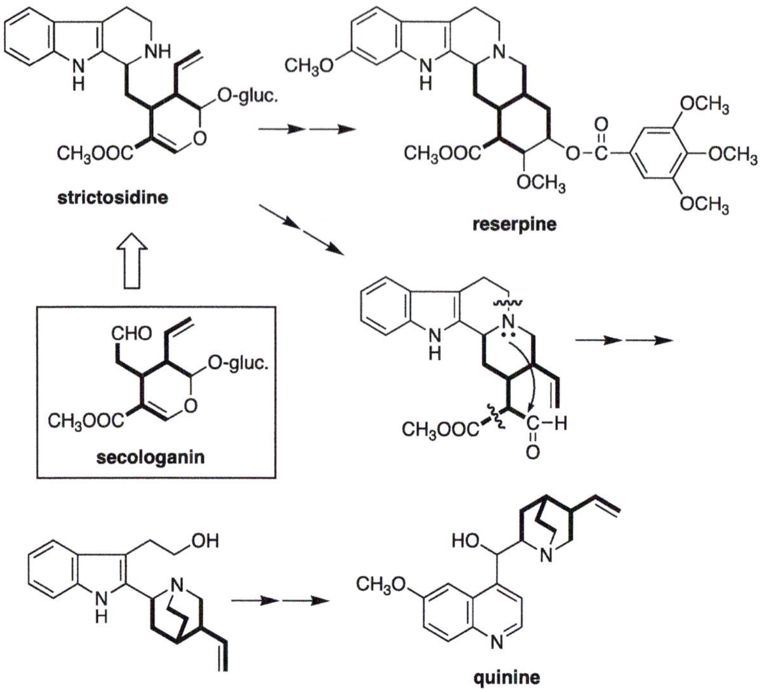

Incorporation Pattern of Secologanin into Reserpine and Quinine

LITERATURE CITED

[1] F. Korte, Chem. Ber 87 (1954) 780.
[2] L. Canonica, F. Pelizzoni, P. Manitto, G. . Jommi, Tetrahedron Lett. 7 (1960).
[3] N.F. Proskurnina, J. Gen. Chem. USSR 14 (1944) 1148 [Chem. Abstr., 40, 72132 (1946)].
[4] N.F. Proskurnina, V.V. Shpanov, R.A. Konovalova, Doklady Akad. Nauk, SSSR 66 (1949) 437 [Chem. Abstr., 44, 159d (1950)].
[5] S. Shibata, M. Fujita, H. Igeta, Yakugaku Zasshi 77 (1957) 116.
[6] T.R. Govindachari, S.S. Sathe, N. Viswanathan, Indian J. Chem. 4 (1966) 201.
[7] H. Inoue, S. Ueda, Y. Nakamura, Chem. Pharm. Bull. 18 (1970) 1856.
[8] H. Inoue, T. Yoshida, Y. Nakamura, S. Tobita, Chem. Pharm. Bull. 18 (1970) 1889.
[9] H. Inoue, S. Ueda, Y. . Nakamura, Chem. Pharm. Bull. 18 (1970) 2043.
[10] G.A. Cordell, Alkaloids 16 (1977) 432.
[11] G.A. Cordell, Alkaloids 52 (1999) 261.

14.3 RHIZOME OF *Nuphar japonicum* AND SESQUITERPENE ALKALOIDS

Nuphar japonicum (Nymphaeaceae) is a perennial herb that grows in shallow running water. The rhizome is called *Nuphar* rhizome (Kohone) and is used in complementary male fertility medicines and in women's diseases.

As chemical components of *Nuphar* rhizome, several sesquiterpene alkaloids, such as nupharidine, (−)-nupharamine, and (−)-deoxynupharidine are known [1–4]. Among them, the absolute configuration of (−)-nupharamine was decided by converting (−)-deoxynupharidine [5–7]into (−)-nupharamine [8]. The racemic (−)-deoxynupharidine was synthesized [9].

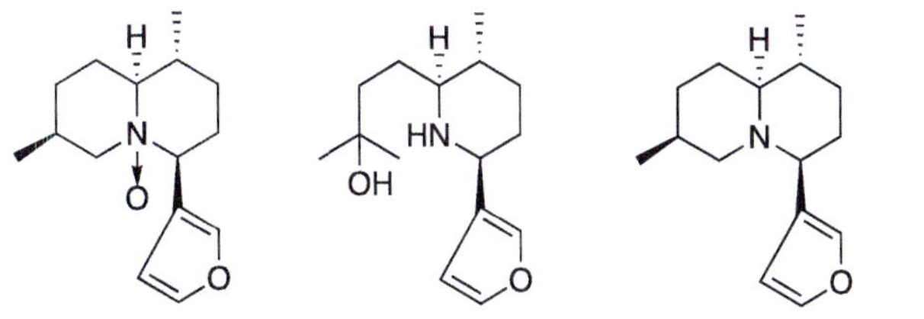

(+)-nupharidine **(-)-nupharamine** **(-)-deoxynupharidine**

These alkaloids possess a sesquiterpene skeleton, and it is considered that they are derived through the isoprenoid pathway, and that a nitrogen atom is incorporated at some point during the pathway. The stage at which this occurs is not known, and no biosynthetic studies have been reported.

LITERATURE CITED

[1] Y. Arata, T. Ohhashi, Yakugaku Zasshi 77 (1957) 236.
[2] T. Ohhashi, Yakugaku Zasshi 79 (1959) 729.
[3] T. Ohhashi, Yakugaku Zasshi 79 (1959) 734.
[4] Y. Arata, T. Ohhashi, M. Yonemitsu, S. Yasuda, Yakugaku Zasshi 87 (1967) 1094.
[5] M. Kotake, I. Kawasaki, T. Okamoto, S. Matsutani, S. Kusumoto, T. Kaneko, Bull. Chem. Soc. Jpn 35 (1962) 1335.
[6] I. Kawasaki, I. Kusumoto, T. Kaneko, Bull. Chem. Soc. Jpn 41 (1968) 1264.
[7] R.T. LaLonde, E. Auer, C.F. Wong, V.P. Muralidharan, J. Am. Chem. Soc. 93 (1971) 2501.
[8] I. Kawasaki, S. Matsutani, T. Kaneko, Bull. Chem. Soc. Jpn 36 (1963) 1474.
[9] Y. Arata, T. Nakanishi, Y. Asaoka, Chem. Pharm. Bull. 10 (1962) 675.

14.4 *Aconitum* AND ACONITINE ALKALOIDS

Aconitum plants (Ranunculaceae) are widely distributed in the subarctic regions and the temperate zones in the Northern Hemisphere. Thus, plants of this genus grow in major regions of Asia and Europe. It has been well known for thousands of years that the tuberous roots of *Aconitum* plants are deadly poisonous. Consequently, there are many stories about aconite poisoning in cultural history.

In Kampo medicine, the young roots of the aconite tuber are known as Bushi, and the mother root is called Uzu, and has been said to act as a diuretic and cardiotonic. It is mainly used for reviving the symptoms generated by the relaxation of various visceral organs, such as recovery of metabolic function, paralysis of the body limbs and joints, recovery from aches, abdominal pain, diarrhea, sperm loss, etc. The aconite tuber that is used in medicine in Japan, known as kako-bushi, has reduced toxicity [1].

There are many species in the genus *Aconitum*, and over 100 kinds of *Aconitum* plants are known in East Asia. For medicinal use, *Actinidia carmichaeli*, *Actinidia triphyllum*, and *Actinidia japonicum* are used in South Korea, the People's Republic of China, and Japan, respectively. Alkaloids characteristic to aconite tuber are known as *Aconitum* alkaloids. The aconite alkaloids are classified, roughly, into two types. One is a toxic alkaloid possessing an acyloxy moiety, as well as methoxyl and/or hydroxyl group. Examples of such alkaloids are aconitine, mesaconitine, and jesaconitine. Another type of alkaloid is the low-toxic alkamine type, which does not have a methoxyl group. The latter group of alkaloids are divided into three types, and as representatives of each group, the structures of atisine [2], napelline [3], and ignavine are shown.

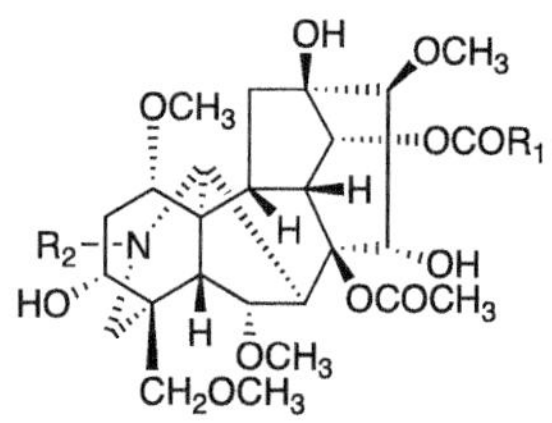

aconitine $R_1 = C_6H_5$, $R_2 = CH_2CH_3$
mesaconitine $R_1 = C_6H_5$, $R_2 = CH_3$
jesaconitine $R_1 = C_6H_4$-(p-OCH_3), $R_2 = CH_2CH_3$

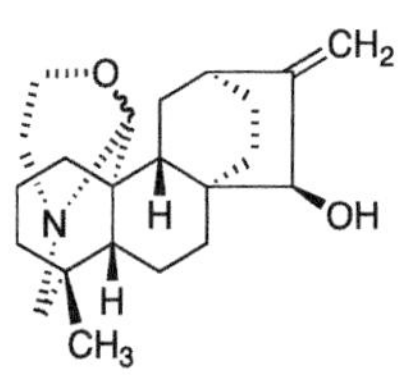

atisine

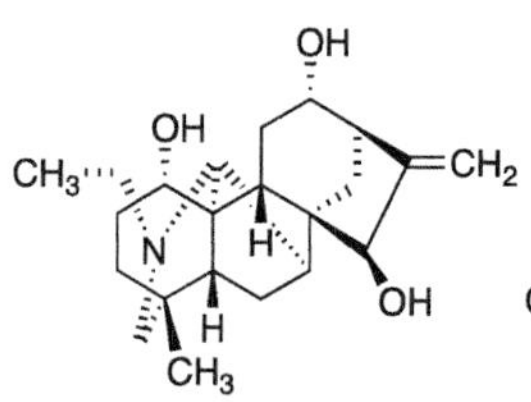

napelline

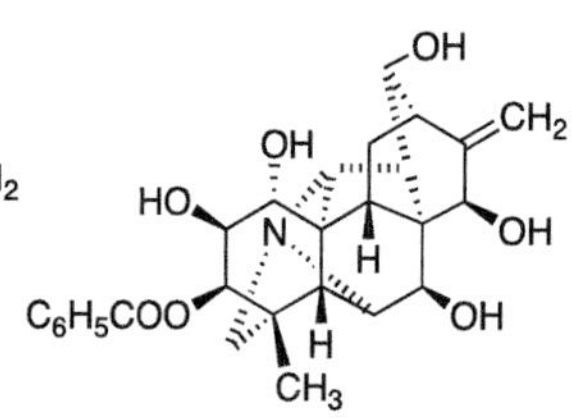

ignavine

Research studies on the aconite alkaloids started at the beginning of the nineteenth century, and aconitine was isolated from *Actinidia napellus* in 1833. As for the *Aconitum* plants of Japan, Shimoyama made the first report in 1882, and the plain chemical structure of aconitine was reported in the 1950s [4]. The stereochemistry, including absolute configuration, of the aconitine skeleton was determined when the total synthesis delphinine, an aconitine-related alkaloid, was achieved in 1972 [5–7].

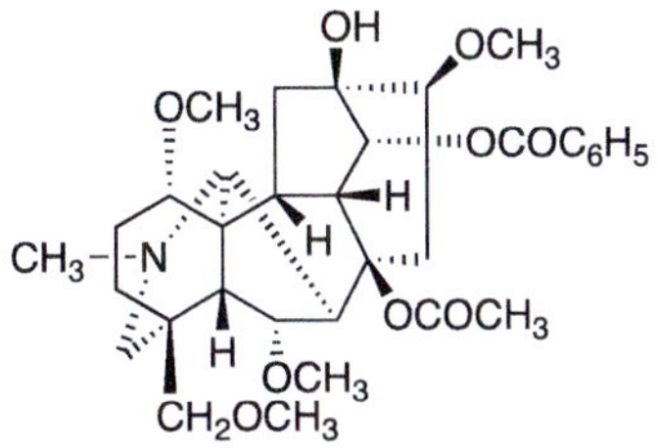

delphinine

Respiratory center paralysis, stimulation after cardiac conduction failure, paralysis of the circulatory system, and perception effects and paralysis of motor nerves are the demonstrated biological activities of the aconitine-type alkaloids in animal models. As for cardiotonic substances, other than higenamine, corynerine was also isolated; and these alkaloids were described in Chapter 1 (phenylalanine- and tyrosine-derived alkaloids).

Though no detailed reports on the biosynthesis of aconitine alkaloids are reported, it seems that these alkaloids may initially have a similar biosynthetic pathway to that of taxol (Section 14.5) and possess the diterpenoid skeleton as the basic skeleton. However, unlike the case of taxol, a nitrogen atom is incorporated into the parent nucleus. The biogenesis for the incorporation of four C_5 units into the aconitine system is shown in the Figure [8].

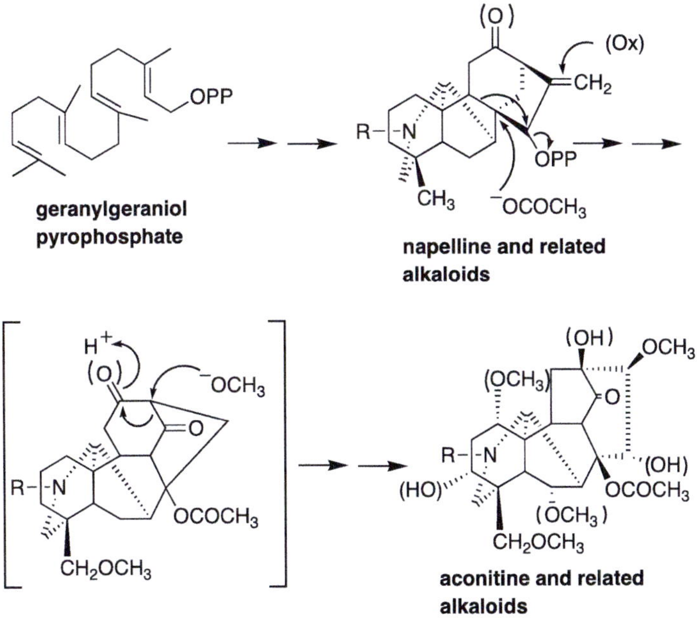

Biogenesis of the Aconitine Skeleton

LITERATURE CITED

[1] J. Bruneton, in: Pharmacognosy, Phytochemistry, Medicinal Plants, Lavoisier Pub. Inc, 1995R, p. 866.
[2] D. Dvornik, O.E. Edwards, Can. J. Chem. 42 (1964) 137.
[3] T. Okamoto, M. Natsume, Y. Iitaka, A. Yoshino, T. Amiya, Chem. Pharm. Bull. 13 (1965) 1270.
[4] E.S. Stern, Alkaloids 4 (1954) 276.
[5] Y. Tsuda, L. Marion, Can. J. Chem. 41 (1963) 1634.
[6] K.B. Birnbaum, K. Wiesner, E.W.K. Jay, L. Jay, Tetrahedron Lett. 12 (1971) 867.
[7] K. Wiesner, E.W.K. Jay, T.Y.R. Tsai, C. Demerson, L. Jay, T. Kanno, J. Krepinsky, A. Vilim, C.-S. Wu, Can. J. Chem. 50 (1972) 1925.
[8] M. Tamura, Y. Ichinohe, in: Bushi No Kenkyuu, Sanwa Shoyaku, 1981R, p. 104.

14.5 *Taxus cuspidata* AND TAXOL

Taxus brevifolia is an evergreen tree of the family Taxaceae that inhabits the American Northwest; and it is also known as "Oregon Yew" or "Pacific Yew Tree." This plant has a close botanical relationship with the *Taxus* species that grows in Japan.

Taxol is a diterpene alkaloid isolated as an anticancer agent from the bark of *T. brevifolia* [1]. Subsequently, it was reported that this alkaloid possessed clinical effects for the treatment of leukemia and had excellent effects on refractory ovarian and mammary cancer. The mechanism of action of this alkaloid is to promote the formation of microtubules, the opposite action to that of the microtubule inhibitors like podophyllotoxin, colchicine, and vinblastine [2,3].

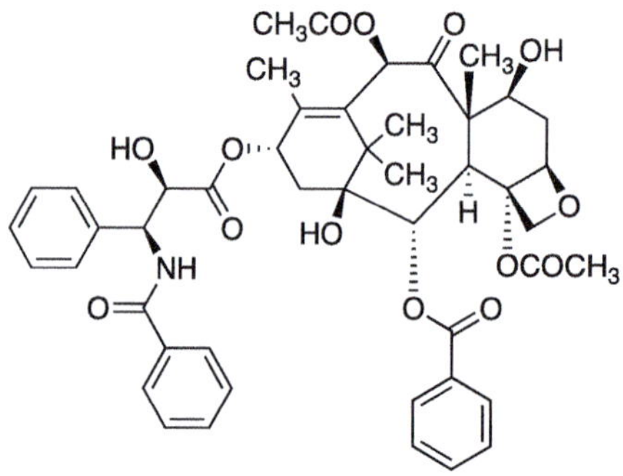

taxol

As described above, taxol is a very desirable compound, but the abundance of this alkaloid in the bark of *T. brevifolia* is only 0.01–0.03%, and thus there was a problem in obtaining an adequate supply for clinical use. The bark harvested from a single mature tree of *T. brevifolia* is about 3 kg, and therefore the taxol isolated from this material is in the range 0.3–1 g. Therefore, a semisynthesis of taxol from 10-deacetylbaccatin III, which is abundantly isolated from the leaf of *T. brevifolia* [4], is the currently preferred route. According to Denis et al., 10-deacetylbaccatin was isolated at a level of about 1 g/kg from the fresh leaves (needles) of *T. brevifolia*.

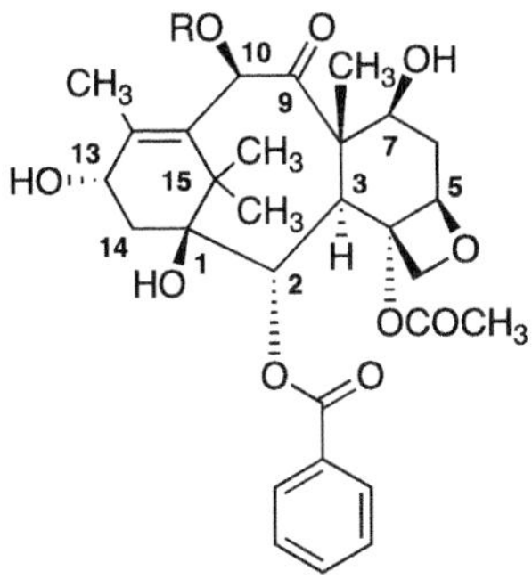

baccatin III **R = COCH$_3$**
10-deacetyl baccatin III **R = H**

In the meantime, the group studying a microorganism that is associated with *T. brevifolia* confirmed that *Taxomyces andreanae*, a kind of mold, which was isolated from the bark of *T. brevifolia*, has the capacity to produce taxol [5]. According to this report, about 200 kinds of microorganisms were isolated from 25 *T. brevifolia* trees in 20 locations, and among them taxol was detected by thin-layer chromatography (TLC) and mass spectrometry (electron ionization mass spectrometry, fast atom bombardment mass spectrometry) in the extract of *Taxomyces andreanae* cultured broth. The output of taxol from *T. andreanae* in culture has been estimated at 24–50 ng/l. Improvement on the output of taxol through manipulation of the culture medium is anticipated because in many cases the metabolite derived from the plant is required for a parasitic microorganism, and this metabolism can be overstimulated.

It is thought that the nucleus of taxol before tailoring is derived from four C$_5$ units, and thus taxol is classified as a diterpene. However, taxol is not always included in the alkaloids, because the position of the nitrogen in taxol is in an esterifying side-chain group. Thus the nitrogen atom is not directly attached to or incorporated into the diterpene nucleus. Taxol is therefore a special example of a diterpene alkaloid included in this chapter.

LITERATURE CITED

[1] M.C. Wani, H.L. Taylor, M.E. Wall, P. Coggon, A.T. McPhail, J. Am. Chem. Soc. 93 (1971) 2325.
[2] P.B. Schiff, J. Fant, S.B. Horwitz, Nature 277 (1979) 665.
[3] S. Rao, N.E. Krauss, J.M. Heerding, C.S. Swindell, I. Ringel, G.A. Orr, S.B. Horwitz, J. Biol. Chem. 269 (1994) 3132.
[4] J.-N. Denis, A.E. Greene, D. Guénard, F. Guéritte-Voegelein, L. Mangatal, P. Potier, J. Am. Chem. Soc. 110 (1988) 5917.
[5] A. Stiele, G. Strobel, D. Stierle, Science 260 (1993) 214.

14.6 *Pachysandra terminalis* AND BUXACEAE ALKALOIDS

The steroids are biosynthetically derived from a triterpenoid nucleus through degradation. Compared with the triterpenoids, these compounds lack the dimethyl moiety at C-4 and a methyl group at the C-14 position. To show the difference between these chemical structures, lanosterol, a representative triterpene, and cholesterol, a representative steroid, are shown [1].

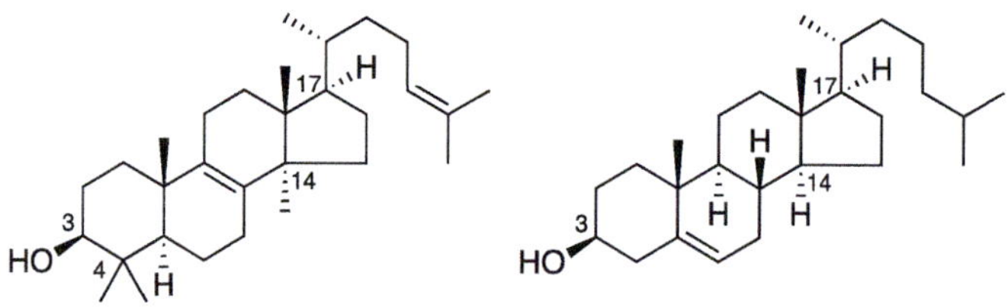

lanosterol **cholesterol**

There are many steroidal compounds that show drug efficacy. Among them are sex hormones, such as estrone, androsterone, and progesterone; adrenocortical hormones, such as cortisone; vitamin D_2, known as the antirachitic vitamin; digitoxin, a cardiotonic glycoside isolated from digitalis (*Digitalis purpurea*) of the family Scrophulariaceae; and bufalin, a cardiotonic steroid isolated from "Senso" (so-called in Kampo medicine) and prepared from the secretion (venom) of *Bufo bufo gargarizans* (Chinese toad) and *Bufo vulgaris formosus* (Japanese toad).

Compounds described in the following sections are the alkaloids derived from steroids and related compounds in which the nitrogen atom(s) have been incorporated into the molecule. The five groups of alkaloids described in this chapter are those of the Buxaceae (14.6), those of the Apocynaceae

(14.7), those derived from the Solanaceae (14.8), the Liliaceae alkaloids (14.9), and those derived from animals (14.10).

The basic skeleton of conessine of the Apocynaceae alkaloids, described in Section (14.7) is very close to that of the Buxaceae alkaloids described in this section.

Pachysandra terminalis (Buxaceae) is a perennial plant that grows in the shade in various locations in Japan and other temperate zones. This plant contains about 0.7% alkaloids in the dried above-ground parts; the main alkaloids are pachysandrine A and pachystermine A [2]. These compounds possess the pregnane skeleton, which has a C_2 unit remaining at the C-17 position, with two nitrogen atoms incorporated at the C_2 side-chain and at C-3.

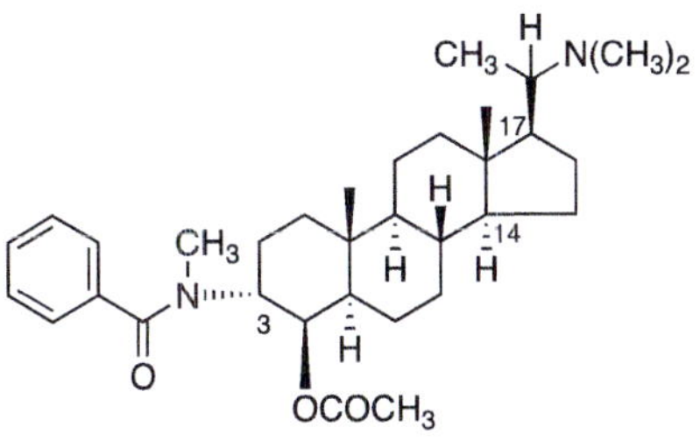

pachysndrine A

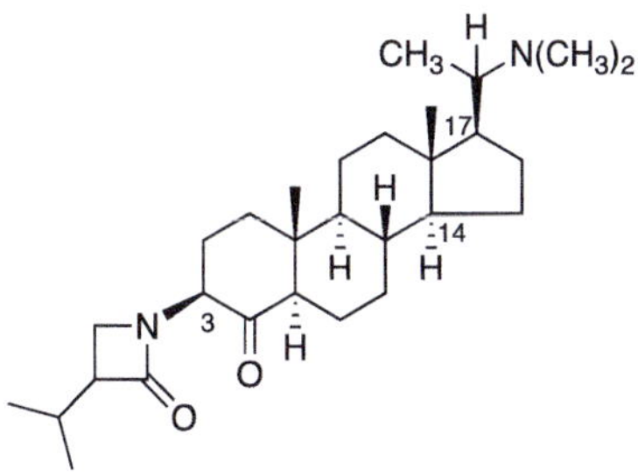

pachystermine A

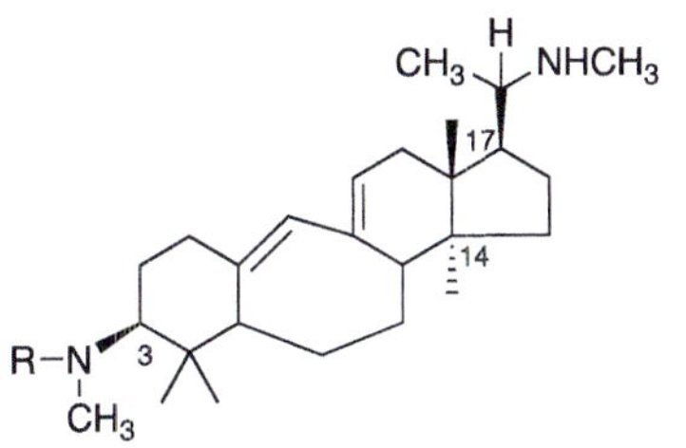

papilamine **R = H**
papiicine **R = CH_3**

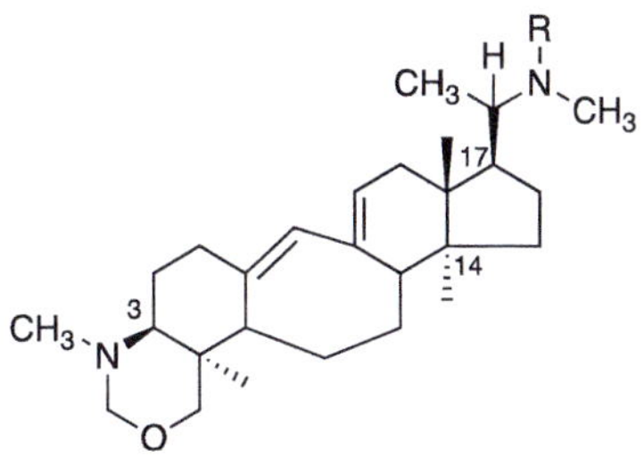

harappamine **R = H**
moenjodaramine **R = CH_3**

Papilamine, papilicine, harappamine, and moenjodaramine were isolated from the leaves of *Buxus papillosa* (Buxaceae). These alkaloids retain the dimethyl moiety at the C-4 position and the methyl group at C-14, as in triterpenes [3]. These alkaloids possess an expanded seven-membered B ring, and the nitrogen atoms are attached similarly to the *Pachysandra* alkaloids. Alkaloids derived from Buxaceae are sometimes known as *Buxus* alkaloids [4].

LITERATURE CITED

[1] (translated by M. Shiota) J.R. Hanson, Introduction to Steroid Chemistry, Kyoritu Shuppan, Tokyo, 1970R.
[2] M. Tomita, T. Kikuchi, S. Uyeo, T. Nishinaga, M. Yasunishi, A. Yamamoto, Yakugaku Zasshi 87 (1967) 215.
[3] M. Atta-ur-Rahman, S. Nisa, T. Farhi, Z. Ismail, Heterocycles 21 (1984) 489.
[4] D.M. Harrison, Nat. Product Rep. 3 (1986R) 443.

14.7 *Holarrhena antidysenterica* AND CONESSINE

Holarrhena antidysenterica (Apocynaceae) is a shrub that grows in India, and the bark of this plant has been used for the treatment of amoebic dysentery. The main component is conessine, a steroidal alkaloid. This drug has an advantage compared with emetine, which possesses activity to treat amoebic dysentery, in that conessine does not produce vomiting as a side effect. Conessine was isolated as wrightine toward the end of the nineteenth century [1], but it was more than 70 years before the plane structure of the alkaloid was determined [2]. The total synthesis of the racemate was completed soon after [3]. This alkaloid contains a pregnane skeleton with a side chain at the C17 position and a nitrogen atom at the C3 position, similar to the alkaloids of *P. terminalis*, described in the preceding section.

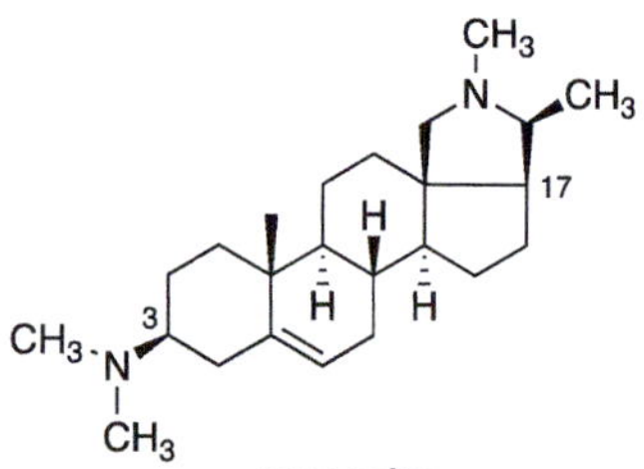

conessine

LITERATURE CITED

[1] H. Warnecke, Chem. Ber 19 (1886) 60.
[2] A. Bertho, M. Goetz, Liebigs Ann. der Chem. 619 (1958) 96.
[3] J.A. Marshall, W.S. Johnson, J. Am. Chem. Soc. 84 (1962) 1485.

14.8 POTATO AND SOLANINE

The toxic alkaloid solanine, having a steroidal skeleton, is present in the sprouts of potato (*Solanum tuberosum*) of the Solanaceae [1]. Subsequently, it was determined that the solanine is a mixture that could be separated into six constituents, namely α-, β-, and γ-solanine and α-, β-, and γ-chaconine [2]. These six alkaloids possess solanidine as a common aglycone, and the differences exist in the sugar moiety. The structure of α-solanine is shown, together with the other five alkaloids. The structures of β- and γ-solanine lack one and two monosaccharide units from that of α-solanine, and the relationship between the structures α-, β-, and γ-chaconine is similar.

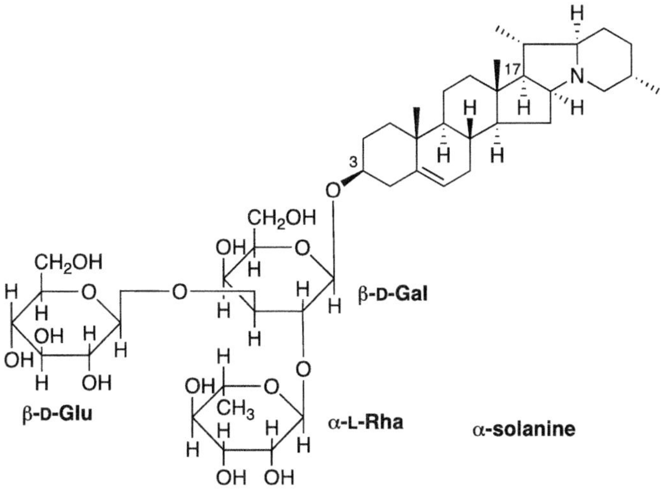

alkaloids	sugar moiety
α-solanine	β-D-Glu(1→3)-β-D-Gal(1→) (1→2)↑ α-L-Rha
β-solanine	β-D-Glu(1→3)-β-D-Gal(1→)
γ-solanine	β-D-Gal(1→)
α-chaconine	α-L-Rha(1→4)-β-D-Glu(1→) (1→2)↑ α-L-Rha
β-chaconine	α-L-Rha(1→4)-β-D-Glu(1→)
γ-chaconine	β-D-Glu(1→)

The steroidal alkaloid solasonine is isolated from the fruit of *Solanum aviculare* and *Solanum sodomeum* [3,4]. The sugar moiety of this alkaloid is equivalent to that of α-solanine, whereas the aglycone is different [5]. The aglycone of solasonine is known as solasodine and possesses an oxo-azaspirodecane structure.

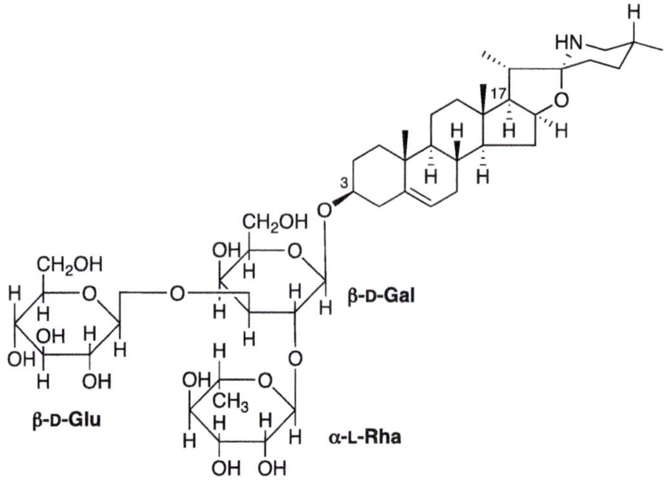

solasonine

Steroidal alkaloids of the Solanaceae can therefore be divided into two categories based on the aglycone; the solanidine- and the solasodine-type alkaloids. The basic skeleton of the former alkaloids is known as solanidane, and that of the latter alkaloids is known as a spirosolane skeleton.

LITERATURE CITED

[1] A. Soltys, K. Wallenfels, Chem. Ber 69 (1936) 811.
[2] R. Kuhn, I. Low, H. Trischmann, Chem. Ber 88 (1955) 1492.
[3] R.C. Bell, L.H. Briggs, J. Chem. Soc. 1942 (1942) 1.
[4] L.H. Briggs, R.P. Newbold, N.E. Stace, J. Chem. Soc. 1942 (1942) 3.
[5] L.H. Briggs, R.C. Cambie, J.L. Hoare, J. Chem. Soc. 1963 (1963) 2848.

14.9 *Fritillaria* AND *Veratrum* ALKALOIDS

Fritillaria verticillata var. *thunbergii* (Liliaceae) is a perennial herb that grows wild in the People's Republic of China. The plant is cultivated in Japan mainly as an ornamental. The dried bulbs are known as "baimo" in Kampo medicine and are used as an antitussive, as an expectorant, for hemostasis and lactogenesis, and as an antipyretic.

Among the alkaloids of *Fritillaria* bulbs, verticine (peimine), peiminoside, and fritillarine (peiminine) have been reported.

Verticine was isolated initially from the bulbs of *F. verticillata* var. *thunbergii* [1,2] and subsequently from *F. roylei* [3–5]. Peimine, another name for verticine, was derived from "pei-mu," the Chinese pronunciation of "baimo."

Peiminoside is the 3-β-D-glucoside derivative of verticine (peimine) [11]. The molecular formula of verticine was clarified in 1944 at the same time as fritillarine, described subsequently. However, it was only in the 1960s that the structures and absolute configurations of verticine and fritillarine were determined [7–9]. The total synthesis of these alkaloids has also been achieved [10].

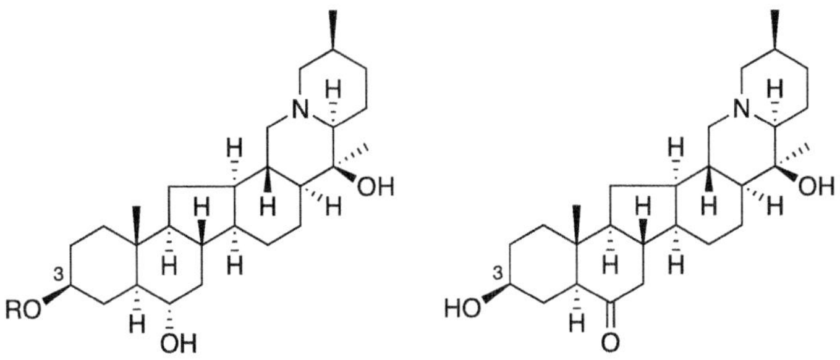

verticine (peimine) **R = H**
peiminoside **R = gluc.**

fritillarine (peiminine)

Fritillarine was first isolated from the bulbs of *F. verticillata* var. *thunbergii* [2], and later from the bulbs of *F. roylei*, under the name peiminine [3–5]. Fritillarine possesses a carbonyl moiety at the C-5 position, where a hydroxy group exists in verticine. Fritillarine can be obtained by oxidation of verticine [12,13].

In pharmacological experiments, remarkable antitussive action was shown for verticine and fritillarine, and hypotension and respiratory movement center paralytic actions were reported for peiminoside. It was reported that the alkaloid fraction showed atropine-like action, including a bronchial smooth-muscle relaxation effect and a mydriatic effect on the pupils [11,14].

The skeleton of verticine, peiminoside, and fritillarine is a C-nor-D-homosteroid, which possesses a five-membered C-ring and a six-membered D-ring. This skeleton is also common to the *Veratrum* alkaloids of the Liliaceae plants.

Veratrum album var. *grandiflorum* (*Veratrum grandiflorum*) and *Veratrum stamineum* (Liliaceae) are perennial herbs that grow wild in the alpine region in the northern center of Japan. It is considered that the rhizome of these plants possesses strong toxicity, and it causes vomiting by stimulating the mucous membrane. *Veratrum japonicum* is a perennial herb observed in various highlands in Japan, and this plant also has similar toxicity. In Japan, almost every year people eat these toxic plants and become poisoned through misunderstanding that they are wild vegetables.

V. album grows from northern Asia to the middle of southern Europe, and the rhizome is used for emetic purposes, as a veterinary medicine, and as an insecticide. The rhizome of *Viride viride* was formerly used as a hypotensive medicine, but it has pronounced side effects, including vomiting. Many alkaloids with various modified steroid nuclei have been isolated from *Veratrum* sp. plants, and these alkaloids are generically known as *Veratrum* alkaloids.

Some *Veratrum* alkaloids were obtained from plants not in the genus *Veratrum*. Zygadenine, isolated from the leaves of *Zygadenus intermedius* (Liliaceae), is an example [15]. Though this alkaloid was obtained at the beginning of the twentieth century, it took almost 50 years to clarify the structure, including its absolute configuration [16,17]. Zygacine is an alkaloid also isolated from the leaves of *Z. intermedius*, where the C-3 position of zygadenine is acetylated [16]. Zygacine was also isolated from the rhizome of *V. grandiflorum* [18].

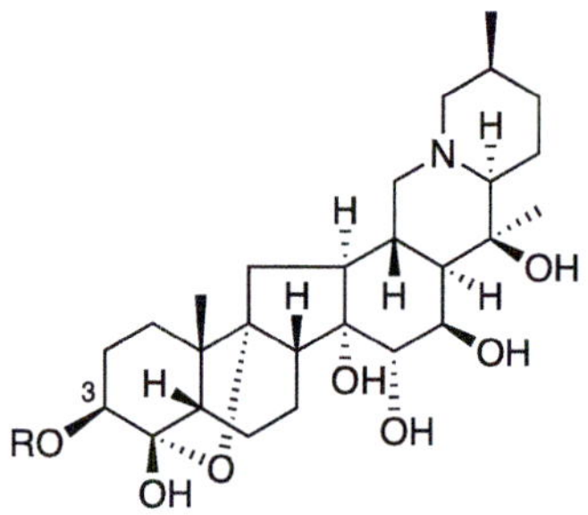

zygacine **R = $COCH_3$**
zygadenine **R = H**

LITERATURE CITED

[1] M. Fukuda, Sci. Reports Tohoku Imperial Univ. Ser. A 18 (1929) 323.
[2] M. Fukuda, Nihon Kagaku Kaishi 50 (1929) 74.
[3] T.-Q. Chou, K.-K. Chen, Chin. J. Physiol. 6 (1932) 265 [Chem. Abstr., 26, 5703(1932)].
[4] Y.-F. Chi, Y.-S. Kao, K.-J. Chang, J. Am. Chem. Soc. 62 (1940) 2896.
[5] T.-Q. Chou, T.-T. Chu, J. Am. Chem. Soc. 63 (1941) 2936.
[6] Y.-H. Wu, J. Am. Chem. Soc. 66 (1944) 1778.
[7] H. Morimoto, S. Kimata, Chem. Pharm. Bull. 8 (1960) 871.
[8] S. Ito, M. Kato, K. Shibata, T. Nozoe, Chem. Pharm. Bull. 11 (1963) 1337.
[9] S. Ito, Y. Fukazawa, T. Okuda, Tetrahedron Lett. 9 (1968) 5373.
[10] J.P. Kutney, C.C. Fortes, T. Honda, Y. Murakami, A. Preston, Y. Ueda, J. Am. Chem. Soc. 99 (1977) 964.
[11] H. Morimoto, S. Kimata, Chem. Pharm. Bull. 8 (1960) 302.
[12] T.-T. Chu, T.-Q. Chou, J. Am. Chem. Soc. 69 (1947) 1257.

[13] M. Fukuda, Nihon Kagaku Kaishi 69 (1948) 165.
[14] K. Kimura, T. Kimura, Genshoku Nihon Yakuyo Shokubutsu Zukan, Hoikusha, 1981R, p. 239.
[15] F.W. Heyl, F.E. Hepner, S.K. Loy, J. Am. Chem. Soc. 35 (1913) 258.
[16] S.M. Kupchan, J. Am. Chem. Soc. 78 (1956) 3546.
[17] S.M. Kupchan, J. Am. Chem. Soc. 81 (1959) 1925.
[18] B. Shimizu, R. Hayashi, Yakugaku Zasshi 79 (1959) 615.

14.10 ARROW TOXINS, TOXIC BIRDS, AND BATRACHOTOXIN

Phyllobates aurotaenia inhabits South America, where it is known as "kokoi." A toxic material is secreted from the skin, and the secretion is utilized as an arrow poison. The toxic material was studied at the National Institutes of Health (U.S.A.), and batrachotoxin was reported as the main toxic component in 1969 [1]. Although there are no reports on the biosynthesis of this compound, it appears that in this alkaloid the nitrogen atom was incorporated into a steroid skeleton. A nitrogen atom has also been introduced into the side-chain moiety in the form of a pyrrole-3-carboxylic acid derivative. The LD_{50} value of batrachotoxin is 2 μg/kg (mouse/subcutaneous injection).

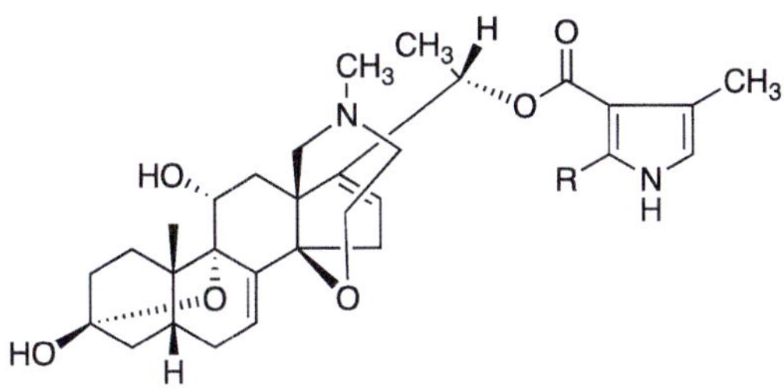

batrachotoxin **R = CH_3**
homobatrachotoxin **R = CH_2CH_3**

It has been known for some time that there are also birds that possess toxic substance(s) in their feathers, skins, muscles, etc. among the birds that inhabit Papua New Guinea. There are three kinds of birds classified as *Pitohui* sp., i.e., "hooded pitohui" (*Pitohui dichrous*), "variable pitohui" (*Pitohui kirhocephalus*), and "rusty pitohui" (*Pitohui ferrugineus*). Among these birds, the hooded pitohui is also known as the "rubbish bird," and it was said that these birds cannot be eaten unless the skin is removed and the meat cooked carefully.

The toxic component of these birds was investigated using gas chromatography–mass spectrometry and TLC by monitoring the toxicity for mice [2]. As a result, the toxic substance was determined to be homobatrachotoxin, which had been reported as one of the toxic principles of *P. aurotaenia* [1,2]. The homobatrachotoxin content in a hooded pitohui (65 g body weight) in the skin and in the feathers was 15–20 μg and 2–3 μg, respectively. These contents are far less than that in the skin of an arrow toxin frog. However, this was the first example of the isolation of a toxic component from birds. The LD_{50} value for the mouse of homobatrachotoxin was reported to be 3 μg/kg (mice/subcutaneous injection).

The word "Chin-Doku" has been used in Japan from olden times, and it was said to be a poison present in the feathers of a toxic bird, "Chin." "Doku" means "poison" in Japanese. Such a description is also shown in "Honzoh-Kohmoku," the famous old Chinese book concerning drugs (published in 1596), indicating that there is a bird known as "Chin." According to this book, the birds inhabit the southern area of China, and they become toxic by eating poisonous snakes in order to accumulate the poison in their bodies [3].

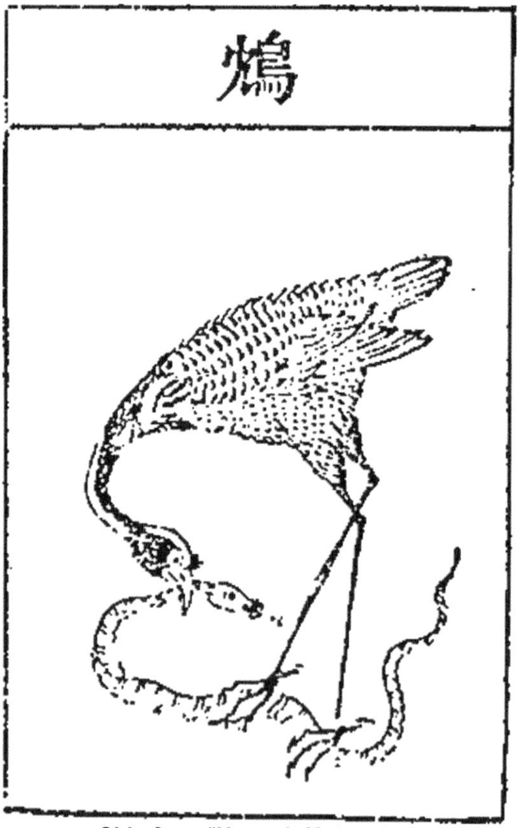

Chin from "Honzoh-Kohmoku"

Until recently, Chin-Doku derived from birds was regarded as a legendary tale. However, although there is no evidence that the poison birds found recently are the Chin itself, and although the large snake pile eagle is described in the attached map of Honzo-Komoku as Chin, the appearance is completely different from the comparatively small poisonous bird of New Guinea. However, it cannot be dismissed as mere legend that there exist poisonous birds [4].

In the meantime, it was shown that the arrow poison frogs are not affected by the above-mentioned poison, since the sodium (Na^+) channel of these frogs does not react to batrachotoxin. However, how the poison birds of the *Pitohui* sp. are protected from their poison remains unsolved. Batrachotoxin has not been detected from these poisonous birds, and the biosynthetic origin of homobatrachotoxin is still unknown [2].

A component that attacks the central nervous system and causes strong convulsions is present in the secretions of the dermal gland of *Salamandra maculosa* (Salamandridae), a type of salamander that inhabits Europe. The main component of the secretion, samandarine, is a steroidal alkaloid. Samandarine comprises 75% of the total alkaloids and it was isolated together with samandarone [5].

According to structure elucidation studies, it was shown that the A ring of the steroid was cleaved, and a nitrogen atom was incorporated. The *cis* configuration between the A and B rings is characteristic [6], and this stereochemistry was also shown in the structure of the batrachotoxins. Samandarone possesses a structure in which the alcohol of samandarine was oxidized to a ketone. The total synthesis of samandarone has been achieved [7].

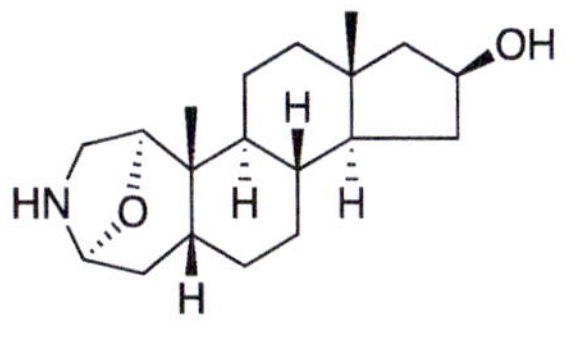

samandarine

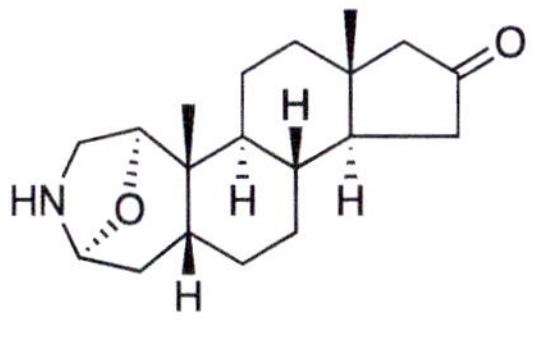

samandarone

LITERATURE CITED

[1] T. Tokuyama, J. Daly, B. Witkop, J. Am. Chem. Soc. 91 (1969) 3931.
[2] J.P. Dumbacker, B.M. Beehler, T.F. Spande, H.M. Garraffo, J.W. Daly, Science 258 (1992) 799.
[3] J.-C. Li, Honzoh-Kohmoku 49 (1596).
[4] S. Funayama, Farumashia 29 (1993R) 1144.
[5] C. Schöpf, W. Braun, Liebigs Ann. der Chem. 514 (1934) 69.
[6] E. Wölfel, C. Schöpf, G. Weitz, G. Habermehl, Chem. Ber 94 (1961) 2361.
[7] S. Hara, K. Oka, J. Am. Chem. Soc. 89 (1967) 1041.

CHAPTER 15

Alkaloids Derived from Polyketides

Conium maculatum (Apiaceae)

A number of alkaloids are biosynthesized without incorporating an amino acid into the basic skeleton. Some examples are those alkaloids derived from a terpenoid nucleus, as described in the previous chapter; coniine, isolated from *Conium maculatum*; and nigrifactin and the piericidins produced by Actinomycetes, which are described in this chapter. The skeleta of these alkaloids are formed through the polyketide pathway, and nitrogen is incorporated subsequently.

$CH_3CO\text{-}SCoA$	$HOOCCH_2CO\text{-}SCoA$	$CH_3CH_2CO\text{-}SCoA$
acetyl CoA	**malonyl CoA**	**propionyl CoA**

15.1 HEMLOCK AND CONIINE

The toxic plant *C. maculatum* (Apiaceae, formerly Umbelliferae) is also known as "hemlock," and the seeds possess strong toxicity. It is for this reason that this plant was formerly used for the execution of criminals (mainly political offenders of the present day) in ancient Greek times. There is a famous story that Socrates (470–399 BC) was executed by an extract of this plant in 399 BC.

Alkaloids
http://dx.doi.org/10.1016/B978-0-12-417302-6.00015-5

(+)-coniine **R = H** **γ-coniceine**
***N*-methylconiine** **R = CH$_3$**

Coniine, the main toxic principle of this plant, was isolated in 1827, and the molecular formula ($C_8H_{17}N$, mw 127) was presented in 1886 [1]. Other than coniine, γ-coniceine and *N*-methylconiine were also isolated from this plant [2].

Coniine possesses a similar chemical structure to that of pelletierine (Section 4.2), isolated from the stem bark and root bark of *Punica granatum*. Thus Robinson [3] estimated coniine to be derived biosynthetically from a lysine moiety and a C_4 unit as in the case of pelletierine, and as shown in the following figure. However, when [2-^{14}C]lysine or its metabolites, [1,5-$^{14}C_2$]cadaverine or [6-^{14}C]-Δ^1-piperideine, were fed to hemlock, none of the alkaloids isolated from this plant were labeled.

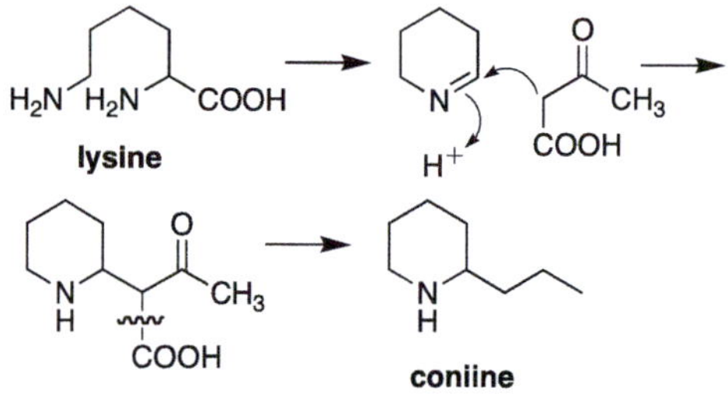

Robinson's Hypothesis for the Biosynthesis of Coniine

On the other hand, when [1-^{14}C]acetate was fed to this plant, labeled coniine was obtained. In addition, through degradation studies, it was found that the even-numbered carbons of coniine were labeled [4]. Thus, through the experiments described above, it was clarified that although the skeleton of coniine is very similar to that of pelletierine, coniine is biosynthesized through the polyketide pathway instead of from lysine, as in the case of pelletierine. These biosynthetic studies on coniine and related alkaloids were reviewed by Leete [5].

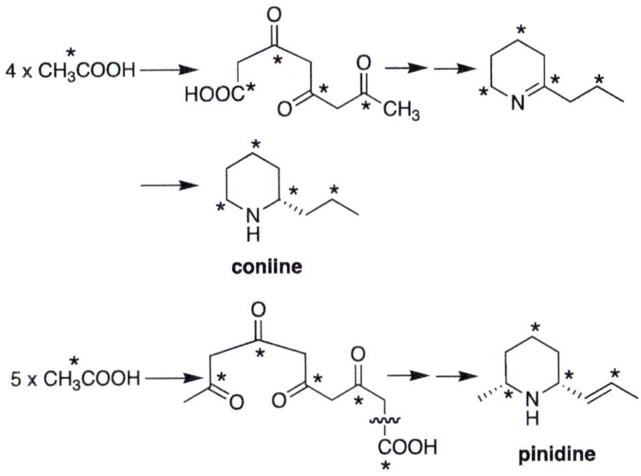

Biogenesis of Coniine and Pinidine

A further example is pinidine, isolated from *Pinus sabiniana* and *Pinus jeffreyi* (Pinaceae), which is derived through the same pathway as coniine [6–8]. It was demonstrated that pinidine was formed in *P. jeffreyi* through the polyketide pathway by feeding experiments with labeled precursors [9].

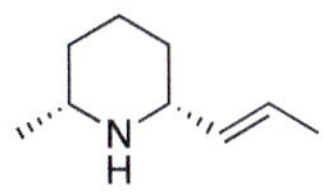

pinidine

LITERATURE CITED

[1] A. Ladenburg, Chem. Ber. 19 (1886) 439.
[2] J.W. Fairbairn, A.A.E.R. Ali, Phytochemistry 7 (1968) 1593.
[3] R. Robinson, The Structural Relations of Natural Products, Clarendon Press, Oxford, 1955, p. 64.
[4] E. Leete, J. Am. Chem. Soc. 86 (1964) 2509.
[5] E. Leete, Acc. Chem. Res. 4 (1971R) 100.
[6] W.H. Tallent, V.L. Stromberg, E.C. Horning, J. Am. Chem. Soc. 77 (1955) 6361.
[7] W.H. Tallent, E.C. Horning, J. Am. Chem. Soc. 78 (1956) 4467.
[8] R.K. Hill, T.H. Chan, J.A. Joule, Tetrahedron 21 (1965) 147.
[9] E. Leete, K.N. Juneau, J. Am. Chem. Soc. 91 (1969) 5614.

15.2 NIGRIFACTIN AND PIERICIDINS

Nigrifactin is a product of *Streptomyces* sp. No. FFD-101, and it possesses antihistaminic effects [1]. The chemical structure of this alkaloid was determined mainly through spectral interpretation [2] and confirmed by chemical synthesis [3]. Through structure elucidation studies, it was clarified that nigrifactin possessed a similar partial structure to those of coniine and pelletierine, described previously.

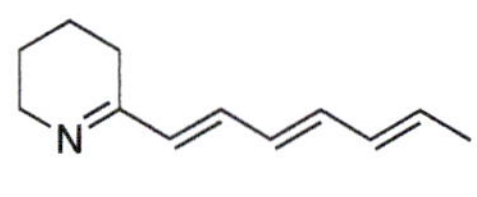

nigrifactin

Feeding experiments using both radioactive and stable isotopes were performed to study the biosynthetic pathway. It was found that in the formation of nigrifactin, [6-^{14}C]DL-lysine was not incorporated, whereas the label from [1-^{14}C]acetic acid was incorporated at the C-6, C-4, C-2, C-2′, C-4′, and C-6′ positions in the labeled nigrifactin.

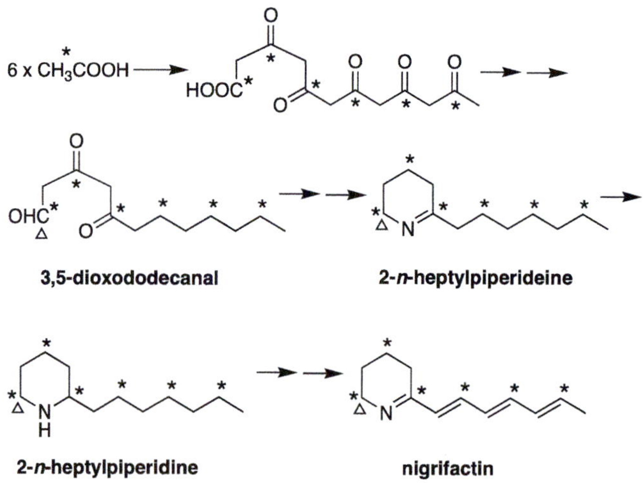

Biosynthesis of Nigrifactin

Thus, it was concluded that nigrifactin is composed of six malonyl CoA units. One nitrogen atom was separately incorporated [4]. The origin of the carbon framework was confirmed by the analysis of the ^{13}C nuclear magnetic resonance spectrum of labeled nigrifactin after feeding [1-^{13}C]acetic acid. In addition, because it was found that [1-^{14}C]-3,5-dioxo-*n*-dodecanol, [6-^{14}C]-2-*n*-heptylpiperideine, and [6-^{14}C]-2-*n*-heptylpiperidine were incorporated into nigrifactin, respectively, the biosynthetic route of nigrifactin was concluded to be as shown in the figure. Also, it became apparent that the double bonds in nigrifactin were formed in the last stages of the biosynthetic pathway because [6-^{14}C]-2-*n*-heptylpiperidine is incorporated into nigrifactin.

Many metabolites of the actinomycetes are biosynthesized through the polyketide biosynthetic pathway. For example, it was clarified that the piericidins, isolated from the fermentation broth of *Streptomyces pactum* and possessing a pyridine nucleus, are produced through the polyketide pathway, as is nigrifactin. The biosynthetic pathway was clarified by stable isotope feeding experiments [5]. Piericidin A_1, the first piericidin derivative, was obtained as an insecticide [6] and the chemical structure, including the absolute configuration, was reported [7]. Other piericidin derivatives, piericidins A_2–A_4, B_1–B_4, C_1–C_4, and D_1–D_4, were also isolated [8].

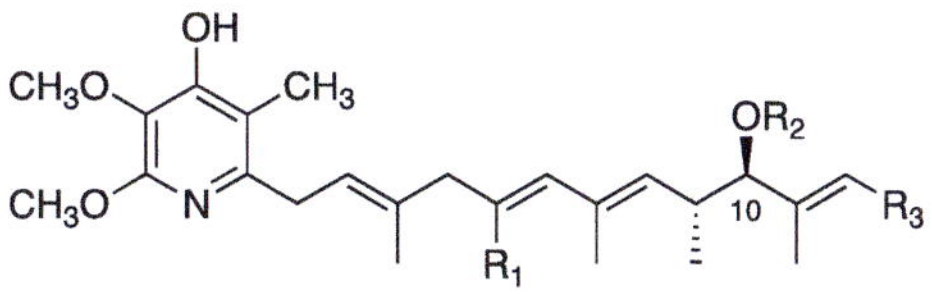

piericidin A_1 — **$R_1 = R_2 = H$, $R_3 = CH_3$**
piericidin A_2 — **$R_1 = R_3 = CH_3$, $R_2 = H$**
piericidin A_3 — **$R_1 = R_2 = H$, $R_3 =$ *i*-Pr**
piericidin A_4 — **$R_1 = CH_3$, $R_2 = H$, $R_3 =$ *i*-Pr**
13-hydroxyglucopiericidin A — **$R_1 = H$, $R_2 =$ gluc., $R_3 = CH_2OH$**

13-Hydroxyglucopiericidin A [9], which possesses a terminal OH moiety and a glucose moiety at the C-10 position, was isolated from the fermentation broth of *Streptomyces* OM-5689 as an alkaloid cytotoxic to B16 melanoma cells.

LITERATURE CITED

[1] Y. Kaneko, T. Terashima, Y. Kuroda, Agric. Biol. Chem. 32 (1968) 783.
[2] T. Terashima, Y. Kuroda, Y. Kaneko, Tetrahedron Lett. 10 (1969) 2535.
[3] H.W. Gshwend, Tetrahedron Lett. 11 (1970) 2711.
[4] T. Terashima, E. Idaka, Y. Kishi, T. Goto, Chem. Commun. 75 (1973).
[5] M. Tanabe, H. Seto, J. Org. Chem. 35 (1970) 2087.
[6] N. Takahashi, A. Suzuki, S. Tamura, Agric. Biol. Chem. 27 (1963) 798.
[7] R. Jansen, G. Höfle, Tetrahedron Lett. 24 (1983) 5485.
[8] S. Yoshida, K. Yoneyama, S. Shiraishi, A. Watanabe, N. Takahashi, Agric. Biol. Chem. 41 (1977) 855.
[9] H. Mori, S. Funayama, Y. Sudo, K. Komiyama, S. Omura, J. Antibiot. 43 (1990) 1329.

CHAPTER 16

Alkaloids Derived from a C_6–C_1 Unit

Ephedra sinica (Ephedraceae)

Ephedrine, isolated from *Ephedra* sp. (Ephedraceae), possesses a C_6–C_2–N skeleton. Therefore, it was initially considered that this alkaloid would be derived from phenylalanine. However, it was established that part of the carbon framework of this alkaloid was derived from a C_6–C_1 unit, and that the nitrogen atom was not provided by retaining the C–N bond of an amino acid.

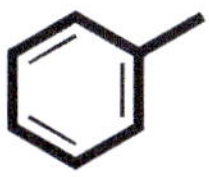

C_6-C_1 unit

In this chapter, ephedrine and related alkaloids, the khat alkaloids (which are related to the ephedrines), and two other types of alkaloids are described. One of them is the naphthalene-isoquinoline alkaloids, which are described in Section 16.3. These alkaloids might also be classified as alkaloids derived from phenylalanine because they possess an isoquinoline skeleton. However, an extra C_1 unit has been joined to the carbon to which the nitrogen is attached, as occurs in the case of ephedrine. Therefore, it is considered that the formation of these alkaloids might involve a biosynthetic pathway not utilizing phenylalanine. The other alkaloids are capsaicin and related alkaloids, which are described in Section 16.4.

Alkaloids
http://dx.doi.org/10.1016/B978-0-12-417302-6.00016-7

As one of the possible biosynthetic pathways of this alkaloid type, a route that includes the C_6–C_1 unit as a biosynthetic precursor of the chromophore was considered. Other possibilities, such that the isoquinoline partial structure would be derived from phenylalanine or a polyketide, etc., are possible, and the partial structure, which includes the isoquinoline nitrogen, might be derived from an amino acid.

On the other hand, it was found that phenylalanine and tyrosine were incorporated into the capsaicins, where it was also clarified that the C–N bond of these amino acids was broken when they were incorporated. Consequently, these alkaloids are not described in Chapter 2, and are described in this chapter (Section 16.4). Through labeling experiments, it was established that a C_6–C_1 unit was comparatively efficiently incorporated into these alkaloids.

16.1 *EPHEDRA* ALKALOIDS

The Chinese herbal drug "Mao" (Japanese name; Chinese name is Ma-huang) is prepared from the above-ground parts of *Ephedra* sp. plants, such as *Ephedra equisetina*, *Ephedra distachya*, and *Ephedra sinica* (Ephedraceae) which are native to China. Mao is used in Chinese traditional medicine prescriptions, and as a material for the preparation of ephedrine hydrochloride used for the treatment of cough.

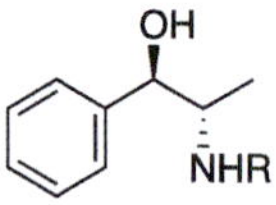

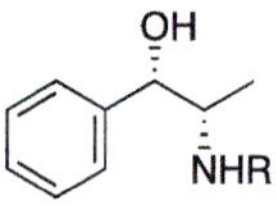

(-)-ephedrine R = CH_3
(-)-norephedrine R = H

(+)-pseudoephedrine R = CH_3
(+)-norpseudoephedrine R = H

In Chinese traditional medicine, this crude drug has been used from ancient times as a cough cure, and as an antipyretic and diaphoretic. On the other hand, the subterranean parts of *Ephedra* plants are used to stop sweating, as this medicine possesses a contradictory action to the above-ground parts of the same plant.

From the ephedrae herba, ephedrine and related alkaloids were isolated as the main components. These alkaloids include (−)-ephedrine and (+)-pseudoephedrine, (−)-norephedrine and (+)-norpseudoephedrine. Among these alkaloids, the *pseudo*-type alkaloids are the diastereomers at the benzylic position of the other alkaloid, respectively.

Research on the constituents of the ephedrae herba was advanced in the Meiji era (1868–1912) by Mototada Yamashina, an assistant engineer of the Tokyo sanitation laboratory, and a crystalline component isolated from this plant material was reported in 1885 by Nagayoshi Nagai (1845–1929), the founder of the Pharmaceutical Society of Japan. However, Yamashina died suddenly, unfortunately. Thus, the first report of the existence of ephedrine was at the Meeting of the Pharmaceutical Society of Japan on July 17, 1885. The isolation of ephedrine as a crystalline form was achieved by Yuzo Hori, of the same laboratory as Yamashina, in 1887 [1]. However, it was not until 1892 that the first report of ephedrine appeared in the literature [2–5].

It was discovered subsequently that this alkaloid was effective for bronchial asthma [6,7]. Ephedrine is a sympathomimetic, and the action is essentially similar to that of adrenaline (epinephrine; Chapter 1) [2]. However, ephedrine differs from adrenaline mainly in its efficacy after oral administration, its much longer duration of action, its more pronounced central actions, and its much lower potency [8,9].

The chemical structure of ephedrine is similar to that of mescaline, a hallucinogen, and it was thought initially that this alkaloid also originated from phenylalanine, as in the case of mescaline. However, when [3-^{14}C] phenylalanine was fed to *E. distachya*, it was established that this unit was incorporated into the aromatic ring of the C_6-C_2-N moiety of (−)-ephedrine [10].

Phenylalanine is not directly introduced into ephedrine in this case, since when [2-^{14}C] phenylalanine was fed, the labeled carbon was not incorporated into ephedrine. Similarly, when [2,3-$^{14}C_2$] phenylalanine was fed, only the C-3 carbon was detected in the resulting ephedrine, and the C-2 carbon was lost. In addition, it was demonstrated that C_6-C_1-type compounds, such as benzoic acid and benzaldehyde, are more efficiently incorporated into ephedrine than phenylalanine [11,12].

From the results described above, it was considered that the origin of the C_6-C_1 moiety and the C_2 + N unit of ephedrine are separate, and that phenylalanine is not introduced intact into ephedrine.

In the meantime, [2,3-$^{13}C_2$] pyruvic acid was fed to the stems of *E. gerardiana* for 5 days, and after cultivation for a further 2 days the aboveground parts were harvested. Ephedrine and related alkaloids were extracted, and it was determined by ^{13}C nuclear magnetic resonance that the proportions of ephedrine, norephedrine, pseudoephedrine, and norpseudoephedrine were 35:3:52:10, respectively.

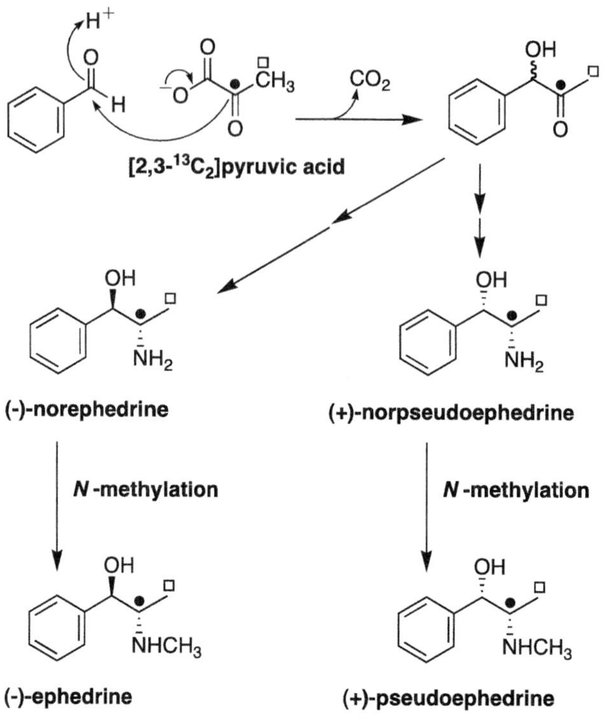

Biosynthetic Route to (-)-Ephedrine and Related Alkaloids

The chemical shifts of the methyl moieties at the C-3 position appeared at $\delta_C = 12.49$, 15.24, 14.39, and 17.36 ppm, respectively, and each of these resonances was ^{13}C–^{13}C coupled with the C-2 carbon, with a coupling constant of $J = 37$ Hz. On the other hand, the signals of the C-2 carbons of these alkaloids appeared at $\delta_C = 62.69$, 54.95, 62.61, and 55.51 ppm, respectively, with a coupling constant of $J = 37$ Hz with C-3. The resonances of C-1 (benzylic carbon) appeared as four singlets at $\delta_C = 74.16$–77.15 [13].

These results showed that the C-2 and C-3 carbons of ephedrine were derived from the C-2 and C-3 carbons of pyruvic acid, respectively. The process of the formation of ephedrine or pseudoephedrine from norephedrine and norpseudoephedrine by the addition of a methyl group from the S-methyl group of methionine was demonstrated by an experiment feeding [methyl-^{14}C]DL-methionine into *E. distachya* [14].

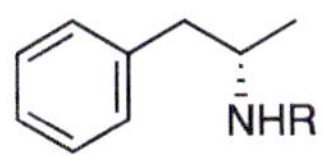

(+)-methamphetamine R = CH_3
(+)-amphetamine R = H

As described in the introductory chapter, methamphetamine and related compounds are known as antihypnotics derived from ephedrine. (+)-Methamphetamine is obtained by reducing (−)-ephedrine or (+)-pseudoephedrine. As well as morphine and heroin (Section 1.10), Lyserg Säure Diäthylamid (LSD) (Section 2.16), and cocaine (Section 3.3), methamphetamine and related alkaloids are causing a variety of social problems in Japan and many other countries at the present time.

LITERATURE CITED

[1] S. Kanao, Nagai Nagayoshi Den, Pharmaceutical Society of Japan, Tokyo, 1960, p. 270.
[2] N. Nagai, Yakugaku Zasshi 12 (1892) 109.
[3] N. Nagai, Yakugaku Zasshi 12 (1892) 832.
[4] N. Nagai, Yakugaku Zasshi 12 (1892) 1186.
[5] N. Nagai, Yakugaku Zasshi 13 (1893) 901.
[6] H. Amatsu, S. Kubota, Kyoto Igaku Zasshi 10 (1913) 301.
[7] H. Amatsu, S. Kubota, Kyoto Igaku Zasshi 14 (1917) 77.
[8] J.G. Hardman, R.W. Ruddon, A.G. Gilman, L.E. Limbird (Eds.), Goodman & Gilman's The Pharmacological Basis of Therapeutics, tenth ed., McGraw Hill Book Co., Inc., New York, 2001, p. 500.
[9] K.K. Chen, C.F. Schmidt, J. Pharmacol. Exp. Ther. 24 (1924) 339.
[10] E. Leete, Chem. Ind. 1958 (1958) 1088.
[11] K. Yamasaki, U. Sankawa, S. Shibata, Tetrahedron Lett. 1969 (1969) 4099.
[12] K. Yamasaki, T. Tamaki, S. Uzawa, U. Sankawa, S. Shibata, Phytochemistry 12 (1973) 2877.
[13] G. Grue-Sfrensen, I.D. Spenser, J. Am. Chem. Soc. 110 (1988) 3714.
[14] S. Shibata, I. Imaseki, M. Yamazaki, Chem. Pharm. Bull. 5 (1957) 71.

16.2 KHAT AND *EPHEDRA* ALKALOIDS

Khat is a medicine prepared from the fresh leaves and branches of *Catha edulis* (Celastraceae) and is mainly used as a stimulant in the Arab world [1]. Khat is also known as "cath," "quat," "chat," "jat," and "tschatt." *C. edulis* grows wild in Ethiopia and is cultivated in the southeast area of the Arabian Peninsula and Eastern Africa.

The active component of khat is ingested by chewing, through the mucosa of the mouth and intestine. It is said that a comparatively large dose of khat is necessary in order to reach the feeling of enhanced awareness. The action of khat is similar to that of amphetamine, and because it produces dependence, it may destroy the normal life of the people who customarily take it. Also, it possesses the action of an appetite suppressant.

The alkaloids contained in khat are phenylalkylamines, and they include khatamine and related alkaloids. In the 1910s, a basic compound with excitatory action was demonstrated in khat, and the isolation of (+)-norpseudoephedrine was the first isolation of an active ingredient in crystalline form [2]. (+)-Norpseudoephedrine is also discussed in Section 16.1.

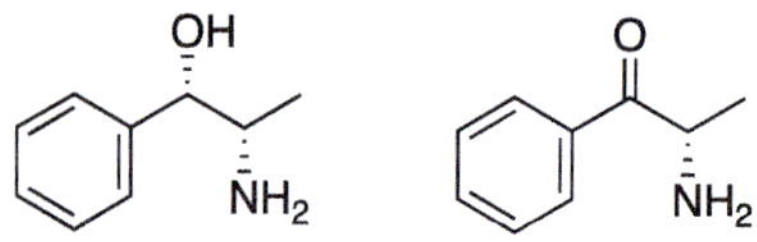

cathine (=(+)-nor-pseudoephedrine) **cathinone**

Initially, (+)-norpseudoephedrine isolated from khat was also called cathine. Subsequently, it was found that a more active alkaloid was present in the fresh plant material, and it was named cathinone [3]. Cathinone corresponds to (−)-α-aminopropiophenone, and the asymmetric carbon has the *S* configuration. Cathinone is the main active ingredient of khat, and its concentration in the phenylalkylamine mixture in khat can be as high as 70%. Cathinone is easily racemized and is dimerized in aqueous solvents to a pyrimidazine (figure). The dimerized component corresponds to that obtained through chemical reaction [4].

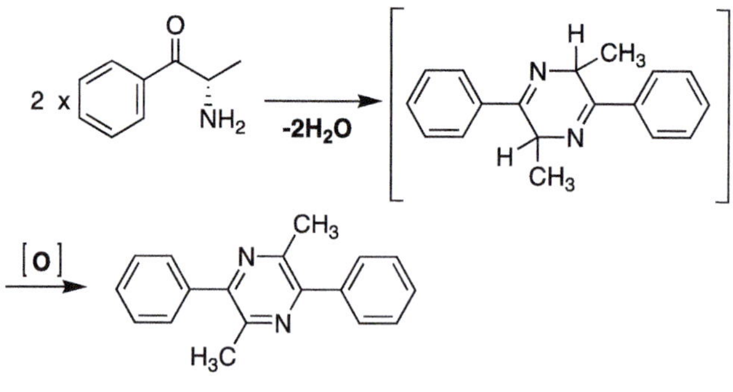

Dimerization of Cathinone

Other minor alkaloids isolated from khat include merucathinone and mercathine [5–8], as well as (−)-norephedrine [1]. The co-occurrence of (−)-norephedrine and merucathinone suggests that phenylalanine is probably not the biosynthetic precursor of these alkaloids.

The total synthesis of optically active cathinone was achieved using racemized norephedrine as the starting material [9]. On the other hand, the total syntheses of merucathinone and merucathine were accomplished using L-alanine as the starting material. Initially, the absolute stereochemistry of merucathine was elucidated to be (3*S*,4*S*) [8], but this was corrected to (3*R*,4*R*) after the completion of its total synthesis [10].

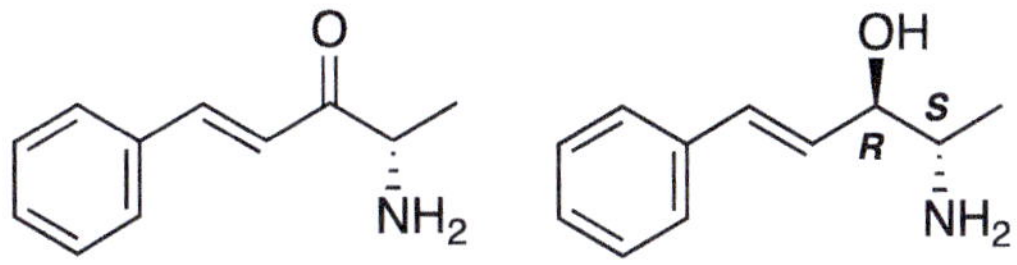

merucathinone **merucathine**

The pharmacologic properties of (−)-cathinone are similar to those of (+)-amphetamine, and both alkaloids possess central nervous system (CNS) excitatory activity and indirect neuronal excitement activity. That is to say, both alkaloids are incorporated at the sympathetic nerve ending and expel noradrenaline (NAdr = norepinephrine, NE) from the amine storage granule, and the expelled NE works as the effector. These activities seem to be the effects caused by chewing khat [11]. Cathine also possesses CNS excitatory activity, but the activity of cathine is less than that of cathinone [12].

Though khat possesses these activities, the number of reports of neurologic disturbance by khat are rare compared with those caused by taking (+)-amphetamine. This may be because of the necessity of taking a large amount of khat to cause such neurologic disturbance, or because of the absence of adequate survey studies.

LITERATURE CITED

[1] L. Crombie, W.M.L. Crombie, D.A. Whiting, in: A. Brossi (Ed.), The Alkaloids, vol. 39, 1990R, p. 139.
[2] O. Wolfes, Arch. Pharm. 268 (1930) 81.
[3] K. Szendrei, Bull. Narc. 32 (1980R) 5.
[4] S. Gabriel, Chem. Ber. 41 (1908) 1127.
[5] R. Brenneisen, S. Geisshüsler, J. Nat. Prod. 50 (1987) 1188.
[6] J.-P. Wolf, H. Pfander, Helv. Chim. Acta 69 (1986) 1498.

[7] X. Schorno, R. Brenneisen, E. Steinegger, Pharm. Acta Helv. 57 (1982) 168.
[8] R. Brenneisen, S. Geisshüsler, X. Schorno, Planta Med. 50 (1984) 531.
[9] B.D. Berrang, A.H. Lewin, F.I. Carroll, J. Org. Chem. 47 (1982) 2643.
[10] J.-P. Wolf, H. Pfander, Helv. Chim. Acta 69 (1986) 918.
[11] P. Kalix, O. Braenden, Pharmacol. Rev. 37 (1985R) 149.
[12] C.E. Johanson, C.R. Schuster, J. Pharmacol. Exp. Ther. 219 (1981) 355.

16.3 NAPHTHALENE-ISOQUINOLINE ALKALOIDS

About 20 species of plant are known in the genus *Ancistrocladus* (Ancistrocladaceae), which grows wild in the tropical areas of Asia, Malaysia, and West Africa. From this genus of plants, as well as two genera of the Dionchophyllaceae family, naphthalene-isoquinoline-type alkaloids have been isolated. This type of alkaloid is quite rare, and only about 20 such alkaloids have been reported [1].

Among them, 10 of the alkaloids possess a junction between C-5 and C-1′, and seven of the alkaloids possess a junction between C-7 and C-1′. The last two alkaloids, ancistrocladidine and ancistrotectorine, possess a junction between C-7 and C-3′ [2]. Ancistrotectorine was isolated from *A. tectorius* in 0.016% yield, and its chemical structure, including the absolute configuration, was reported [2].

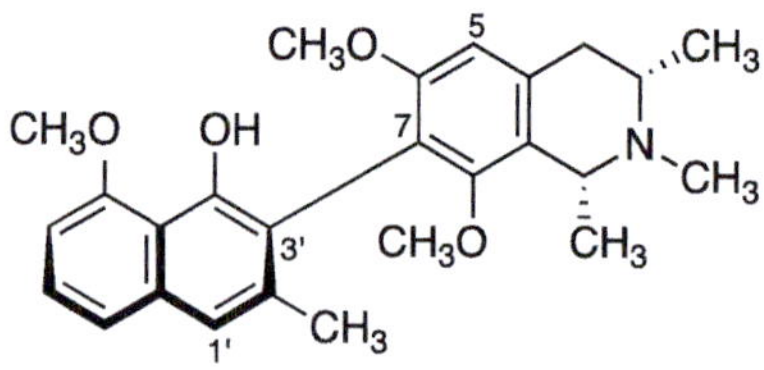

ancistrotectorine

There are no reports on the biosynthesis of this type of alkaloid. It is possible that the origin of these alkaloids is phenylalanine, as for other isoquinoline-type alkaloids. However, because this type of alkaloid possesses a methyl moiety attached to the carbon alpha to the nitrogen atom, like ephedrine and related alkaloids, it is also possible that these alkaloids are derived from a C_6-C_1 unit, such as benzoic acid or benzaldehyde, as in the case of ephedrine and related alkaloids. That is why these alkaloids are included in this chapter. Elucidation of the biosynthetic pathway for these alkaloids is awaited.

LITERATURE CITED

[1] G. Bringmann, F. Pokorny, in: G.A. Cordell (Ed.), The Alkaloids, vol. 46, 1995R, p. 127.
[2] N. Ruangrungsi, V. Wongpanich, P. Tantivatana, H.J. Cowe, P.J. Cox, S. Funayama, G.A. Cordell, J. Nat. Prod. 48 (1985) 529.

16.4 RED PEPPER AND CAPSAICIN

The origin of *Capsicum annuum* (Solanaceae) is said to be South America, and it is widely cultivated from the tropical zone to the temperate zone for the use of its fruits as red peppers. It was introduced into Spain in 1494, and rapidly spread worldwide, being introduced into Japan from Portugal in 1542.

Red peppers, or the matured fruits of its several varieties, are used as a spice or as a raw material for a spice, such as Shichimi-Togarashi (seven kinds of spices including red pepper) and Ra-yu (its hot oil), in many systems of cooking. It is also used as a hot-taste stomachic salve in Chinese herbal medicine, and the alcoholic extract of red pepper is mixed in a rubefacient.

The pungent principle of *Capsicum*, capsaicin, was isolated by Thresh et al. in 1876 [1]. Afterward, several synthetic methods for capsaicin were reported [2,3]. According to reports concerning the biosynthesis of capsaicin [4], when tritium-labeled precursors were administered to immature red pepper (*C. annuum*), it was demonstrated that DL-[3′-^{3}H]phenylalanine, L-[3′,5′-$^3H_2$]tyrosine, [5′-^{3}H]vanillylamine, and various cinnamic acids (labeled in the aromatic ring) were incorporated. It was also shown that the rate of uptake of phenylalanine was better than that of tyrosine.

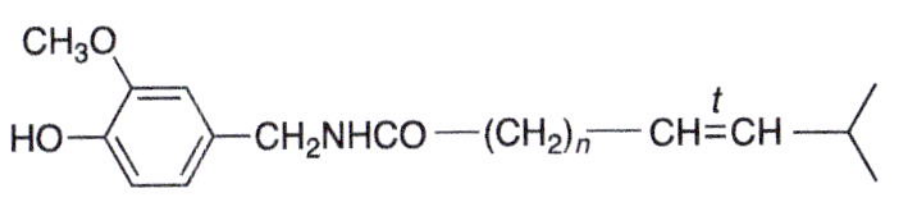

capsaicin *n* = 4
homocapsaicin *n* = 5

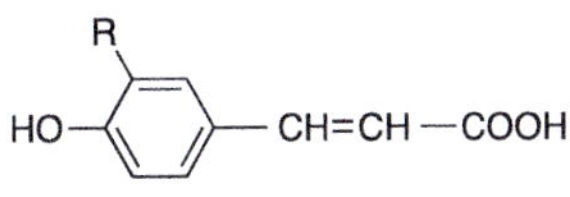

***p*-coumaric acid** R = H
ferulic acid R = OCH_3
caffeic acid R = OH

Cinnamic acid derivatives, such as *p*-coumaric acid, ferulic acid, and caffeic acid, were incorporated better than tyrosine, but the incorporation rate was inferior to that of phenylalanine [4]. However, since the water solubility of these phenylpropanoids is low, except for caffeic acid, these compounds might not be taken up by plants well.

On the other hand, it was shown that vanillylamine was comparatively well incorporated; thus, vanillylamine might be a precursor that is closer to the terminal biosynthetic stages. However, whether the nitrogen atom of vanillylamine is retained into capsaicin is uncertain at present.

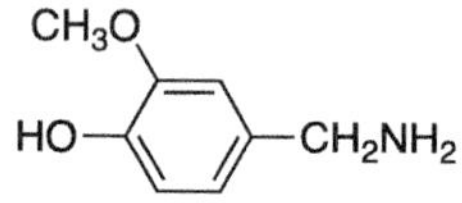

vanillylamine

From these experimental results, it is clear that phenylalanine may be a precursor in capsaicin biosynthesis, and that a C_6-C_1 unit, like vanillylamine, might be a more direct precursor. Even if phenylalanine is introduced, it is clear that the C–N combination is not retained. Based on these experimental observations, capsaicin is described in this chapter.

Though there are many varieties of red pepper (*C. annuum)*, so-called "Takano-tsume" (*C. annuum* var. *parvo-acuminatum*) and "Yatsu-busa" (*C. annuum* var. *parvo-acuminatum* f. *erectum*) are mainly used and cultivated in Japan.

Among the material used for the spice, other than *C. annuum* described above [5], which contains 0.2–0.3% of capsaicin, the so-called African chilies (African red pepper), which possess a more potent hot taste, are also known. It is indicated that the capsaicin content of the material of African chilies (*Capsicum frutescens*) may reach 0.6–0.9%.

Ninety percent of the pungent principle of red pepper is present in the pericarp, and the remaining 10% is in the seed. On the other hand, other than capsaicin (69%) as a main pungent principle, dihydrocapsaicin (22%), nordihydrocapsaicin (7%), homocapsaicin (1%), and homodihydrocapsaicin (1%) are known as minor pungent constituents [4].

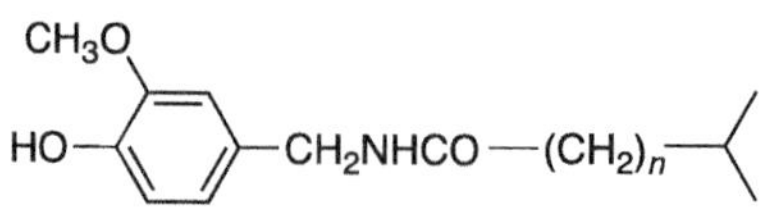

nordihydrocapsaicin $n = 5$
dihydrocapsaicin $n = 6$
homodihydrocapsaicin $n = 7$

As for the capsaicin derivatives, it was shown that only the *trans* isomer is naturally occurring. Biologically, it was reported that purified capsaicin depletes substance P stores in sensory neurons and blocks further synthesis of this neuropeptide. Therapeutically, capsaicin has been used successfully to treat several painful conditions (e.g., rheumatoid arthritis, osteoarthritis, and various peripheral neuropathic disorders). To date, there are no capsaicinoids that have proven to be as potent as the parent compound capsaicin for neuropeptide activity [6].

LITERATURE CITED

[1] J. Molnár, Arzneimittel-forsch. 15 (1965R) 718.
[2] E. Späth, S.F. Darling, Chem. Ber. 63 (1930) 737.
[3] L. Crombie, S.H. Dandegaonker, K.B. Simpson, J. Chem. Soc. 1955 (1955) 1025.
[4] D.J. Bennett, G.W. Kirby, J. Chem. Soc. (C) 1968 (1968) 442.
[5] M. Fujita, T. Furuya, A. Kawana, Yakugaku Zasshi 74 (1954) 766.
[6] G.A. Cordell, O.E. Araujo, Ann. Pharmacother. 27 (1993R) 330.

INDEX

Note: Page numbers with "f" and "t" denote figures and tables, respectively.

D

Q

R

S

T

CPI Antony Rowe
Eastbourne, UK
May 14, 2020